Oak
Forest
Ecosystems

Oak
Forest
Ecosystems

Ecology and Management
for Wildlife

EDITED BY
WILLIAM J. MCSHEA
AND
WILLIAM M. HEALY

THE JOHNS HOPKINS UNIVERSITY PRESS
BALTIMORE AND LONDON

The book has been brought to publication with the generous assistance of the Smithsonian Institution.

© 2002 The Johns Hopkins University Press
All rights reserved. Published 2002
Printed in the United States of America on acid-free paper
9 8 7 6 5 4 3 2 1

The Johns Hopkins University Press
2715 North Charles Street
Baltimore, Maryland 21218-4363
www.press.jhu.edu

Library of Congress Cataloging-in-Publication Data

Oak forest ecosystems : ecology and management for wildlife / edited by William J. McShea and William M. Healy.
     p. cm.
Includes bibliographical references (p.   )
     ISBN 0-8018-6745-2 (hardcover)
     1. Forest ecology—North America.   2. Oak—ecology—North America.
3. Forest management—North America.   4. Wildlife conservation—North America.   I. McShea, William J.   II. Healy, William M.   III. Title.
     QH102 .O24 2002
     577.3'0973   dc21

00-012804

A catalog record for this book is available from the British Library.

Contents

Part III **Management of Oaks for Wildlife**

Contributors

Marc D. Abrams
School of Forest Resources
Pennsylvania State University
University Park

Patrick H. Brose
U.S. Department of Agriculture
  Forest Service
Forestry Sciences Laboratory
Irvine, Pennsylvania

John P. Buonaccorsi
Department of Mathematics
  and Statistics
University of Massachusetts
Amherst

Daniel Dey
U.S. Department of Agriculture
  Forest Service
North Central Research Station
Columbia, Missouri

Joseph S. Elkinton
Department of Entomology
University of Massachusetts
Amherst

George A. Feldhamer
Department of Zoology
Southern Illinois University
Carbondale

Peter F. Ffolliott
University of Arizona
Tucson

Lee E. Frelich
Department of Forest Resources
University of Minnesota
St. Paul

Cathryn H. Greenberg
U.S. Department of Agriculture
  Forest Service
Southern Research Station
Bent Creek Experimental Forest
Asheville, North Carolina

William M. Healy
U.S. Department of Agriculture
  Forest Service
University of Massachusetts
Amherst

Roy L. Kirkpatrick
Department of Fisheries
  and Wildlife Sciences
Virginia Polytechnic Institute
  and State University
Blacksburg

Johannes M. H. Knops
School of Biological Science
University of Nebraska
Lincoln

Walter D. Koenig
Hasting Reservation
University of California
Carmel Valley

Nelson W. Lafon
Virginia Department of Game
    and Inland Fisheries
Verona

Andrew M. Liebhold
U.S. Department of Agriculture
    Forest Service
Northeast Forest Experimental
    Station
Morgantown, West Virginia

William J. McShea
National Zoological Park
    Conservation and Research
    Center
Front Royal, Virginia

William H. McWilliams
U.S. Department of Agriculture
    Forest Service
Radnor, Pennsylvania

Gary W. Norman
Virginia Department of Game
    and Inland Fisheries
Verona

Steven W. Oak
U.S. Department of Agriculture
    Forest Service
Southern Region Station
Asheville, North Carolina

Renee A. O'Brien
U.S. Department of Agriculture
    Forest Service
Forestry Sciences Laboratory
Ogden, Utah

Richard S. Ostfeld
Institute of Ecosystem Studies
Millbrook, New York

Bernard R. Parresol
U.S. Department of Agriculture
    Forest Service
Southern Research Station
Asheville, North Carolina

Peter J. Pekins
Department of Natural
    Resources
University of New Hampshire
Durham

Gordon C. Reese
Department of Fisheries
    and Wildlife
Colorado State University
Fort Collins

Peter B. Reich
Department of Forest
    Resources
University of Minnesota
St. Paul

Peter D. Smallwood
Department of Biology
University of Richmond
Richmond, Virginia

Christopher C. Smith
Division of Biology
Kansas State University
Manhattan

Richard B. Standiford
Integrated Hardwood Research
  Management Program
University of California
Berkeley

Martin A. Stapanian
Ohio Cooperative Fisheries
  and Wildlife Research Unit
Columbus

Michael A. Steele
Department of Biology
Wilkes University
Wilkes-Barre, Pennsylvania

David E. Steffen
Department of Game and Inland
  Fisheries
Vinton, Virginia

David H. Van Lear
Department of Forest Resources
Clemson University
Clemson, South Carolina

Michael R. Vaughan
Virginia Cooperative Fisheries
  and Wildlife Research Unit
Virginia Polytechnic Institute
  and State University
Blacksburg

Karen L. Waddell
U.S. Department of Agriculture
  Forest Service
Forestry Sciences Laboratory
Portland, Oregon

Acknowledgments

We thank the authors of this volume for tolerating our repeated requests and vague instructions. Liz Reese helped pull all the chapters together into a coherent form. Melissa Songer did an excellent job of bringing continuity of form to the diverse chapters, worked to ensure that each figure and citation was properly verified, and kept us all on schedule. The following individuals contributed greatly to the quality of this volume by reviewing multiple chapters: Matthew Knox, David deCalesta, Michael Bowers, Mario Castellanos, Jerry Wolff, Michael Pelton, Jack Cranford, James Rieger, William M. Block, Robert D. Childs, Matthew J. Kelty, William A. Patterson III, Charles M. Ruffner, and David W. Smith. Terry Sharik gave a critical read to the entire manuscript and helped improve many sections. At the Johns Hopkins University Press, Sam Schmidt kept faith with the project and Anne Whitmore fixed problems we could not see.

This volume is dedicated to Lucy Braun, whose descriptions of bygone forests still resonate with today's ecologists.

Oak
Forest
Ecosystems

Chapter 1

Oaks and Acorns as a Foundation for Ecosystem Management

WILLIAM J. MCSHEA AND WILLIAM M. HEALY

Acorns are the most important wildlife food in the deciduous forests of North America, the ecological equivalent of manna from heaven. The pattern and abundance of acorns, and of their parent trees (*Quercus* spp.), are of critical importance to most wildlife species that reside in the temperate forests of North America. Understanding the function and role of oaks within forested ecosystems will help natural resource managers understand the dynamics of wildlife populations. The genus *Quercus* is distributed almost worldwide in more than 500 species (Smith 1993). Although there must be similarities between the dynamics of all oak systems, we have limited this volume to North American oaks. There has been work on masting species in Europe (Shaw 1968a,b) and New Zealand (King 1983), but we feel only the literature from North America is extensive enough to be the basis for specific management recommendations. Our emphasis is also on working landscapes, where at least part of the area is used for commodity production, usually timber or livestock, because most oak forests are on private rather than protected land (see Chapter 2).

This volume emphasizes wildlife and its interactions with oak ecosystems. There are many advantages to oak ecosystems (e.g., timber, recreation, and aesthetics), but we share the belief of many wildlife biologists that the true value of oak forests rests in their contribution to wildlife. With increasing pressure on landholders to generate capital from their lands, we want to offer guidelines that will profit wildlife, as well as the bottom line.

Oaks are one of the most diverse group of tree species in North America, with approximately 50 species found from the deciduous forests of the Appalachians to the open savannas of California. North American oaks occupy temperate climes, where moisture is slightly xeric and fires are common but not necessarily annual. Although the seedlings are usually shade-intolerant, oaks can form old-growth stands because of their longevity, resistance to fire, and their ability to recolonize following disturbance. Among the hardwood species, oaks are at present the most economically important group within North America. There is concern for their health and continued productivity (Healy et al. 1997) because of the stresses caused by the introduction of insect pests (Houston 1981, Gottschalk 1989), recent increases in deer populations (Healy 1997a), suppression of fire (Abrams et al. 1995), and widespread oak decline (Oak et al. 1988). Conservationists are concerned with the loss of forest to agricultural and suburban development (Robinson et al. 1995) and the industrial conversion of deciduous forest stands into pine plantations (Palik and Engstrom 1999). For many wildlife species, the loss of deciduous forests is compounded by the loss of seed crops in the autumn.

This volume is intended to bring together the knowledge of a diverse group of scientists and managers who work with oak forests. We hope this knowledge will form the basis for better management of this critical resource and allow managers, scientists, and students to see oaks within the holistic field of ecosystem management. With so many species dependent on oaks for shelter and food, and with so many consumer uses of oak forests, it is futile to deal with wildlife issues on a species-by-species basis. Management for one species, or one consumer use, impacts a myriad of other resources. Many of this book's chapters deal with specific species or ecological processes, because these are the stepping stones toward viewing oak forests and their wildlife as a complex ecosystem. Managing oak forests in a sustainable manner for wildlife involves understanding all the parts and recognizing the true scope of management actions.

We firmly believe that the management of oak forests will only succeed within the context of ecosystem management, but exactly what does ecosystem management entail? Recent approaches to forest and wildlife management have been characterized by terms such as *ecosystem management, biodiversity management, ecological forestry, new forestry, landscape management,* and *sustainable forestry.* Within recent years, major federal land management agencies, including the U.S. Forest Service, Bureau of Land Management, National Park Service, Fish and Wildlife Service, Environmental Protection Agency, and the National Oceanic and Atmospheric Administration, have made a commitment to ecosystem management (Rauscher 1999). Each agency defines and implements ecosystem management in a way that is tailored to meet its particular mission and goals, but all emphasize large spatial scales, long time frames, system dynamics, and sustaining ecosystem processes.

Much of the controversy concerning an ecosystem approach to management has actually been about which values ought to prevail in the management of public lands, rather than the technical merits of the approach (Maguire 1999). Yaffee (1999) recently published a useful classification of the different meanings of ecosystem management which organizes the definitions into three broad perspectives that form a continuum of management. The first category includes environmentally sensitive, multiple-use management that aims at satisfying a diverse set of human needs and values. This view is close to traditional sustained-yield, multiple-use management, in which the emphasis is on sustaining goods and services. The second grouping includes an ecosystem-based approach to resource management which is based on the principles contained in the ecosystem management literature. The emphasis is on a systems approach to management rather than a specific ecosystem. This approach promotes ecological integrity while allowing sustained human use. The third category includes ecoregional management that incorporates the principles of ecosystem-based approaches but emphasizes landscape-scale management as a fundamental goal. The emphasis is on real geographic units and restoring and maintaining ecosystem function while allowing sustained human use.

Yaffee's topology provides insights into the diversity of values, experiences, and goals that underlie different management paradigms. We agree with Yaffee that professionals should learn to recognize the different meanings of ecosystem management and continue working toward the goal of ecoregional management. In contrast to the multiple definitions of ecosystem management, several common themes emerge in the literature on ecosystem management (Grumbine 1994, Yaffee 1999) and its application to forested landscapes (Boyce and Haney 1997).

Reviews of ecosystem management theory identify from 6 to 10 elements that are common to ecosystem management approaches to natural resource management (Grumbine 1994, Yaffee 1999). Recent forestry literature incorporates most of these elements into new management and silvicultural approaches to sustaining forest diversity (Boyce and Haney 1997, Hunter 1999). Eastern forests and western woodlands have received less attention in the forestry literature on ecosystem management than old-growth, coniferous forests in the Pacific Northwest (e.g., Franklin 1997). This is unfortunate because the extent of oak forests dwarfs that of other forest types (see Chapter 2), and although these forests are rarely old-growth (but see Chapter 8), their mast production is abundant at a relatively early age (see Chapter 10). If the goal of ecologically managed forests is to maintain timber harvest while preserving biodiversity at the local or landscape scale (Seymour and Hunter 1999), then in oak forests this is accomplished by attention to the pattern and severity of disturbance and to mast production. For protected oak forests the goal is to develop management techniques that mimic the natural disturbance regimes that created the oak habitat in the first place, yet to maintain mast production at the landscape level.

The development of oak forests is not a recent phenomenon; pollen records from the Midwest and the East show that oaks have been a significant component of the forest for the last 10,000 years (Delcourt 1979, Abrams 1992). This abundance appears to wax and wane with cycles of moisture and temperature (Grimm 1983, Winkler et al. 1986). Fire, both natural and man-made, had a role in these fluctuations (Abrams 1992). In eastern North American forests, oaks are part of a rich diversity of deciduous species. With the loss of the American chestnut (*Castanea dentata*) as a significant forest species, oaks have increased in abundance and importance (Woods and Shanks 1959), but probably not beyond that experienced in past periods, when the climate was drier. In the Midwest, oaks form the boundary between rich deciduous forests and prairie grasslands. This boundary has fluctuated with the weather cycles and their companion fire regimes (Baker et al. 1996). With human dominance of the landscape and the suppression of fire in the last century, this dynamic edge has become stable and the management challenge is to maintain the transitional ecotone. In the West and Southwest, we do not have a good record of how persistent oaks were, because the forest types, primarily savanna or shrub, do not provide good paleologic records. However, based on our knowledge of present systems, we can assume that fire played a major role in maintaining oak species within these habitats.

Placing the current importance of oaks within a historical context is difficult because of the widespread loss, or change in abundance, of wildlife and plant species due to the actions of humans. These species losses include both seed predators, such as the passenger pigeon (*Ectopistes migratorius;* Bucher 1992), and co-dominants in the canopy, such as the American chestnut (Woods and Shanks 1959, Diamond 1989). The passenger pigeon, with population densities in the billions and mass migrations across eastern forests, must have created a dynamic that is not repeated by the present suite of seed predators (Bucher 1992). There is also no precedent for the age structure of our present forests, which contain such a large proportion of trees in early maturity. The original preponderance in old growth will not be duplicated under present economic conditions. Restoring historical balance is not as important as understanding the dynamics of current forests and maintaining their potential to support biodiversity (McShea et al. 1997). Oaks, with their long history of numerical dominance, their high diversity, and their widespread distribution, create a matrix within which wildlife and other species persist.

We cannot discuss oaks without focusing on their seed crop, acorns. Oak trees can serve as denning sites, forage for vertebrate and invertebrate herbivores, and provide the structure of the forest itself, but these functions can be found in any deciduous forest. It is oaks' production of a seed crop that is large, both in number and mass, but also annually variable that makes the genus unique and leads to its impact on wildlife. A survey by Martin et al. (1961) lists 96 North American vertebrate species that consume acorns, and acorn production figures prominently in habitat suitability models for several wildlife species (Schroeder and Vangilder 1997). Although many species depend on acorns, their consumption of the resource does not fit most predator/prey or herbivore/plant systems, as the acorn crop is beyond the short-term regulation of seed predators. The production of acorns is a function of the basal area of oaks and weather conditions (Goodrum et al. 1971, Sork et al. 1993). The longevity of oaks is an order of magnitude longer than that of their seed predators; a hypothetical system in which predators removed every seed would only result in reduced seed production decades later, when the trees produced by those seeds would have reached maturity. Oak-dominated forests are not a system characterized by an equilibrium between the oaks and their seed predators but rather a system where the density of the predators is driven by the abundance of the seed resource. The tie between seed production and seed predator populations is best demonstrated by measuring predator reproduction. As in xeric systems, where plants and animals respond rapidly to pulses in rainfall (Brown and Heske 1990), many small seed predatory mammals reproduce following large seed falls (Gashwiler 1979, Wolff 1996). The magnitude of the annual pulses for oak seeds is unpredictable (Sork et al. 1993), and therefore reproduction does not anticipate, but rather tracks, changes in seed productivity.

The terms *mast* and *masting* are commonly used when discussing oaks, but they have multiple meanings. *Mast* can refer collectively to the fruit of *masting* species, those that engage in the intermittent production of seeds. Ecologists tend to use these meanings. Deriving from the variable abundance as well as intermittency of this masting, *mast* can also refer to seed crops and crop years of particular abundance—for example, "mast years." Wildlife and timber managers will refer to a stand of oaks as "masting" if it is producing a large quantity of seeds that year. Because this volume contains contributions from many fields, the meaning of *mast* and *masting* will vary depending on the context of the discussion.

Seeds of masting species differ in their utility to wildlife species. Soft mast consists of seed crops where the seed is surrounded by, or imbedded in, a fleshy pulp (e.g., *Cornus* sp., *Rubus* sp., *Vaccinium* sp., and *Vitis* sp.). These species produce valuable seed crops that provide energy and nutrients for late-season reproduction of smaller mammals, premigratory fattening for many bird species, and fat deposition for large mammals, such as bears (Martin et al. 1961). However, these pulpy seeds cannot be externally stored. Hard mast is made up of seed crops that possess a hard outer seed coat that maintains the seed's moisture and enables the seed to be stored for extended periods. Many of the dominant trees in North American forests produce hard mast (*Pinus* sp., *Quercus* sp., *Juglans* sp., and *Fagus americana*). In the eastern United States, the American chestnut was a dominant producer of hard mast prior to its demise.

Endothermic vertebrates rely on relatively high intakes of energy to maintain their body temperature (Schmidt-Nielson 1984). Nonpredatory mammals and birds in temperate forests have few means for surviving periods of low productivity (e.g., winter); they can migrate to more productive areas, hibernate, or eat hard seeds. Most volant vertebrates use the first option, but, for nonvolant vertebrates, migration is an option only for large animals in situations where relatively short movements, (e.g., along elevation gradients) result in large changes in habitat productivity. Larger animals can also support a thick insulative covering and therefore store sufficient fat to hibernate, or at least persist at a negative energy balance for extended periods (Eisenberg 1981, Reynolds 1993). Sufficient fat deposits can be obtained by feeding on hard and soft mast crops or vegetation. The high surface-to-volume ratio of small birds and mammals does not permit them to develop sufficient insulation or to store sufficient body fat to maintain their body temperature through extended periods of low food productivity, and they have evolved alternative strategies (Eisenberg 1981, Peters 1983). For nonvolant small mammals, it is essential that usable energy be packaged so that it can be cached, or at least can persist in the environment, without degradation. Most hard mast seeds mature in the autumn, but their germination is delayed until the following spring; a life history trait advantageous for all nonvolant small vertebrates. Without hard mast, it is difficult to believe that temperate forests could sustain more than a handful of vertebrate species.

Although most of our focus in this volume is on granivorous or omnivorous species, many generalist predators do consume acorns (Martin et al. 1961), and many predators are impacted indirectly by variations in acorn crops, because their principal prey are mast-dependent small mammals (King 1983, McShea 2000). Without a prey base of small mammals, most small predators would not be able to persist until the arrival of migrant bird species and the emergence of insect populations.

That so many species rely on hard mast crops to survive winter and to maximize reproduction attests to the evolutionary importance of the seeds within forest ecosystems. Many of our forest ecosystems have undergone profound changes in the last 200 years, including the loss of major plant and animal species. Luckily, the species shift in eastern deciduous forests from American beech (*Fagus americana*) and American chestnut to oaks has not resulted in the loss of hard mast, but it has potentially changed the pattern of hard seed production.

Plant species that produce mast crops are not adapted to feed wildlife; their production of large quantities of seeds is to ensure their own effective reproduction. The interrelationship between seed trees and their predators is an important subject and is covered extensively in this volume. This relationship is probably what has driven the boom or bust phenomenon so often seen in mast crops. Annual fluctuations in acorn crops eliminate seed predators during a year of scarcity and overwhelm the remaining seed predators during the next year of abundance. This scenario only works if individual trees act in synchrony, at least within the home range or dispersal radius of the seed predators. Local and regional synchrony of acorn production is explored in this volume by Koenig and Knops, and Greenberg and Parresol. What needs to be examined is the degree of synchrony between tree species, as most oak forests are composed of several hard mast species, including pines, beech, and *Carya* sp., as well as soft mast–producing species. For larger mammals, the loss of hard mast following timber management may be compensated for by an increase in soft mast or herbaceous cover, but there is little supporting evidence for this hypothesis. The lack of specialization for most seed predators indicates the necessity, and opportunity, of seed predators to switch between tree species as mast crops for individual species wax and wane. We don't know how animal species differ in their ability to process and handle a variety of hard seeds; we can only assume that seed size and predator size must play a role (Aizen and Patterson 1990).

Oaks exist within a matrix of species, whose interactions compose a food web with complex feedback loops across trophic levels (Elkinton et al. 1996, Ostfeld 1997, Jones, Ostfeld, and Wolff 1998, McShea 2000). Two types of those interactions are the focus of this volume: first, species and events that impact the persistence and abundance of oaks and their seeds; second, how variability in seed production impacts the persistence and abundance of wildlife populations. Forest management focuses on the first set of interactions, and the results can be measured through the second set of interactions.

To accomplish ecosystem management goals we need to know enough about ecosystem function to identify critical species or interactions. Oaks forests are one of the few systems about which we have built a sufficient data set to begin following the principles of ecosystem management. The first section of this volume contains chapters on distribution, disease, disturbance, and silviculture of oaks. These chapters are the basis for understanding the ecology of this dominant group. The purpose of the second section is to identify interactions we know exist, and the authors cover the rich history of research on wildlife species and their dependence and interaction with oaks. Scientists are just beginning to see how these interactions come together within complex food webs. The third section focuses on selected oak systems for which we do possess enough data to start making predictions at the landscape scale of how forestry practices will impact wildlife communities. We cannot provide examples from all oak systems in this volume, so we offer three systems (i.e., oak woodlands of California, southwestern oak forests, and northen Appalachian oak forests) selected for their completeness and with an eye toward representing different regions of North America. We end this section with basic goals and guidelines to manage wildlife in oak forests.

For wildlife management to succeed in oak forests, foresters and wildlife managers must work together at a scale that is meaningful to the species involved. Within the forestry community there is a maxim that good forest management is good wildlife management, and good forest management means the active manipulation of stand composition and basal area. Although there may be more exceptions to this rule than examples of it, with oaks one sees the maxim in action. Oak forests are the product of disturbance; and ensuring adequate mast production is a process of managing the timing and scale of that disturbance. Management conflict arises when the optimal timing and scale for mast production do not match the optimum for economic return. However, the economics of forest management rarely include the value of wildlife (Lippke and Bishop 1999), and as go acorns, so go the wildlife within oak-dominated forests.

Because of their dominance in both biomass and numbers, oaks do not usually qualify as keystone species, which are those species whose impact on the ecosystem is disproportionate to their low abundance (Power et al. 1996). However, this volume will highlight that oaks are far more critical than even their numbers indicate, because their production of acorns is the linchpin upon which much of the forested ecosystem hinges. Effective management of oaks, and by that we mean the retention of sufficient mast-crop potential and the effective regeneration of harvested stands, is the key to maintaining wildlife populations and practicing true ecosystem management.

Part I

Patterns and Processes of Oak Forests

Chapter 2

Distribution and Abundance of Oaks in North America

WILLIAM H. MCWILLIAMS, RENEE A. O'BRIEN,

GORDON C. REESE, AND KAREN L. WADDELL

It is common to hear oak (*Quercus* L.) referred to as the most important tree genus in North America. The significance of oaks results from their commercial value for timber and as food and cover for wildlife. The classic finding of Van Dersal (1940) that 186 different kinds of birds and mammals feed on oaks and that the distribution of a number of animals coincides with or is dependent on the range of oaks underscores their importance in sustaining North American fauna. (In this chapter, North America is defined as north of Mexico: the continental United States and Canada.)

Except for the region to the north of the oak biome and a few other locations, species of oak are found from coast to coast in North America. In the East, oaks are typically members of broadleaf deciduous forest communities. Oaks often are the dominant species in the central and southern regions and in mixed-oak forests to the north and east. In the interior West, oaks are common in shrub communities, where their importance to wildlife is magnified because other sources of food are rare. On the Pacific Coast, various evergreen and deciduous oak species are common as shrubs and trees. Oaks dominate in a number of communities, for example, the oak woodlands of California.

A member of the beech family (*Fagaceae*), the oak genus comprises from 500 to 600 species worldwide (Smith 1993). The exact number of species in North America varies depending on the author and life form considered; however, oak clearly is the largest genus of native trees. The Plants National Database of the U.S. Natural Resources Conservation Service (1998) lists 93 species of oak and about twice as many varieties and hybrids. Kartesz's checklist of vascular flora (1994) includes 86 species of oak. Little's widely accepted checklist of "trees" for the United States (1979) documents 58 tree species, one naturalized tree, and about 10 native shrubs. Little lists 10 oaks in Canada. There are no recognized oak species we are aware of that are unique to Canada.

The native range of oak in North America extends from Nova Scotia across the continent to the Pacific coast (Figure 2.1). Native oaks are essentially absent in the northern Rocky Mountains of Idaho, western Montana, and western Wyoming. Although the Great Plains contain little forest area and native oaks are relatively uncommon, oaks are found on some sites in this region. Oaks also are found along the southern border of the United States. The southernmost oak, live oak (*Q. virginiana* Mill.), is found at the tip of Florida (at about 24° south latitude). To the

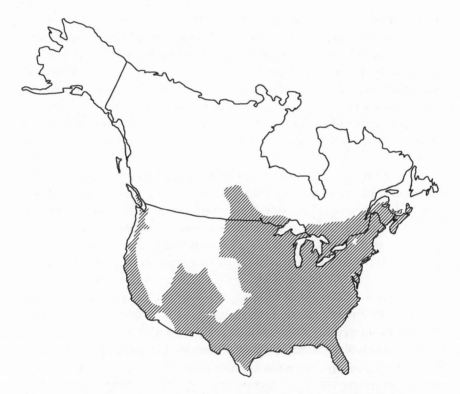

FIGURE 2.1. General distribution of oak (*Quercus* L.) in North America. (Adapted from Burns and Honkala 1990, Clary and Tiedemann 1992, Miller and Lamb 1985, and Peterson and Boyd 1998.)

north, the range of oak stretches into southern Canada, with the northern limit at about 54° north latitude. The range of bur oak (*Q. macrocarpa* Michx.) stretches the farthest north, extending just north of Lake Winnipeg in the Boreal Plains of Manitoba. In the West, the northern limit is reached by Oregon white oak (*Q. garryana* Dougl. Ex Hook), which is found in southeastern Vancouver Island. Oregon white oak is notable as the only oak in Canada west of southeastern Saskatchewan. Most of the oaks in Canada are found in southern Ontario and Quebec. The most widely distributed of these are white oak (*Q. alba* L.), swamp white oak (*Q. bicolor* Willd.), bur oak, pin oak (*Q. palustris* Muenchh.), chestnut oak (*Q. prinus* L.), and northern red oak (*Q. rubra* L.). The 58 species of oak listed by Little (1979) include 33 eastern species and 25 western species. Many of the western oaks occur as shrub or small tree forms and rarely form a dominant forest canopy. Burns and Honkala (1990) list 20 "commercially" important oaks in the East. In the West, four commercially important oaks are listed.

The diversity of oak species in North America increases significantly from north to south. The Plants Database lists 9 species of oak for Minnesota but 55 for Alabama (U.S. Natural Resources Conservation Service 1998). The greater diversity to the south also is evident in the West. For example, Gambel oak (*Q. gambelii* Nutt.) is the only oak species found at the northern extent of oak's distribution in Utah and southern Wyoming. By contrast, at least 16 oak species and numerous hybrids are found in the southwestern United States, excluding west Texas and California (Miller and Lamb 1985). This point is further emphasized by the incredible diversity of oaks that occur south of the study area, in Mexico. The exact number of oak species in Mexico has not been confirmed, but it ranges from Little's description (1979) of 125 to Miller and Lamb's description of 251 (1985).

Because of the large number of oak species and the myriad of communities of which oaks are major associates, our analysis considers oak in three broad regions: the East, the interior West, and the Pacific Coast. These regions generally follow the distribution of major oak communities within the oak biome and together represent the major socioeconomic forces that affect the character and status of oak in North America. Much of the information presented here is derived from regional forest-inventory data collected by the U.S. Forest Service's Forest Inventory and Analysis (FIA) units, which have traditionally focused on tree life forms. As a result, some of the source data reflect information that is limited to tree species. Other sources are used where the existing FIA data are scant, for example, for shrub communities of the Rocky Mountains.

THE EAST

Oaks have been a major component of eastern forests for more than 6,000 years (Lorimer 1993). Today's eastern mixed-oak forests evolved in response to a complex mix of anthropogenic and ecological factors, and their interactions (see later chapters). For example, the importance of oaks for eastern wildlife has increased in the twentieth century as American chestnut (*Castanea dentata* Marsh. Borkh.) and American beech (*Fagus grandifolia* Ehrh.) have declined (Healy et al. 1997). It has been estimated that 49 species of birds and mammals utilize oak mast in the East (Miller and Lamb 1985). An analysis of their status is timely, in light of heavy cutting of oak for timber products, infestations of gypsy moth (*Lymantria dispar* L.) (Gansner et al. 1993), oak decline syndrome (Millers et al. 1989), and regeneration failure (Lorimer 1993, Smith 1993).

Existing maps of forest-type groups containing oak, such as that presented by Powell et al. (1993), lack details on the actual degree of oak occupancy. To overcome these limitations, we applied spatial statistical techniques to data contained in FIA's Eastwide Forest Inventory Database (Hansen et al. 1992) to examine where oak is most prevalent across the East. Figure 2.2 shows the distribution of timberland with at least 25% of total basal area in oaks species based on indicator kriging. The occurrence of oak in forest stands throughout the eastern states is readily apparent. The largest block of high-density oak is located in the Ozark Plateau of Missouri and northern Arkansas. Other areas with high oak densities are the central lowlands of Minnesota and Wisconsin, the Appalachian Mountains from central Pennsylvania to northern Alabama, the Nashville Basin of Tennessee, and the western side of the Florida Peninsula.

Of the major forest-type groups of the eastern United States, oaks are common associates in all but the northernmost groups: spruce-fir and aspen-birch. Oak is most prevalent in the two upland oak groups: oak-hickory and oak-pine. The area of upland oak groups covers 63.7 million hectares, or 43% of eastern timberland (Table 2.1). Upland oak forests compose half or more of the timberland in the central, Gulf, mid-Atlantic, and Atlantic states. Oaks also are members—though sporadic—of the

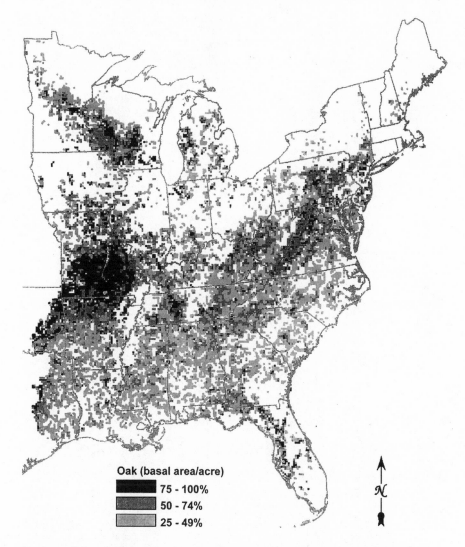

Oak (basal area/acre)
75 - 100%
50 - 74%
25 - 49%

FIGURE 2.2.   Estimated oak occupancy on timberland with at least 25% of total basal area in oak species, eastern United States. (Available data excludes Connecticut, Massachusetts, and Rhode Island.)

other upland deciduous forest types, primarily maple-beech-birch. The other upland deciduous groups make up 19% of eastern timberland. Lowland deciduous forests of the Atlantic and Gulf states account for 12% of the eastern timberland and occupy sites on a continuum from moist to wet. Lowland deciduous forests (commonly referred to as bot-

Table 2.1
Area of timberland by region and broad forest type, eastern United States (in thousands of hectares)

Region	Total	Upland oak[a]	Other upland deciduous	Lowland oak[b]	Other lowland deciduous	Conifer	Untyped
Lake states[c]	19,845.8	2,441.6	10,819.7	—	1,798.8	4,705.1	80.6
Central states[d]	10,758.3	6,599.9	2,252.8	12.5	1,442.0	394.2	56.9
New England states[e]	12,675.6	1,668.1	6,087.9	—	335.0	4,573.9	10.7
Mid-Atlantic states[f]	22,427.7	11,039.2	8,167.1	56.8	1,025.9	2,113.0	25.7
Atlantic states[g]	39,419.2	19,531.5	421.0	1,748.2	3,918.1	13,800.4	—
Gulf states[h]	41,532.6	22,437.5	45.1	2,290.8	4,480.5	12,171.0	107.7
Total	146,659.2	63,717.8	27,793.6	4,108.3	17,108.6	37,757.6	281.7

Source: Eastwide Forest Inventory Database (Hansen et al. 1992).

[a]Includes oak-hickory and oak-pine forest-type groups.
[b]Includes swamp chestnut oak–cherrybark oak, sweetgum–Nuttall oak–willow oak, and overcup oak–water hickory forest types.
[c]Michigan, Minnesota, and Wisconsin.
[d]Illinois, Indiana, Iowa, Kansas, Missouri, Nebraska, North Dakota, and South Dakota.
[e]Connecticut, Massachusetts, Maine, New Hampshire, Rhode Island, and Vermont.
[f]Delaware, Maryland, New Jersey, New York, Ohio, Pennsylvania, and West Virginia.
[g]Florida, Georgia, Kentucky, North Carolina, South Carolina, and Virginia.
[h]Alabama, Arkansas, Louisiana, Mississippi, east Oklahoma, Tennessee, and east Texas.

tomland hardwoods) are regarded as perhaps the most productive fish and wildlife habitats in North America (Barrett 1995). Oaks are common members of moist-site forest types, which are a subset of these lowland forests and contain some of the most valuable timber found in the southern United States. The eastern coniferous forest-type groups (26% of timberland) often contain significant overstory and understory oak components.

As defined by the FIA, oak forest-type groups include 9 detailed types within the oak-hickory group, 8 oak-pine types, and 3 lowland oak types (see Table 2.2). The oak-hickory and oak-pine groups are distinguished by the amount of oak stocking relative to other species. The oak-hickory group includes stands where 50% or more of the stocking is contributed by oak or oak-dominated stands. For the oak-pine group, stocking of oaks and other deciduous species is from 25% to 50%. The information in Table 2.2 is presented by stand-size class to provide insight into the distribution of oak forest types by age or stage of stand development. There has been concern that existing oak forests are aging and that their distribution is skewed toward older stands with a relative shortage of young stands (Abrams and Nowacki 1992, Healy et al. 1997, and Lorimer 1993). The FIA stand-size variable provides a coarse surrogate for stage of development, that is, seedling-sapling, poletimber, and sawtimber; but it is somewhat limited for discussion of stand age, because it is based solely on tree diameter. The overall distribution of oak timberland by stand-size class is 25% seedling-sapling, 27% poletimber, and 48% sawtimber.

Roughly three-fourths of the oak timberland in the East is classified as oak-hickory. The distribution of oak-hickory by stand-size class is 22% seedling-sapling, 28% poletimber, and 50% sawtimber. The most prevalent type within this group is the white oak–red oak–hickory type, which accounts for about one-third of the oak-hickory timberland. This type is particularly common in the Appalachian and Ozark Mountains. One-fourth of the timberland in this type is in seedling-sapling stands. Oak-hickory forest types with below average amounts of young seedling-sapling stands include chestnut oak (7%), northern red oak (9%), yellow-poplar–white oak–northern red oak (10%), and white oak (12%). Forest types with above average proportions of seedling-sapling stands are southern scrub oak (69%) and sweetgum–yellow-poplar–oak (41%). The sweetgum–yellow-poplar–oak type is often a transition type that temporarily occupies southern pine timberland following harvest of merchantable overstory trees.

To examine the distribution of oak-dominated timberland spatially,

Table 2.2
Area of oak timberland by forest type and stand-size class, eastern United States (in thousands of hectares)

Forest type	Total	Seedling-sapling	Poletimber	Sawtimber	Untyped
Upland oak					
Oak-hickory					
Post oak–black oak–bear oak	2,284.9	64.8	896.2	922.6	1.3
Chestnut oak	1,870.9	127.1	590.2	1,153.6	—
White oak–red oak–hickory	17,479.5	4,454.5	5,137.3	7,877.3	10.4
White oak	2,311.2	274.1	690.6	1,345.0	1.5
Northern red oak	950.5	84.7	247.2	618.6	—
Yellow-poplar–white oak–northern red oak	3,095.0	323.1	684.2	2,083.8	3.9
Southern scrub oak	725.4	498.2	135.4	79.8	12.6
Sweetgum–yellow-poplar–oak	2,059.3	836.9	426.7	775.7	20.0
Other oak-hickory[a]	19,269.5	4,069.5	5,216.5	9,947.4	36.1
Total oak-hickory	50,046.2	11,132.9	14,024.3	24,803.8	85.2
Oak-pine					
White pine–northern red oak–white oak	945.9	82.6	297.6	565.7	—
Eastern red cedar–hardwood	1,039.1	410.1	400.0	228.3	1.5
Longleaf pine–scrub oak	574.1	334.6	106.5	131.0	2.0
Shortleaf pine–oak	2,097.2	379.9	726.0	991.3	—
Virginia pine–southern red oak	1,054.3	242.7	338.2	473.4	25.1
Loblolly pine–hardwood	6,533.0	2,882.8	1,151.7	2,473.4	25.1
Slash pine–hardwood	754.1	331.5	157.6	257.7	7.3
Other oak-pine[a]	673.1	187.2	207.4	277.3	1.2
Total oak-pine	13,671.6	4,851.4	3,385.0	5,398.1	37.1
Total upland oak	63,717.8	15,984.3	17,409.3	30,201.9	122.3
Lowland oak					
Swamp chestnut oak–cherrybark oak	181.7	16.0	19.5	145.1	1.1
Sweetgum–Nuttall oak–willow oak	3,480.3	671.7	563.5	2,215.6	29.5
Overcup oak–water hickory	446.1	38.7	58.6	347.9	0.9
Total lowland oak	4,108.1	726.4	641.6	2,708.6	31.5
Total all oaks	67,825.9	16,710.7	18,050.9	32,910.5	153.8

Source: Eastwide Forest Inventory Database (Hansen et al. 1992).

[a] The "other oak-hickory" and "other oak-pine" categories include timberland that is not classified into the specific oak forest types. All of the oak-hickory and oak-pine timberland in the lake states, Illinois, and Indiana is included in these categories.

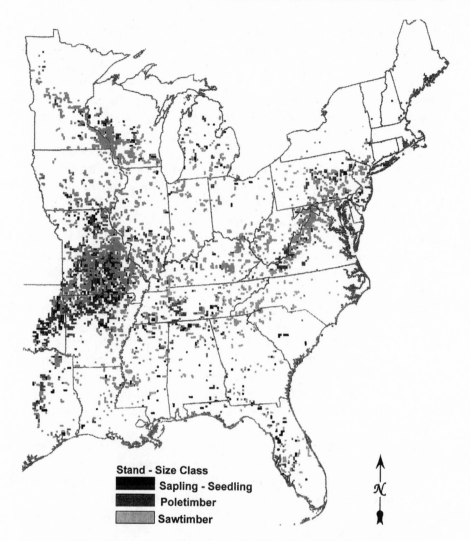

FIGURE 2.3.   Distribution of timberland with at least 50% of total basal area in oak species by stand-size class, eastern United States. (Available data excludes Connecticut, Massachusetts, and Rhode Island.)

we processed the stand-size class variable for FIA sample locations where oak composed at least 50% of total basal area per acre (Figure 2.3). It is immediately apparent that young successional stands are rare and that older sawtimber-size stands predominate across much of oak's native range in the East. The areas with the most interspersion of successional stages are in Missouri and Arkansas. The current distribution of oak-dominated timberland by successional stage arose from a complex mix of factors. Abrams and Nowacki (1992) described repeated logging and burning as factors in the development of mixed-oak forests in Pennsylvania following European settlement. A subsequent decline in heavy logging and elimination of wildfire has accelerated canopy dominance of red maple (*Acer rubrum* L.), sugar maple (*A. saccharum* Marsh.), and black cherry (*P. serotina* Ehrh.). In some areas, the problem of securing adequate oak regeneration is further exacerbated by heavy browsing by white-tailed deer (*Odocoileus virginianus* L.) (Marquis and Brenneman 1981). Oak regeneration is particularly problematic on mesic sites where more-tolerant and faster-growing species outcompete young oak trees (Hodges and Gardiner 1993). A study of heavily disturbed mixed-oak stands in Pennsylvania found that 92% of the stands studied were adequately stocked with woody species, but only 16% had adequate stocking of oak (McWilliams et al. 1995).

The oak-pine forest-type group composes 20% of the timberland classified as oak. Most of the oak-pine timberland (81%) consists of oak growing in combination with southern pines, such as the shortleaf pine–oak, loblolly pine–hardwood, and Virginia pine–southern red oak types. These types contain a high proportion of young successional stands made up of small pine, oak, and other deciduous species. These young stands commonly convert to pine stands as pine species outgrow their competitors.

Although it occupies a comparatively small portion of the oak-pine timberland (0.9 billion acres), the white pine–northern red oak–white oak type is among the most economically important oak-pine types, particularly in New England. Currently, 60% of the timberland in this type is sawtimber size and only 9% is seedling-sapling size. New sources of white pine–northern red oak–white oak forests are rare, and the type is decreasing in the more populated areas of New England as forest land is converted to other land uses.

The pattern of ownership for oak timberland has important implications for public policy and other conservation initiatives. In the East, roughly 9 out of 10 hectares of oak timberland are controlled by private owners (Table 2.3). Nearly all of the privately owned oak timberland is held by the "other private" owner group. The other private owners are a diverse group, ranging from individuals owning small woodlots for a variety of reasons to corporations (other than the forest industry) managing large tracts for financial gain. Forest industry owners are defined

Table 2.3
Area of oak timberland by region and ownership group, eastern
United States (in thousands of hectares)

Region	Total	National forest	Other public	Forest industry	Other private
Lake states	2,441.6	125.2	442.7	35.3	1,838.4
Central states	6,612.3	574.7	336.1	93.0	5,608.5
New England states	1,668.0	8.2	228.0	30.7	1,401.1
Mid-Atlantic states	11,096.2	267.9	1,184.3	454.2	9,189.8
Atlantic states	21,279.8	1,456.8	818.3	1,593.7	17,411.0
Gulf states	24,728.0	1,335.0	1,170.4	3,917.2	18,305.4
Total	67,825.9	3,767.8	4,179.8	6,124.1	53,754.2

Source: Eastwide Forest Inventory Database (Hansen et al. 1992).

as individuals or corporations that operate a primary wood-using plant. The forest industry owns 9% of the oak timberland. Public ownership accounts for only 12% of the oak timberland. Publicly held oak timberland is split between National Forest (52% of the public ownership) and other public owners (48%).

Ranking the FIA sample using numbers of trees highlights overall dominance traits and the role of the oaks in relation to other eastern tree species. Deciduous species dominate eastern forests, accounting for roughly 8 of 10 stems (Table 2.4). Within the deciduous component, oak species make up 14%, or 29.3 billion stems. The FIA sample includes 25 oak species, including 2 commonly accepted varieties. Red oak species (subgenus *Erythrobalanus*) are somewhat more abundant than white oak species (subgenus *Leucobalanus*). The red and white oak groups contribute 60% and 40% of the oak trees, respectively. White oak is the most common oak in eastern forests, with 20% of the oak stems, followed by water oak (*Q. nigra* L.) with 15%, northern red oak and post oak (*Q. stellata* Wangenh.), 10% each, and black oak (*Q. velutina* Lam.), 9%. These 5 oak species make up nearly two-thirds of the total oak component of eastern forests. Other important oak species (5% or more) include southern red oak (*Q. falcata* Michx.), laurel oak (*Q. laurifolia* Michx.), and chestnut oak (*Q. prinus* L.).

The importance of oaks in eastern forests is underscored by their contribution to total forest volume. Oak volume totals 3.1 billion cubic meters, or 23% of the eastern forest resource (see Table 2.5). The regional breakdown of oak inventory percentage is as follows: lake states, 7; cen-

Table 2.4
Number of trees (in millions) measuring at least 2.5 cm in diameter, by FIA species group and detailed oak species, eastern United States

Coniferous		
Loblolly–shortleaf pine	*Pinus taeda* L.-*Pinus echinata* Mill.	20,322.8
Spruce-fir	*Picea* spp. A. Dietr.-*Abies balsamea* (L.) Mill.	13,346.7
Other coniferous[a]		21,786.5
Total coniferous		55,456.0
Deciduous		
Tupelo-blackgum	*Nyssa* spp. L.	48,838.7
Red oaks	*Quercus* spp. L.	
Water	*Q. nigra* L.	4,510.2
Northern red	*Q. rubra* L.	2,811.0
Black	*Q. velutina* Lam.	2,641.9
Southern red	*Q. falcata* Michx.	2,115.8
Laurel	*Q. laurifolia* Michx.	1,594.3
Scarlet	*Q. coccinea* Muenchh.	1,068.3
Blackjack	*Q. marilandica* Muenchh.	911.7
Willow	*Q. phellos* L.	745.6
Cherrybark	*Q. falcata* var. *pagodafolia* Ell.	335.2
Live	*Q. virginiana* Mill.	305.1
Shingle	*Q. imbricaria* Michx.	127.0
Northern pin	*Q. ellipsoidalis* E. J. Hill	109.7
Pin	*Q. palustris* Muenchh.	86.6
Shumard	*Q. shumardii* Buckl.	68.9
Nuttall	*Q. nuttallii* Palmer	51.4
Total red oak		17,482.7
White oaks	*Quercus* spp. L.	
White	*Q. alba* L.	5,707.9
Post	*Q. stellata* Wangenh.	3,063.3
Chestnut	*Q. prinus* L.	1,949.3
Bur	*Q. macrocarpa* Michx.	384.4
Chinkapin	*Q. muehlenbergii* Engelm.	268.1
Overcup	*Q. lyrata* Walt.	188.8
Swamp chestnut	*Q. michauxii* Nutt.	159.8
Swamp white	*Q. bicolor* Willd.	58.1
Durand	*Q. durandii* Buckl.	4.8
Delta post	*Q. stellata* var. *paludosa* Sarg.	1.9
Total white oak		11,786.4
Total oaks		29,269.1
Soft maple	*Acer rubrum* L. (and small numbers of *Acer* spp. L.)	21,940.7
Sweetgum	*Liquidambar styraciflua* L.	12,571.9
Hard maple	*Acer saccharum* Marsh. (and small numbers of *Acer* spp. L.)	9,119.3

(*continued*)

Table 2.4 (*Continued*)

Hickory	*Carya* spp. Nutt.	8,171.8
Cottonwood-aspen	*Populus* spp. L	7,533.5
Ash	*Fraxinus* spp. L.	7,402.8
Other deciduous[a]		68,394.0
Total deciduous		213,241.8
Total all species		268,697.8

[a]For brevity, FIA species groups constituting less than 3% of the total are grouped into the "other" category.

tral states, 10; New England states, 4; Mid-Atlantic states, 19; Atlantic states, 32; and Gulf states, 28. Given the significance of oak, a review of overall health and sustainability will highlight expected future trends. The FIA statistics include estimates of change in the inventory due to mortality, net growth, and removals.

Mortality estimates are the primary indicator of health from the regional forest inventories and are expressed on an annual basis to provide information that can be compared among species. The mortality estimates in Table 2.5 are indexed as a percent of volume of the species. The overall rate of oak mortality is 0.75%. The rates for red and white oak are 0.95% and 0.49%, respectively. To put these into perspective, the highest mortality rates occur for species with well-established stressors. For example, a study of mortality in the mid-Atlantic region found that American elm (*Ulmus americana* L.) had the highest mortality rate, at 3.28, resulting from a wilt fungus (*Ceratocystis ulmi* (Buisum.) C. Moreau) that is the agent of Dutch elm disease (McWilliams et al. 1997). Balsam fir (*Abies balsamea* [L.] Mill.) had a rate of 2.64 due to outbreaks of spruce budworm (*Christoneura fumifera* (Clemens)). The highest oak mortality rates are for scarlet oak (*Q. coccinea* Muench.) at 1.52, laurel oak (*Q. laurifolia* Michx.) at 1.51, and shingle oak (*Q. imbricaria* Michx.) at 1.49. It should be noted that these regional estimates mask the impact of some important local mortality events. For example, the most recent inventory of Pennsylvania indicated a high mortality among chestnut oak (*Q. prinus* L.) and white oak (*Q. alba* L.) that resulted from gypsy moth infestations of the 1980s (Alerich 1993).

Oak sustainability can be gauged by the relationship between average annual net growth (gross growth minus mortality) and removals. Removals volume includes removals due to loss of forest land, and so the growth-to-removals ratio is a useful indicator of the ability of the oak in-

Table 2.5
Inventory volume of oak growing stock, annual averages 1983–1997
(in millions of cubic meters), eastern United States

Species	Volume	Mortality	Mortality percentage	Net growth before removals	Removals	Growth-to-removals ratio
Red oaks						
Water	196.0	2.1	1.07	9.0	7.2	1.25
Northern red	550.1	3.6	0.65	13.4	8.1	1.65
Black	335.5	3.4	1.01	8.0	5.5	1.45
Southern red	165.0	1.4	0.85	6.7	6.4	1.05
Laurel	92.6	1.4	1.51	3.4	2.9	1.17
Scarlet	177.2	2.7	1.52	4.6	3.3	1.39
Blackjack	3.0	0.0	—	0.1	0.0	—
Willow	75.3	0.9	1.20	2.8	2.6	1.08
Cherrybark	61.8	0.4	0.65	2.8	2.2	1.27
Live	14.6	0.0	—	0.3	0.1	3.00
Shingle	6.7	0.1	1.49	0.3	0.0	—
Northern pin	10.8	0.1	0.93	0.3	0.1	3.00
Pin	18.8	0.2	1.06	0.5	0.3	1.67
Shumard	10.3	0.1	0.97	0.4	0.3	1.33
Nuttall	15.7	0.1	0.64	0.6	0.3	2.00
Total red oak	1,733.5	16.4	0.95	53.0	39.3	1.35
White oaks						
White	716.8	2.6	0.36	19.8	12.8	1.55
Post	148.1	1.1	0.74	4.0	4.0	1.00
Chestnut	335.4	2.1	0.63	6.9	3.4	2.03
Bur	40.3	0.1	0.25	0.8	0.3	2.67
Chinkapin	16.7	0.1	0.60	0.4	0.2	2.00
Overcup	34.4	0.3	0.87	1.0	0.6	1.67
Swamp chestnut	21.2	0.2	0.94	0.6	0.7	0.86
Swamp white	11.3	0.0	—	0.3	0.1	3.00
Durand	0.4	0.0	—	0.0	0.0	—
Delta post	0.4	0.0	—	0.0	0.0	—
Total white oak	1,325.1	6.5	0.49	33.82	2.2	1.52
Total oak	3,058.6	23.0	0.75	86.9	61.5	1.41

Source: Eastwide Forest Inventory Database (Hansen et al. 1992).

ventory to expand. A ratio of 1.00 indicates a balance between growth and removals. Ratios less than 1.00 indicate overcutting and ratios greater than 1.00 indicate inventory expansion. The ratio for all oak species combined is 1.41. The ratios for red and white oak are 1.35 and 1.52, respectively. The tightest relationships are for swamp chestnut oak (*Q. michauxii* Nutt.) at 0.86, post oak (*Q. stellata* Wangenh.) at 1.00, southern red oak (*Q. falcata* Michx.) at 1.05, and willow oak (*Q. phellos*

L.) at 1.08. (The low ratio for swamp chestnut oak may be discounted, because of a sparse sample.) Many other oak species have relatively tight ratios and should be watched closely because of their economic importance and the likelihood that future increases in demand will further tighten the ratios. This is particularly important because, as the oak resource matures, increases in net growth will come more from accretion on existing trees and less from ingrowth of new oak stems.

In summary, the eastern oak resource is immense and colossally important to wildlife populations and the timber economy. There have been no consequential decreases in acreage of oak forest types in recent years, although a scarcity of sources for new oak forest and pressure from competing land uses will limit future expansion. Also, it should be remembered that clearing of lowland oak forest for agriculture was common in the not too distant past, making it important to keep policies and programs that limit these activities in place. The finding that the distribution of oak timberland is skewed toward older, sawtimber-size stands for several oak types should be considered along with current knowledge that successional trends tend to move oak stands toward dominance by more shade intolerant species and that oak regeneration is lacking in some areas.

Current estimates of mortality rates and growth-to-removals relationships, while favorable, should be tempered with the understanding that pests have had considerable impacts, some local areas have undergone significant oak decline, and that increased demand for oak timber and competing land uses will likely increase future removals of oak. Therefore, it is not unreasonable to characterize the oak biome as being at the brink of a time when monitoring will be critical for ensuring that management, policies, and other programs will foster a future for oak in the East and will mollify the factors that led Healy et al. (1997) to note that oak may well be reduced to a minor forest component over the next century.

THE INTERIOR WEST

About 25% of the interior West is classified as forest; however, this varies widely throughout the region. For example, Nevada is only about 11% forested, but northern Idaho and northwestern Montana are roughly 78% forested. Oak-dominated woodland occupies about 1.2 million hectares, or about 2% of the total forested area. Oak-dominated woodland includes stands in which oak species constitute the majority of stocking and usually attain tree size. About 46% of oak-dominated woodland is privately owned. Oaks also present as shrubs in many other forest and woodland types. Oak-dominated woodland, savannas, and shrublands occupy as many as 9.8 million hectares in the interior West and can be divided into three very general groups: Gambel oak, interior chaparral, and sand shinnery. The most widely occurring of these groups is dominated by Gambel oak, a deciduous white oak. The Gambel oak group encompasses varieties whose vegetative habits range from a shrub about 1 meter tall found in dense patches, to a small tree or tall shrub interspersed with other species, to a tree in widely spaced stands. Clary and Tiedemann (1992) estimated that Gambel oak is an important component on about 3.8 million hectares. The distribution of Gambel oak is almost entirely within Utah, Colorado, Arizona, and New Mexico, at elevations ranging from 990 to 3,110 meters (Harper et al. 1985).

According to Clary and Tiedemann, while there is little information on which to project long-term changes in the density or distribution of Gambel oak, there is some evidence that Gambel oak clones (expanding as rhizomes) may have increased in number and size since the turn of this century. Gambel oak is very important as a source of food and winter cover for wildlife. In recent years, suburban home development has been encroaching into some of these areas (Barrett 1995). In some areas, fire, herbicides, and mechanical removal have been used to offset the dominance of Gambel oak by expanding forage and accessibility for livestock and large game. Because Gambel oak is highly valued for fuelwood, fuelwood sales are combined with oak stand management in some areas. Management aimed specifically at Gambel oak is not common.

The second group, interior chaparral, is a combination of two types described by the Society for Range Management (Shiflet et al. 1994): (1) Arizona interior chaparral and an evergreen oak zone that occurs in the transition between the oak-juniper and the mahogany-oak associations, and (2) the Arizona oak woodlands described by McPherson (1992). This grouping consists primarily of 10 shrub-form evergreen oaks that are found with a variety of other trees and shrubs: Arizona white oak, (*Q. arizonica* Sarg.), Emory oak (*Q. emoryi* Torr.), gray oak (*Q. grisea* Liebm.), silverleaf oak (*Q. hypoleucoides* A. Camus), chinkapin oak (*Q. muehlenbergii* Engelm.), Mexican blue oak (*Q. oblongifolia* Torr.), netleaf oak (*Q. rugosa* Nee), Toumey oak (*Q. toumeyi* Sarg.), turbinella oak (*Q. turbinella* Greene), and wavyleaf oak (*Q. x. pauciloba* Rhyb.).

Estimates of the area inhabited by interior chaparral range from 1.6

to 3.2 million hectares. This type usually is found on south-facing moderate to steep slopes in shallow, often rocky soils; but it can extend to moderately deep to deep soils on mesas and in valleys. The chaparral oak merges with pinyon-juniper and ponderosa pine (*Pinus ponderosa* Dougl. Ex Laws.) at the upper limits of its elevation range, and with oak-juniper, desert shrub, and desert grassland at lower elevations. Many stands are dense and nearly impenetrable by livestock and large game. Historical data indicate that stands of the past were mostly open shrub-savanna interspersed with dense stands of grass.

The third group, sand shinnery, consists primarily of shin oak (*Q. harvardii* Rydb.). Sand shinnery is common in New Mexico, western Texas, and western Oklahoma (Peterson and Boyd 1998) and there are scattered populations in southern Utah and Arizona (Welsh et al. 1993). This type is found on sandy soils and is typically only about 1.2 meters in height. Estimates of area vary widely, but an approximate figure is 2.0 to 2.8 million hectares (Peterson and Boyd 1998). Roughly one-third of this area has only a light or scattered oak canopy.

Much of the sand shinnery is on private lands that are used primarily for livestock grazing and hunting. Many consider this type undesirable for agriculture and livestock use, because its leaves are toxic to cattle for several weeks in the spring, and they use fire, chemical, and mechanical means to kill or reduce its cover. Therefore, it is unlikely that its distribution is expanding.

THE PACIFIC COAST

Oaks occupy a diverse forest habitat along the Pacific Coast. These hardwoods are found at low elevations in temperate rain forests of the Northwest and in dry, open forests on rocky slopes in southern California. Oaks are located on both productive commercial "timberland" and less productive "woodland." The majority of oak types are found on private lands. On timberland, 70% of the area occupied by oaks is privately owned; on woodland, 82%.

There are an estimated 23.6 million hectares of unreserved timberland forests outside of wilderness lands in California, Oregon, and Washington and about 5.2 million hectares of woodland. Although the forests on this timberland are dominated by coniferous forest types (84% of the area), 3.1 million hectares are classified as hardwood types (see Table 2.6). Hardwoods of the Pacific Coast include both evergreen and decid-

Table 2.6
Area of unreserved timberland by forest type and state, Pacific Coast,
United States (in thousands of hectares)

	Total	California	Oregon	Washington
Oaks				
California black	357.2	327.7	29.5	—
Canyon live	205.7	157.5	48.2	—
Oregon white	139.5	69.0	65.3	5.2
Oak-madrone	205.9	72.5	128.0	5.4
Other	63.1	63.1	—	—
Total oak types	971.4	689.8	271.0	10.6
Other deciduous	2,176.9	552.4	845.0	779.5
Coniferous	19,756.2	5,275.9	8,398.7	6,081.6
Nonstocked	677.4	399.5	128.6	149.3
Total all types	23,581.9	6,917.6	9,643.3	7,021.0

Source: Adapted from Woudenberg and Farrenkopf 1995.

uous species. Oak forest types are found on 1.0 million hectares of timberland, just 4% of the total timberland area. The picture is different on
woodland, where more than 50% of the land area is occupied by oak
types, primarily in the oak woodlands of California (see Table 2.7).

On timberland, three species of oak reach densities sufficient to form

Table 2.7
Area of unreserved woodland by forest type and state, Pacific Coast,
United States (in thousands of hectares)

Species	Total[a]	California	Oregon[a]	Washington[a]
Oaks				
Blue	—	1,209.7	—	—
Interior live	—	407.2	—	—
Coast live	—	355.5	—	—
Canyon live	—	248.2	—	—
Oregon white	—	153.0	—	—
California black	—	66.4	—	—
Valley	—	52.4	—	—
Engelmann	—	25.5	—	—
Total oak types	2,702.9	2,517.9	160.7	24.3
Other deciduous	149.0	141.0	0.8	7.2
Coniferous	2,396.9	933.5	1,133.8	329.6
Total all types	5,248.8	3,592.4	1,295.3	361.1

Note: Columns may not sum to totals due to rounding.

[a]Details on specific oak types are not available for Oregon and Washington because trees were
not sampled on unreserved woodlands.

pure forest types: California black oak (*Q. kelloggii* Newb.), Oregon white oak (*Q. garryana* Dougl. Ex Hook.), and canyon live oak (*Q. chrysolepis* Liebm.). Most oak types on timberland are found in California (about 0.7 million hectares or 71% compared to only 1% in Washington). About 80% of the area of oak types has developed into large, sawtimber-size stands. Oaks are components of at least six conifer-forest types and five non-oak hardwood types. California black oak is one of the more abundant oak types in the Pacific Coast states (nearly 0.4 million hectares) and is the most common hardwood species harvested for lumber (Bolsinger 1988). Of the three oak species mentioned, canyon live oak

Table 2.8
Number of trees (in millions) at least 2.5 cm in diameter, by species, Pacific Coast, United States

Coniferous		
Douglas-fir	*Pseudotsuga menziesii* (Mirb.) Franco	5,138.2
Pine	*Pinus* spp. L.	3,697.7
True firs	*Abies* spp. Mill.	3,414.8
Western hemlock	*Tsuga heterophylla* (Raf.) Sarg.	2,238.7
Cedars[a]		1,333.6
Other coniferous[b]		868.7
Total coniferous		16,691.7
Deciduous		
Red oaks	*Quercus* spp. L.	
Canyon live oak	*Q. chrysolepis* Liebm.	518.6
California black oak	*Q. kelloggii* Newb.	385.2
Interior live oak	*Q. wislizeni* A. DC.	46.7
Coast live oak	*Q. agrifolia* Nee	15.2
Total red oaks		965.7
Oregon white oak	*Q. garryana* Dougl. ex Hook.	147.2
Other oaks[b]	*Q.* spp. L.	8.2
Total, all oak species		1,121.1
Tan oak	*Lithocarpus densiflorus* (Hook. & Arn.) Rehd.	1,183.3
Red alder	*Alnus rubra* Bong.	943.0
Pacific madrone	*Arbutus menziesii* Pursh	416.9
Bigleaf maple	*Acer macrophyllum* Pursh	278.5
California laurel	*Umbellularia californica* (Hook. & Arn.) Nutt.	118.6
Other deciduous[b]		447.4
Total deciduous		4,508.8
Total all species		21,200.5

[a]Cedars include incense cedar (*Libocedrus decurrens* Torr.), Port-Orford-cedar (*Chamaecyparis lawsoniana* (A. Murr.) Parl.), and western redcedar (*Thuja plicata* Donn ex D. Don).
[b]For brevity in the table, species with less than 3 percent are grouped into the "other" categories.

accounts for most of the trees across all three states One reason may be that this species is more tolerant of shade and drought than its associates (Burns and Honkala 1990) and canyon live oak may have greater survival ability as forest canopies close. Both California black oak and canyon live oak grow from Lane County in Oregon down to southern California. Oregon white oak extends from Vancouver Island and British Columbia south through the valleys and foothills of Washington, Oregon, and California (Burns and Honkala 1990). Oregon white oak is the only native oak that grows east of the Cascades in Oregon and throughout the state of Washington. (See Table 2.8.)

On the 5.2 million hectares of woodland, 8 oak forest types have been found in FIA inventories: blue oak (*Q. douglasii* Hook. & Arn.), interior live oak (*Q. wislizenii* A. DC.), coast live oak (*Q. agrifolia* Nee), canyon live oak, Oregon white oak, California black oak, valley oak (*Q. lobata* Nee), and Engelmann oak (*Q. engelmannii* Greene). These oak types occupy 2.7 million hectares of land, primarily in scattered open stands in the oak woodlands of California. Valley and blue oaks are the only oaks endemic to California (Griffin and Critchfield 1972). Coast and interior live oaks are limited to California and northern Baja (Little 1971). The most abundant oak forest types on woodland in the Pacific Coast states are

Table 2.9
Percent of growing-stock volume for deciduous species of the oak woodlands of California

Oaks		
Blue oak	*Quercus douglasii* Hook. & Arn.	27
Coast live oak	*Q. agrifolia* Nee	18
Canyon live oak	*Q. chrysolepis* Liebm.	15
Interior live oak	*Q. wislizeni* A. DC.	9
California black oak	*Q. kelloggii* Newb.	7
Oregon white oak	*Q. garryana* Dougl. ex Hook.	7
Valley oak	*Q. lobata* Nee	4
Engelmann oak	*Q. engelmannii* Greene	1
Total oak		88
Other deciduous		
California laurel	*Umbellularia californica* (Hook. & Arn.) Nutt.	3
Pacific madrone	*Arbutus menziesii* Pursh	3
Eucalyptus	*Eucalyptus* spp.	2
Bigleaf maple	*Acer macrophyllum* Pursh	1
Black cottonwood	*Populus trichocarpa* Torr. & Gray	1
Other deciduous		2
Total other deciduous		12

blue oak (1.2 million hectares) and interior live oak (0.4 million hectares). Ranking species by volume of growing stock shows the relative importance of blue oak (27% of the total), coast live oak (18%), and interior live oak (15%) (see Table 2.9).

The oak woodlands are typically not harvested for timber but are valued for other uses. In early years, hardwoods were considered weeds, and considerable areas were cleared for range improvement and firewood. In recent times, these same ecosystems have become critically important for wildlife habitat, watershed protection, and recreation, in addition to grazing. Fortunately, most species are regenerating well, except for coast live oak, blue oak, and valley oak. Current concern centers on suburban development that competes with nonconsumptive forest-land uses (Barrett 1995).

Chapter 3

The Postglacial History of Oak Forests in Eastern North America

MARC D. ABRAMS

Temperate hardwood forests dominate much of the eastern United States, from central Maine to northern Minnesota and eastern Texas to north central Florida (Barnes 1991, Abrams and Orwig 1994). Within the eastern deciduous biome, oak represents one of the dominant forest types (Abrams 1996). Approximately 30 oak species occur in the eastern United States, of which white oak (*Quercus alba*), red oak (*Q. rubra*), black oak (*Q. velutina*), scarlet oak (*Q. coccinea*), post oak (*Q. stellata*), and chestnut oak (*Q. prinus*) are among the most dominant. Oaks are an important component of many forest associations in the eastern United States, including northern hardwood-conifer, maple-beech-basswood, mixed mesophytic, oak-hickory, oak-pine, and southern evergreen (Braun 1950, Abrams and Orwig 1994).

Oak species have a long and important history in the forests of the eastern United States and southern Ontario (Abrams 1992). In the recent past, oak forests changed dramatically in response to profound changes in land-use history. Following European settlement, the dominance of oak species increased in many parts of the eastern United States. Contemporary forests have exhibited various developmental pathways. These pathways include an expansion of oak in the tallgrass prairie region in response to fire suppression, an increase within northern hardwood-conifer forests after the original cutting and burning, and heightened dominance in the original mixed-oak forests following cutting and burning associated with the charcoal-fueled iron industry, lumbering, and the chestnut blight. It has been argued that a variety of disturbance factors, including changes in the frequency and intensity of fire, Native American and Euro-American land-use practices, and insect and disease outbreaks, have played a major role in the historical development and long-term dynamics of eastern oak forests (Russell 1980, Lorimer 1985, Crow 1988, Abrams 1992, Clark and Royall 1995). However, a decline in oak canopy recruitment has occurred during the middle and later twentieth century throughout most of the eastern forest biome, which has been attributed to a suite of factors such as fire suppression, deer and small mammal browsing and acorn predation, insects and diseases, and competition from later successional tree species (Abrams 1992, 1998).

The expansion of oak forests following European settlement and the recent decline in oak regeneration and recruitment in eastern North America has been extensively reviewed by scientists during the last decade (see reviews by Lorimer 1985, 1989, Crow 1988, Abrams 1992, 1996, 1998, Whitney 1994). The purpose of this chapter is to synthesize the literature on paleoecology, Native American land-use history, and witness trees mapped at the time of European settlement and thereby to describe the distribution and dynamics of oak forests during the Holocene (10,000 years B.P.). I will attempt to elucidate the environmental conditions and land-use practices associated with oak forest development during the 10,000 years before European settlement. By studying the natural and cultural environment under which oak forests developed, we may better understand the ecological requirements needed to manage and conserve our existing oak forests.

PALEOECOLOGICAL HISTORY OF OAK FORESTS

Pollen and charcoal preserved in undisturbed sediment in lakes and bogs can be concentrated, dated, and analyzed; the resulting data provide an invaluable source of information on the long-term dynamics of vegetational assemblages. Major spatial and temporal patterns and changes in the prevailing vegetation (inferred from pollen composition) can be used to infer climatic changes and disturbance history, including fire history (from sediment charcoal abundance), Native American activity, and insect and disease outbreaks. Eighteen thousand years ago, prior to the Holocene, the eastern United States was dominated by pine forests in the southeastern and mid-Atlantic regions, and *Picea* (spruce) and *Cyperaceae* (sedge) dominated communities in the mid-Atlantic, midwestern, and central regions; glaciers covered most of the northeastern and Great Lakes regions (Webb 1988). During this time, oak occurred in the South and Southeast, but it represented only 1–5% of the pollen percentages, compared with 20–40% for pine. By the beginning of the Holocene, oak abundances started to increase dramatically and this genus apparently dominated most forests in the eastern United States, excluding the northern tier which was primarily pine, spruce, and birch (*Betula*) (Webb 1988).

Using a network of sites in central Appalachia and the New Jersey section of the Coastal Plain, Watts (1979) reported on the expansion of pine that occurred 10,000 to 6,000 years B.P. Further expansion of pine was attributed to wetter climate, the increased formation of bogs and swamps, and the possible replacement of deciduous trees (e.g., birch and alder). On several of his study sites (e.g., Szabo Pond, N.J., and Potts Mountain Pond, Va.), a change from pine to oak domination occurred, and this coincided with an increase in charcoal abundance. The increase in charcoal associated with the change from pine to oak domination is somewhat counterintuitive, given the higher flammability of pine foliage, but it may be explained by the change from a swamp conifer to drier upland oak forest types as the climate became warmer and drier. Upland oaks, as well as pines, have a suite of physiological and morphological adaptations for dealing with drought conditions (Abrams 1990, 1996).

A number of paleoecological studies from the eastern United States provide long-term data on oak forest distribution and dynamics in specific regions. For example, at Mirror Lake, New Hampshire, in the northern hardwood-conifer forest region, peak domination of white pine and oak occurred in the early Holocene (9,000–7,000 years B.P.) when the climate was warmer and drier than at present; charcoal was most abundant 8,000–7,000 years B.P. (Davis 1985). When the climate cooled thereafter, hemlock (*Tsuga*), beech (*Fagus*), and birch (*Betula*) increased and charcoal abundances decreased. Charcoal levels at Mirror Lake are low relative to those at other sites in the northeastern and north central United States, and Clark (1997) has suggested that local fire may not be an important ecological factor. Nonetheless, Clark and Royall (1995) reported on the transformation of northern hardwood (beech-maple) forests to white pine (*P. strobus*)–oak forests in response to Iroquois cultivation and burning during the 1400s at Crawford Lake in southern Ontario. Indeed, local forest fires in the northeast United States were historically associated with Indian settlements and land-use practices (Russell 1983, Patterson and Sassaman 1988, Fuller et al. 1998). The natural fire return interval (not related to Indian activity) for northern hardwood-conifer forests in the Northeast may be on the order of 800 to 1,200 years, due to the cool and moist conditions, despite the fact that lightning is a potential ignition source for the region (Henry and Swan 1974, Lorimer 1977, Fahey and Reiners 1981, Foster and Zebryk 1993).

Analysis of sediment pollen and charcoal evidence from the last 2,000 years for northern hardwood-conifer forests across a longitudinal gradient in the northeastern and north central United States and southern Ontario reveal regional patterns of oak abundance and fire frequency (Clark and Royall 1996). The most western forests in Minnesota and Wisconsin were dominated by oak, birch, and pine, and evidence of past fires was most abundant in these forests. Further east, the sites were increasingly dominated by beech, hemlock, yellow birch, and spruce, and amounts of sediment charcoal decreased. One small catchment area in western New York, named Devil's Bathtub, has been dominated by oak (surrounded by beech-hemlock forest) since 8,300 years B.P., but the site lacks the high frequency input of charcoal that is indicative of local fires. Low levels of charcoal led Clark and Royall (1996) to question the necessity of fire for oak recruitment and maintenance at Devil's Bathtub and other sites in the Northeast.

However, Devil's Bathtub is not representative of more expansive oak forests in the region, and a general lack of knowledge about charcoal input from low-intensity surface fires in hardwood forests makes interpretation of the data ambiguous (Abrams and Seischab 1997). Simply stated, charcoal particles exist in a wide range of sizes, each of which has different dispersal distances. Fine particles can travel great distances and therefore may settle in lakes or bogs far from the fire in which they originated. Large charcoal particles, on the other hand, tend to stay on the site where they are formed. Moreover, many paleocharcoal studies use techniques that average sediment over many years and do not detect individual fires (Clark 1997). For these reasons, the presence of small-sized charcoal in sediment profiles may not prove that a fire occurred at a particular site. Nonetheless, many paleoecology papers show an intrasite increase in regional or local charcoal with increases in oak abundance, and newer papers using improved microsectioning techniques or particle-size distribution analysis reveal the occurrence of localized fires on oak sites during the Holocene (e.g., Clark and Royall 1995, Delcourt and Delcourt 1997).

A paleoecological study of the vegetation dynamics and disturbance history of a hemlock forest in north central Massachusetts reported spruce domination of the site in the early Holocene, followed by pine-oak, hemlock–northern hardwoods, and finally hemlock–chestnut–northern hardwoods at the time of European settlement (Foster and Zebryk 1993). Repeated, infrequent disturbances at the site (including fire), coupled with changing climate, led to a decrease in northern hardwoods and an increase in pine, oak, and chestnut. In addition, pollen and charcoal data spanning the past 1,000 years from 11 small lakes in the same region indicate that oak, chestnut, and hickory were abundant at lower (warmer) elevations, whereas hemlock, beech, sugar maple, and yellow birch were common at higher (cooler) elevations (Fuller et al. 1998). Fires were more frequent and/or intense at lower elevations, probably due to the relatively high Native American populations, and maintained a high abundance of oak. Northern hardwood-conifer forests in two small lake basins in southern Ontario were dominated by pine (white pine, red pine, and jack pine), hemlock, beech, birch, and oak during the Holocene (Fuller 1997). Long-term vegetation dynamics at the sites were affected by climate change, soil development, species migration, biotic and abiotic disturbances (including fire), and competition. The ecological factors related to the persistence of low to moderate levels of oak throughout the Holocene were not discussed. One possible explanation is that oak can persist in northern hardwood-conifer forests by occasional gap capture following small-scale disturbance (cf. Henry and Swan 1974).

Pollen records indicate that the precolonial forests in northern New Jersey were dominated by chestnut and oak in the uplands and by alder in the swamps; other major genera in the region were hemlock, birch, and pine (Russell 1980). Following settlement, tree cutting for charcoal promoted the expansion of oak and birch. Pollen data from four lakes in northern New Jersey and southeastern New York indicate that forests were dominated by oak, chestnut (*Castanea*), and hickory (*Carya*) prior to 1900 (Loeb 1989). After land clearing and subsequent field abandonment and the chestnut blight, pollen percentages of oak, ragweed (*Ambrosia*), and grasses (*Gramineae*) increased. After 1900, oak and birch had continued importance. An abundance of oak pollen was also reported for the early European settlement period in coastal forests of southeastern Long Island, New York, coupled with historical and field evidence of fire (Clark 1986). A paleoecological study of the Holocene vegetation, climate and fire history of the Hudson Highlands in southeastern New York revealed that oak dominated forests were invaded by white pine 10,000 years B.P., by hemlock 9,650 years B.P., by beech 8,100 years B.P., by hickory 6,200 years B.P., and by chestnut 3,600 years B.P. (Maenza-

Gmelch 1997). Large oak pollen percentages were associated with continuous charcoal influx, and the author concluded that fire may have played a large role in the development and maintenance of oak forests in the region. However, while oak pollen was abundant and exhibited high constancy throughout the Holocene, the charcoal accumulation rate was highly variable, with peak values having been reached between 5,500 and 4,500 years B.P. Following European settlement, the pollen abundances of oak, pine, and hemlock decreased, probably due to the domestic and industrial use of the species. Charcoal influxes increased during the postsettlement period, from land clearance and industrial wood burning (Maenza-Gmelch 1997). Continuous charcoal input in the sediment during the last 9,000 years was also reported for oak, pine, and birch forests in the Shawangunk Mountains in southeastern New York (Laing 1994). Charcoal input increased with the increase of pitch pine and white pine pollen after 3,500 years B.P., according to that study.

A study of vegetation history for the eastern Highland Rim and adjacent Cumberland Plateau of Tennessee reported that mixed mesophytic forest taxa (e.g., oak, ash, ironwood, hickory, birch, walnut, elm, beech, sugar maple, basswood, and hemlock) were most abundant during the early Holocene (Delcourt 1979). A warming trend in the region from 8,000–5,000 years B.P. was reflected in higher pollen influx values for oak, ash, and hickory and lower pollen influx for the mixed-mesophytic species. Domination by oak continued for the remainder of the Holocene on several of the study sites. A paleoecological investigation in eastern Tennessee revealed that the study area was dominated by oak, pine, and ragweed during the last 1,500 years (Delcourt 1987). Charcoal influx was continuous during the period and has increased in response to growing human populations and impacts in the valley. In the Little Tennessee River Valley, oak, pine, chestnut, and hickory were the most abundant arboreal pollen species during the last 3,000 years (Delcourt et al. 1986). These species also represent most of the wood charcoal remains of hearth fires and habitations from direct human use during that period. Moreover, hickory, oak, and walnut were the most frequent contributors to native mast tree nutshell remains, indicating that these species were an important food source of the Native Americans during the Holocene. A study of a bog in the southern Blue Ridge of North Carolina indicates that oak, chestnut, pine, and birch were the dominant tree taxa during the last 4,000 years (Delcourt and Delcourt 1997). There was constant influx of charcoal to the bog during the period, including small, medium, and large charcoal particles from regional, watershed, and local fires, respectively. Delcourt and Delcourt state that fire suppression since 1930 is leading to a demise in the historically dominant tree species and that periodic fire must be restored to the region to maintain the fire-adapted pine and oak.

The temporal dynamics of boreal forests, oak forests, and grassland have been described in paleoecological studies of the Central Plains and north central regions. In southeastern Minnesota, prairie development began about 7,200 years B.P., but prairie was replaced by oak forests about 5000 years B.P., presumably due to moister climatic conditions (Wright et al. 1963). Pollen diagrams from Iowa and south central Minnesota describe a change in forest communities from spruce-fir and spruce-larch to pine-oak-elm to prairie-oak during the Holocene (Wright 1964, Cushing 1965, Durkee 1971). A study in northeastern Iowa reported that late-glacial spruce forest was replaced by oak and elm in the early Holocene; these forests were replaced with sugar maple–basswood–*Ostrya–Carpinus* forests by the mid-Holocene (Baker et al. 1996). After 5,500 years B.P. the forest was replaced by prairie. Baker et al. state that the vegetational changes were primarily driven by climatic changes, particularly seasonal changes in temperature and precipitation. In addition, the arrival of prairie species coincided with evidence of fire in the region.

In the Big Woods of southern Minnesota, oak woodlands began invading prairie about 5,000 years B.P.; oak woods persisted until about 300 years ago, when elm, hop hornbeam, basswood, and sugar maple became dominant (Grimm 1983). The change from prairie to oak to northern hardwoods presumably required a reduction in fire frequency, caused by increased precipitation and possibly decreased temperature. Sediment cores from south central Wisconsin indicate a change in vegetation from rich mesophytic forest before 6,500 years B.P. to oak savanna between 6,500 and 3,500 years B.P., associated with warmer and drier conditions and the increased abundance of charcoal (Winkler et al. 1986). After 3,500 years B.P., closed oak forests formed and charcoal levels decreased, suggesting a cooler and wetter climate. The upland vegetation surrounding a marsh in northeastern Indiana was dominated by pine, oak, elm, and hickory from 10,000 to 5,700 years B.P. (Singer et al. 1996). After 5,700 years B.P., the closed, mixed-species forest was replaced by oak savanna in response to drier climate and periodic fires. Between 3,000 and 2,000 years B.P., the uplands shifted to more mesic vegetation, although oak remained dominant on the site. A study of the postglacial history of oak savanna on dry and well-drained soils in southern Ontario reported that jack pine forest dominated between 10,000 and 9,000 years B.P., followed by white pine domination until 6,000 years B.P. with relatively low charcoal-to-pollen ratios (Szeicz and MacDonald 1991). Between 6,000 and 4,000 years B.P., white pine was replaced by oak savanna, which remained intact until European settlement in the 1850s. A high charcoal-to-pollen ratio suggests that fire was an important component of the oak savanna environment.

OAK DISTRIBUTION DURING THE EURO-AMERICAN SETTLEMENT PERIOD

During the protracted period of Euro-American settlement of the eastern United States, from the early 1600s to the late 1800s, land surveys were conducted by recording witness (warrant or bearing) trees in both metes and bounds surveys of property corners and General Land Office surveys of section lines. These data are an invaluable source of information about the species composition and distribution of forest types at the time of Euro-American settlement.

Table 3.1 summarizes examples of the published witness tree accounts throughout the eastern forest biome in which oak was an important component. Oak was widely distributed in the presettlement forest, although its degree of dominance varied. The major oak regions were southern New England, the mid-Atlantic region, the Southern Appalachians, southeastern parts of the Coastal Plain, and the Midwest and southern Lake States. Oak occurred in northern hardwood-conifer forests of New England and the Lake States, but it was a minor component compared to beech, sugar maple, birch, hemlock, and white pine. However, Native American land clearing and use of fire were apparently responsible for increasing the local distribution of oak within the northern hardwood-conifer forest, including the formation of oak savanna in northeastern Wisconsin (Dorney and Dorney 1989, Fuller et al. 1998). Temporal and spatial increases in oak abundance in New England are also related to warmer and drier climate and sites (Foster, Motzkin, and Slater 1998). On the southeastern coastal plain, loblolly pine, slash pine, and longleaf pine were much more important than oaks in the original forests. The Central Plains had bur oak, black oak, white oak, post oak, and blackjack oak growing in savannas and gallery forests (Abrams 1986, Nuzzo 1986), but this region was primarily dominated by prairie vegetation.

At the time of European settlement, white oak, black oak, red oak,

Table 3.1
Presettlement oak forest types in eastern United States, by region

Forest type	State	Reference
Northeast		
White oak–black oak–pine	MA	Whitney and Davis 1986
Pine–white oak–chestnut–black oak	MA	Whitney 1994
Oak-pine	MA	Foster, Motzkin, and Slater 1998
White oak–black oak–hickory	NY	Glitzenstein et al. 1990
White pine–white oak–black oak	WI	Nowacki et al. 1990
Mid-Atlantic region		
Beech–sugar maple–basswood–white oak	NY	Seischab 1990
White oak	PA	Abrams and Downs 1990
White oak–white pine–hickory	PA	Nowacki and Abrams 1992
Chestnut oak–white oak–pitch pine	PA	Nowacki and Abrams 1992
Black oak–white oak–chestnut–hickory	PA	Mikan et al. 1994
White oak–chestnut–red maple–white pine	PA	Abrams and Ruffner 1995
White oak–chestnut oak–white pine– Virginia pine	PA	Abrams and Ruffner 1995
White oak–black oak–chestnut	NJ	Russell 1981
Red oak–chestnut	NJ	Ehrenfeld 1982
White oak–black oak–chestnut–hickory	NJ	Ehrenfeld 1982
White oak–chestnut–hickory–pine	WV	Abrams et al. 1995
White oak–pitch pine	WV	Abrams and McCay 1996
White oak–red oak	VA	Orwig and Abrams 1994
Chestnut–red oak	VA	Braun 1950
White oak–red oak	VA	Spurr 1951
White oak–red oak–chestnut oak– hickory–pine	VA	Spurr 1951
Southeast		
White oak–black oak	NC	Braun 1950
Oak-chestnut	NC	Keever 1953
Chestnut–chestnut oak–red oak	NC	Woods and Shanks 1959
Oak-hickory-pine	GA	Nelson 1957
Oak–red maple–sweetgum	GA	Nelson 1957
Longleaf pine–slash pine–turkey oak	FL	Myers 1985
Midwest and Central Plains		
White oak–hickory	OH	Whitney 1994
White oak–black oak–hickory	OH	Whitney 1994
Beech–sugar maple–white oak	OH	Whitney 1994
White oak–black oak–bur oak	IL	Rodgers and Anderson 1979
Post oak–hickory	IL	Fralish et al. 1991
White oak–black oak	IL	Fralish et al. 1991
White oak–white ash–beech	IL	Fralish et al. 1991
White oak–bur oak–hickory	MO	Howell and Kucera 1956
White oak	MO	Pallardy et al. 1988
Post oak–shortleaf pine–juniper	MO	Guyette and Cutter 1991
Blackjack oak–post oak	OK	Rice and Penfound 1959
White oak–red oak–pin oak	TX	Schafale and Harcombe 1983

(continued)

Table 3.1 (*Continued*)

Forest type	State	Reference
North central		
Bur oak–red oak	MN	Grimm 1984
Bur oak savanna	WI	Cottam 1949
White oak–black oak	WI	Dorney and Dorney 1989
White oak–black oak	MI	Dodge and Harman 1985
Red pine–white oak–white pine	MI	Kilburn 1960

Source: Data reconstructed from witness tree data recorded in land surveys during the early European settlement period in various regions of the eastern United States (adapted from Abrams 1992, 1996, 1998).

chestnut oak, chestnut, and hickory were most often recorded as witness trees throughout the East, and among these white oak was usually the most prevalent (see Table 3.1). The abundance of white oak in these records is probably an indication of its concentration on valley floor sites, most favored by settlers for farming. Thus, the valleys are where most of the metes and bounds surveys were conducted. This contrasts with unfavorable ridges and side slopes, where little farming or settlement took place, which were dominated by chestnut oak, red oak, and chestnut; these species appear somewhat less frequently in witness tree records.

CONCLUSIONS

The paleoecological and witness tree literature reviewed here indicates that oak species have had a long history of domination in much of the eastern forest biome. Some regions have had a high constancy of oak domination, while other regions show the dynamic relationship of the oak with other vegetation types during the Holocene. Paleocharcoal studies suggest that the importance of oak was affected by fire and by changes to a warmer and drier climate in most of the eastern United States. Significant levels of charcoal influx occur almost routinely with oak pollen influx throughout the Holocene. However, several studies show a high constancy of oak pollen input for long periods of the Holocene, marked by highly variable levels of sediment charcoal or peaks in charcoal abundance that do not match the peaks in oak pollen input. Moreover, charcoal inputs for oak forests are often much lower than those for the more flammable pine and prairie vegetation. Indeed, closed oak forests were probably maintained by low-intensity surface fire, in contrast to the higher-intensity surface or crown fires associated with prairie, savanna, and certain pine forests (Abrams 1992). Northern hardwood or mixed-mesophytic forests likely had lower fire frequency and intensity than oak forests.

The occurrence of fire and oak during the Holocene was not independent of other biotic and abiotic factors. Clearly, the historic fire frequency in oak forests was linked to climatic changes that resulted in intermediate levels of temperature and moisture. During the Holocene, oak forest replaced the white pine, northern hardwood, or mixed-mesophytic forests associated with warmer and drier climate and, apparently, with increased fire. Conversely, when climate became cooler and moister and fire frequency or intensity was reduced, closed oak forests replaced oak savanna, prairie or some pine forests. Oak species were themselves replaced by other vegetation types when a change in the climatic and fire environment dictated.

In some locations fire frequency and oak distribution are also linked to Native American population density and land-use practices. Throughout the Holocene, Indian populations generally increased in much of the eastern United States, as did their use of fire (Pyne 1983). In addition, the native people may have locally increased the dominance of oak and other nut trees via anthropogenic fires and by caching and cultivating the seeds of these species (Delcourt et al. 1986, Delcourt 1987, Wykoff 1991). Indeed, regardless of the climate, Native American land use transformed local patches of northern hardwood-conifer or mixed-mesophytic forests into forests dominated by oak, chestnut, and hickory.

The direct dating of fire scars on trees within old-growth forests provides further confirmation that periodic fire (in the range of every 5–25 years) occurred in eastern North American oak forests (Buell et al. 1954, Abrams 1985, Guyette and Cutter 1991, 1997, Guyette and Dey 1995, Sutherland 1997). Strong indirect evidence for this idea is found in the fact that when fire has been suppressed in twentieth-century oak forests, they have been rapidly invaded by successional, fire-sensitive species, such as red maple, sugar maple, beech, birch, and black gum. These species are now replacing oak and becoming dominant trees. I believe that the suppression of fire and the successional replacement of oak forests is a cause-and-effect relationship. Nonetheless, additional studies on sediment charcoal input from low-intensity oak understory fires, paleoecological studies specifically addressing the hypothesized link between fire and oak replacement, long-term burning studies in oak forests, and the direct dating of fire scars in old-growth oak forests are needed to resolve some of the remaining issues about the ecological role of fire in the development of oak forests. Moreover, fire is not the only factor that has influenced the ecology of oak forests during th twentieth century. The impacts of acid rain, elevated CO_2, potential global warming, deer browsing, and gypsy moth outbreaks need to be studied in relation to the decline in oak recruitment and the increase in successional tree species in eastern oak forests (Foster et al. 1997). Throughout the eastern United States, interactions among climate, vegetation, soils, topography, large- and small-scale disturbances, anthropogenic factors, and wildlife created an ecological dynamic for oak forests that can only be understood from a long-term, multifactor perspective.

ACKNOWLEDGMENTS

I wish to thank E. W. B. Russell and D. A. Orwig for their review of an earlier draft of the chapter.

Chapter 4

Fire History and Postsettlement Disturbance

DANIEL DEY

Over the past 300 years, forests have undergone dramatic changes, in response to major shifts in disturbance regimes in eastern North America (Orwig and Abrams 1994). Past disturbances were important in the expansion of oaks throughout this region (Nowacki and Abrams 1992). Oak dominance across such a diversity of site types is a legacy of this disturbance history and oak's ability to prosper during periods of disturbance and environmental stress (Hicks 1998). Today's oak forests originated at a time when logging, fuelwood cutting, charcoal production, woods burning, grazing, agriculture, and other anthropogenic disturbances occurred extensively and frequently across the landscape (Nowacki et al. 1990, Abrams 1992, Abrams and McCay 1996). The forest landscape was indelibly altered by a human population that grew exponentially, as did its voracious appetite for food, shelter, and the other amenities of life.

Fire is one of the oldest tools humans have used to manipulate vegetation for the production of food and other plant materials, provision of browse and forage, and to prepare fields for agriculture. Although few fire histories have been documented for eastern oak forests, what we do know is derived from dendrochronological studies of fire scars, pollen and charcoal sediment studies, analyses of early surveyor notes, and reviews of early explorers' and settlers' journals. Disturbance histories are better documented after European settlement.

This chapter highlights what is known about fire history from the late Native American period (circa early 1600s) through European settlement and into the twentieth century. Also covered is the change in human disturbances that occurred when Europeans settled in eastern North America. The nature of human disturbances changed dramatically during the mid-1800s and continues to do so today. The role of fire abruptly diminishes from the 1930s to the 1950s, and fire is replaced by a disturbance regime that is causing the successional replacement of oak-dominated forests in many eastern ecoregions.

USE OF FIRE BY NATIVE AMERICANS

For 10,000 to 20,000 years, Indians used fire to influence forest composition and structure and the extent of grasslands in North America (Pyne 1982, Cronon 1983, M. Williams 1989, Delcourt et al. 1993). Indian fires created a mosaic of plant communities quite different from today's landscape. Early explorers and settlers observed a complex, quilt-like pattern of old-growth, open oak, and pine woodlands, oak and pine savannas, prairies, barrens, bald ridges, oak openings, meadows, grasslands, and scrub oak forests throughout eastern North America.

Indian uses of fire favored the widespread dominance of oaks and pines. Repeated burning gave rise to the longleaf pine and mixed southern pine forests; the oak-hardwood forests; the eastern prairies, glades, and barrens; the oak and pine savannas; and the northern pineries. Many forests had an open, parklike appearance because of their long history of fires (Bartram [1791] 1955, Lorimer 1993). In the Cross Timbers region and throughout the Midwest, the cycle of fire determined the balance between tallgrass prairie, oak savanna, oak woodlands, and pine forests (Gleason 1913, Curtis 1959, Grimm 1984, Engle et al. 1996). However, in some regions fire was less common, due to low human population or cultural characteristics of the resident population (e.g., hunter-gatherer versus agrarian societies); and this lower incidence of fire affected forest type by permitting succession to more shade-tolerant and fire-sensitive species. Elements of a fire regime that influence succession include frequency, intensity, seasonality, spatial extent, and type of fire.

During the Native American period, annual fires were common in regions of prairies and oak-pine savannas (M. Williams 1989, Whitney 1994). Fires burned extensively, favoring the eastern extension of grass-dominated ecosystems. Elsewhere, Indians burned berry- and nut-producing areas on a 3-to-5-year basis and woodlands every 10 to 15 years, although there was variation across regions (Table 4.1). In the

Table 4.1

Fire frequency before 1850 for oak-pine-dominated forests in eastern North America

Region	Mean fire interval (years)	Reference
Missouri Ozarks	2–6	Cutter and Guyette 1994, Guyette and Dey 1997
New Jersey Piedmont	14	Buell et al. 1954
Southeastern United States	2–10	Wright and Bailey 1982
Northeastern United States	1–15	Little 1974
Southern New England	1–10	Niering et al. 1970
Central Ontario	10–17	Dey and Guyette 2000

southeastern Coastal Plain, fires every 2 to 3 years favored longleaf pine forests, but less frequent burning (e.g., every 10 years) led to loblolly pine dominance (Garren 1943).

Frequent fires reduced the density and size of woody species in the understories of oak-dominated and pine-dominated forests, increased the diversity in ground flora, and favored the growth of grasses, legumes, and other herbaceous plants (Garren 1943, Wright and Bailey 1982, Reich et al. 1990, Kruger and Reich 1997a,b). The overall effect was to promote the accumulation and growth of advance reproduction of the fire-adapted oaks by reducing understory competition and causing occasional overstory mortality, thus increasing light at the forest floor. Oaks often persisted as "grubs" in the understories of forests and savannas, and in prairies, glades, and barrens.

The length of fire-free intervals varied substantially, ranging from 1 to 70 years throughout eastern North America (Harmon 1982, Abrams 1985, Cutter and Guyette 1994, Dey and Guyette 2000). This variability in fire frequency played a critical role in the regeneration and recruitment of species into the overstory, because pines and oaks require a fire-free period to develop sufficient bark thickness to minimize burning injury during the next fire. Species such as oaks that can persist as grubs or stool sprouts during periods of frequent fire are able to produce rapid shoot growth during fire-free periods and thus increase their fire resistance and probability of surviving subsequent fires. Pines can grow rapidly, produce thick bark early, and develop fire resistance quickly during fire-free periods when they grow in low to moderate competition conditions.

Before fire suppression became prevalent, low-intensity fires were common in eastern hardwood forests, occurring every 5 to 15 years (Heinselman 1973, Little 1974, Whitney 1994, Sutherland 1997). Fires of moderate and greater intensity happened less frequently (e.g., every 40–50 years) and coincided with regional or subcontinental drought (Heinselman 1973, Cwynar 1977, Guyette et al. 1995, 1999). Catastrophic fires, intense enough to induce stand replacement, were least common in eastern hardwood forests. The periodicity of such fires is not well documented. Frequent low-intensity fires decreased the probability of stand replacement by fragmenting fuels and limiting fuel loads.

Native Americans often set brush-clearing fires in the spring and fall seasons when dry weather and fine fuels such as grasses, leaf litter, and other herbaceous plant litter made burning easy (Little 1974, Pyne 1982, Cronon 1983). Many of these fires burned at a time of year when plants were dormant. An abundance of fine fuels such as the warm-season grasses, which responded well to frequent burning, made it easier for fires to ignite, and increased the length of fire seasons. In southern regions, where snow cover is intermittent or absent during the winter, burning was done whenever weather and fuel conditions permitted, which might be in any month. In years of normal or above average precipitation, fires were less common during the growing season. In dense forests, summer fires were rare except in years of severe drought, because fires do not spread well in the understory of closed-canopy forests after leaf-out. The chance of growing-season fires increased, however, in drought years and in areas that had extensive stands of warm-season grasses and other herbaceous plants that created a continuous matrix of fine, flashy fuels.

Once ignited, fires usually burned extensively, because there was little to no effort by man to extinguish them. They burned out when they ran into natural fire breaks, the weather changed to rain or snow, or they encountered a less combustible fuel type. Periodically, during subcontinental droughts, fires would burn simultaneously throughout eastern North America. Such a year was 1780, when fires burned large tracts of oak, pine, and mixed hardwoods in Algonquin Park, Ontario, and in the Missouri Ozarks (Guyette and Dey 1995, Guyette et al. 1999).

Fires were common in bottomland areas, and even wetlands burned during drought periods (Whitney 1994, DeBano et al. 1998). Europeans found large prairies and openings along the valleys of rivers such as the Potomac, Rappahannock, Shenandoah, Missouri, and Mississippi, which were the result of Indian fires (M. Williams 1989). George Catlin (1844), who left St. Louis in 1832, recorded extensive fires and expansive prairies and savannas in the bottomlands and surrounding countryside as he traveled the Missouri River to Fort Union.

During the Native American period, most fires in eastern hardwood forests were surface fires (Whitney 1994). In the northeastern northern hardwoods, surface fires were common and caused injury rather than death to mature trees (Stickel 1935). Surface fires also were common in Ontario oak-pine forests (Dey and Guyette 2000) and in the Great Lakes region (Heinselman 1973), where they led to the development of multi-aged overstories. Before European settlement, the disturbance regime in oak forests included surface fires that caused noncatastrophic, small-scale mortality of mature trees (Abrams et al. 1995). More severe and even catastrophic fires did occur, however, though less frequently. Fires were more likely to induce stand replacement when they occurred in drought years, or, later in North American history, when they burned after logging on the site.

Fire history in eastern North America has been linked to human history (Pyne 1982, Cronon 1983, M. Williams 1989). Diseases introduced by Europeans decimated Indian populations, often killing upwards of 80% of the people (Delcourt et al. 1993). The frequency of fire was greatly reduced during this period, which may have lasted from 100 to 200 years (Dey and Guyette 2000). Migrating Indian tribes, displaced from their ancestral lands in the East, carried fire into areas of low population as did the vanguards of European settlement. Thereafter, fire's presence and role would be ever changing in eastern North America.

FIRE AND EUROPEAN SETTLEMENT

European settlers continued Indian burning practices but often increased fire frequency and carried fire into more remote areas (Pyne 1982, Cronon 1983, Batek et al. 1999). In general, during the period 1850 to 1930 (somewhat earlier in New England and Atlantic coastal areas), when Europeans were busy converting a North American wilderness into farms and villages, fires were the most frequent in the region's history. At the beginning of the twentieth century, fires burned on average every 5 to 6 years in Ontario oak-pine forests (Howe and White 1913, Dey and Guyette 2000), every 3 to 4 years in southeastern Ohio oak forests (Sutherland 1997), every 10 years in northeast Kansas oak gallery forests (Abrams 1985) and in southern New England (Niering et al.

1970), and every 13 years in oak-pine forests of the Great Smoky Mountains (Harmon 1982). Similar fire frequencies and trends in fire history have been reported for oak woodlands and glades in the Missouri Ozarks (Guyette and McGinnes 1982, Guyette and Cutter 1991, Cutter and Guyette 1994). These estimates of fire frequency are conservative, because not all fires resulted in fire scars on surviving trees.

Fires, and now other forest disturbances such as grazing, logging, and fuelwood cutting, maintained the open, parklike character of eastern forests, with understories dominated by grasses and herbaceous plants (Komarek 1974, Wright and Bailey 1982). Frequent burning, grazing, and logging created forests of sprout origin dominated by oaks. In fact, when much of the area of southern New England, New York, and New Jersey was mapped by Hawley and Hawes (1912), they named it the "Sprout Hardwoods Region."

Fires burned extensively following wholesale logging of the Great Pineries of eastern North America (Howe and White 1913, Kittredge and Chittenden 1929). In the early 1900s, many of our most famous fires burned millions of acres of cutover forests and took the lives of many people in the Lake States. The cycle of logging and burning also greatly reduced the extent of pine forests in eastern North America. After the mature, seed-bearing pines were harvested, intense slash fires and repeated burnings eliminated or greatly reduced the abundance of pine reproduction. In addition, frequent burning (e.g., every 3 years on average) for up to 100 years before the pine logging era may have eliminated much of the pine advance reproduction (Record 1910, Guyette and Dey 1997). Aspen, white birch, and jack pine replaced the white and red pine forests in the Great Lakes Region (Heinselman 1973). Elsewhere, oaks succeeded the pines and shade-tolerant mesophytic hardwoods (Abrams 1992). Oaks expanded their dominance on mesic, highly productive sites through frequent and widespread burning. The stage was set for a great expansion of oak throughout eastern North America.

The widespread suppression of wildland fires began in the 1930s and 1940s in most regions of eastern North America; however, in the Ozark Highlands and in the South, wildland fires were common until the 1950s (Pyne 1982, Pyne et al. 1996). The occurrence of wildland fires has dropped drastically over the past 100 years. At the beginning of the twentieth century, the fire rotation period was 90 years in Michigan and 50 years in Pennsylvania (Whitney 1994). Before European settlement in the Missouri Ozarks, mean fire intervals averaged about 3 years (Guyette and Cutter 1991, Cutter and Guyette 1994), and the fire rotation period

Table 4.2
Estimated fire rotation periods for the modern period for oak-pine-dominated forests in eastern North America

Region	Fire rotation period (years)	Reference
Missouri Ozarks	715	Westin 1992
Pennsylvania	910	Whitney 1994
Lower Michigan	1,400 to 2,000	Whitney 1986, 1994
Upper Michigan	1,273[a]; 4,545[b]	Frelich and Lorimer 1991
Smokey Mountains	> 2,000	Harmon 1982
Southern Illinois	900	Haines et al. 1975
Monongahela National Forest, West Virginia	6,000	Haines et al. 1975

[a]For surface fires.
[b]For stand replacement fires.

in the Smoky Mountains before 1940 was less than 10 years (Harmon 1982). Now, the fire rotation period is estimated to be significantly longer (Table 4.2).

At the outset of the fire suppression period, modern oak forests developed rapidly across eastern North America, expanding their range by replacing savannas, barrens, and prairies and dominating old fields and cutover lands. Oak advance reproduction in these systems grew quickly into closed-canopy forests once fires were suppressed (Cottam 1949, Curtis 1959, Grimm 1984).

As fires continue to be suppressed throughout eastern North America, oak forests are being replaced by more shade-tolerant, mesophytic species, such as the maples (Lorimer 1984, Pallardy et al. 1988, Hix and Lorimer 1991, Abrams 1992, 1998). Where fires have been absent, oak forests have increased in structural complexity, and this has been accompanied by the growth of a midstory of shade-tolerant trees and an understory dominated by shade-tolerant shrubs and advance reproduction. In the heavy shade, large oak advance reproduction is unable to develop, and other species dominate following overstory disturbances. However, on xeric pine sites in the southeastern Coastal Plain succession in the absence of fire is to oaks and other southern hardwoods (Garren 1943, Wright and Bailey 1982).

Although wildland fires are less extensive and less frequent than they were just 60 years ago, humans continue to be the most significant source of wildland fires in eastern North America (Guyette et al. 1999). For example, arsonists ignited 47% of all wildland fires that occurred in

Arkansas from 1991 to 1996 (Garner 1999). These fires accounted for 64% of the total acreage of burned wildland. Another 27% of Arkansas's wildland fires were started by humans burning debris. Nationally, arson and debris burning accounted for 74% of wildland fires from 1983 to 1987 (Garner 1999). On the other hand, lightning caused less than 2% of all fires in Arkansas (1991–96) (Westin 1992, Garner 1999). In the East, most lightning occurs during thunderstorms and is accompanied by rain, which reduces the likelihood of fire ignition or spread. Outside the southeasternmost portion of the United States, lighting causes fewer than 5 fires per million acres in eastern North America. Throughout the Mid-Atlantic and Midwest less than 1 fire per million acres results from lightning.

AGRICULTURAL AND INDUSTRIAL TRANSFORMATION PERIOD

Forest clearing for agriculture was the primary cause of deforestation in eastern North America (Whitney 1994). Only 10%–30% of the forest remained in southern New England by the mid-1800s; coastal forests had been cleared much earlier (Niering et al. 1970). As settlers moved into the Ohio Valley, they established farms until, by 1900, most (85%) of the land had been cleared (Sutherland 1997). Similar events played out in the Lake States (Pyne et al. 1996) and the Southeast (Martin and Boyce 1993). From 1860 to 1910, the rate of conversion to agriculture accelerated such that more forests (190 million acres) were converted during this period than in all previous time since European arrival in North America (M. Williams 1989, Powell et al. 1993). Since the 1850s, forest clearing and drainage of wetlands for agriculture have caused the loss of 70%–98% of bottomland forests in the United States (Sharitz and Mitsch 1993).

During this agricultural revolution in North America, fire was used by settlers to clear forests, maintain farm and pasture, and improve the woods for grazing, as the Indians had done but on a much grander scale. Forests were further disturbed by cattle and hogs, who grazed freely everywhere Europeans settled (Cronon 1983, M. Williams 1989, Whitney 1994). Grazing on open range was practiced in many regions (e.g., Ozark highlands) as late as the 1950s and 1960s. Livestock grazing affected forest regeneration, often eliminating it in the understory of woodlands and savannas.

Locally throughout the range of the oak, the production of charcoal for the iron industry affected the surrounding forests. Forest harvesting for charcoal peaked in the late eighteenth century in the Piedmont and Coastal Plain regions (Orwig and Abrams 1994) and later in the nineteenth century in the Midwest and Northeast (Sutherland 1997). Forests were harvested on short rotations (e.g., 20–30 years) and fires fueled by logging slash burned with greater intensity (Cronon 1983, M. Williams 1989, Whitney 1994). These practices created coppice forests dominated by oaks (Clatterbuck 1991, Orwig and Abrams 1994).

Industrial logging transformed the forests of eastern North America during the period from 1850 to 1930. Forest harvesting had previously been a local activity to supply the timber and fuelwood needs of nearby villages and specialty products sought by foreign sovereigns (e.g., masts for British sailing ships). Small populations and primitive transportation systems limited the need and the ability, respectively, to lumber extensively. Timber was transported by log drives down rivers that had connections to ocean ports. Not until railroads were built into the more remote forested regions were many forests linked to population centers. This set the stage for the wholesale logging and "destruction" of eastern North America forests documented in Ontario (Howe and White 1913), the South (Martin and Boyce 1993), the Lake States and Northeast (Cronon 1983, M. Williams 1989), the Ozarks (Cunningham and Hauser 1989), and the central hardwood region (Hicks 1998). Beginning in 1850, the annual production of forest products increased markedly; in 1910 an estimated 13 billion cubic feet of timber were harvested (Powell et al. 1993). Repeated and often catastrophic fires burned the cutover forests. The forests had never before experienced disturbances of such extent or severity. Far from being destroyed, the forests renewed themselves, although they changed in character. This logging and fire history favored the oaks, which rose like the phoenix out of the ashes to assume widespread dominance throughout eastern North America.

Much of the eastern forests have developed on agricultural land abandoned during the late 1700s in the Northeast, Mid-Atlantic, Piedmont, and Coastal Plain, and later in the Midwest, Lake States, and other interior regions. The regrowth of forest on agricultural lands accelerated during the Great Depression. Often, these forests were dominated by pine and oak (Orwig and Abrams 1994). Oak seedlings and grubs sprouted in pastures and, in the absence of mowing or grazing, grew to maturity. Elsewhere, stands dominated by pine provided ideal conditions for oaks to establish and grow in the understory, setting the stage for oak dominance after the pine was harvested. As agricultural land was abandoned throughout the Midwest, Lake States, and Southeast, the amount of forest land increased.

In some places, forest cover has increased to levels approaching that seen by the first European settlers. Current forest acreage in eastern North America equals about two-thirds of the forest area estimated to have existed in the 1600s. In the unglaciated plateau in Ohio, forest cover has increased from 15% to 65% over the past 100 years (Sutherland 1997), and Vermont has gone from being 65% cropland in 1850 to 77% forested today (Powell et al. 1993). Since 1907, the extent of forest land in the United States has stabilized at 730–760 million acres. The change in proportions is due largely to a cessation of forest clearing for agriculture (Powell et al. 1993). Deforestation in the lower Mississippi River bottomlands continues today, however (Sharitz and Mitsch 1993).

Europeans brought to the New World insects and diseases that have affected forest composition and succession over large areas. One of the most notable examples was the introduction of the chestnut blight (*Endothia parasitica* [Murrill] P. J. Anderson and H. W. Anderson), which eliminated chestnut as a major overstory species in eastern North America by the early 1900s. Oak dominance increased after the loss of chestnut in the Appalachian and mid-Atlantic regions (Hinkle et al. 1993, Abrams and McCay 1996, Hicks 1998). Chestnut oak, red oak, and scarlet oak replaced chestnut. The introduction of exotic species and the spread of invasive species will continue to modify natural forest succession and alter native forest character (see Chapters 6 and 7).

Forest product manufacturing in the United States declined from 1910 until after World War II, when a boom in the housing market sparked an increase in harvesting (Powell et al. 1993). Since then, production has increased (see Figure 4.1), resulting in historic levels of harvesting, but annual growth of hardwoods in the East still exceeds timber removals (e.g., by 80% in 1991). In the eastern United States, most (94%) of the 380 million acres of forest land is timberland; this proportion has been relatively constant since the early 1950s (Powell et al. 1993). About half of the eastern forest land is classified as oak-pine (32.2 mill. a.), oak-hickory (129.7 mill. a.), or oak-gum-cypress (29.2 mill. a.) forest type, and the largest areas of highly productive (> 120 cu. ft. per acre per year) forest lands occur in the oak-hickory and loblolly-shortleaf pine types. Most of the timberlands in the East are owned by private

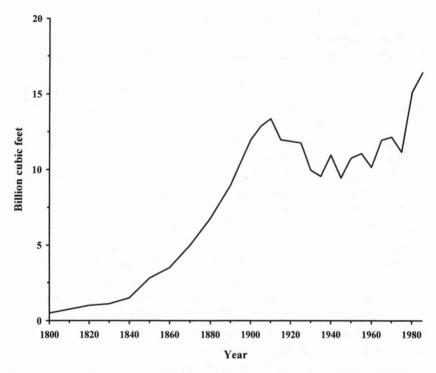

FIGURE 4.1.   Domestic production of forest products from 1800 to 1985 in the United States. (Adapted from Powell et al. 1993.)

industry (16%) and nonindustrial private individuals (70%), and they are primarily responsible for the hardwood harvest (e.g., 90% of the hardwood production in 1991).

Unfortunately, written management plans have been used by only an estimated 5% of private forest land owners, who control 39% of the private forest lands (Birch 1996). More than half of all private forest lands have no management plans. Although it is difficult to quantify the amount of timber harvested by regeneration method, common harvesting techniques on unmanaged private land include selective cutting or high grading and diameter limit cutting. These rogue harvesting practices create small gaps in the overstory canopy, which usually do not favor oak development, especially on the more productive sites. Harvesting by these methods often results in understocked stands of reduced quality and value.

WIND DISTURBANCES

Wind has contributed to the overall level of damage to forests, acting in concert with fire, logging, grazing, and other disturbance factors. Since the suppression of wildland fires, wind is the most common natural force capable of altering forest character on a landscape basis (Greenberg and McNab 1998). Humans have had little direct effect on the extent and intensity of wind storms. They have, however, substantially affected the importance of wind disturbances, by altering the type and distribution of forests. For example, much of the original old-growth forest lands that stood at the time of European settlement have been converted to agricultural uses or been harvested to produce younger forests, which are more resistant to damage by wind (Oliver and Larson 1990, Kozlowski et al. 1991). In contrast, high-grade logging and reckless woods burning have increased the amount of decay in hardwood forests (e.g., 37% of the live volume in Missouri is cull) rendering trees more susceptible to windthrow (Powell et al. 1993). Wind storms have been and continue to be an important forest disturbance, altering forest dynamics by creating small gaps in the canopy by the windthrow of a single or several overstory trees or by causing catastrophic extensive loss of the overstory (Lorimer 1980, Myers and Van Lear 1998).

Storms that cause large-scale windthrows generally are rare (Table 4.3). Because large-scale catastrophic wind disturbances are less frequent than those that cause small canopy gaps, much emphasis has been placed on the importance of small-scale disturbances in determining forest succession, structure, and diversity in mesophytic forests of the East. Large infrequent disturbances such as catastrophic wind storms and fires, however, are receiving more recognition for their role in shaping the regional landscape and their ability to influence ecosystem processes for centuries (Lorimer and Frelich 1994, Foster, Knight, and Franklin 1998).

CONCLUSION

The abundance, diversity, and extent of eastern forests, savannas, and prairies are intricately linked to human disturbances, demography, and culture. Humans have been one of the most dominant and influential agents of forest disturbance over the past millennium. Fire is one of humankind's oldest and most reliable tools for culturing vegetation. Fre-

Table 4.3
Rotation periods for wind disturbances in common forest types of the eastern United States

Rotation period (years)	Forest type	Disturbance type[a]	Region	Reference
110–125	Old-growth mesophytic hardwoods	Small canopy gaps	S.W. Ohio	Runkle 1990
50–200	Old-growth mesophytic hardwoods	Small canopy gaps	Eastern U.S.	Runkle 1985
50–250	Old-growth mesophytic hardwoods	Small canopy gaps	S. Appalachian	Runkle 1982
541	Presettlement; hemlock–northern hardwoods	—	U.P. Michigan	Zhang et al. 1999
1,200	Presettlement; hemlock–white pine–northern hardwoods	Catastrophic wind	Michigan	Whitney 1986
1,300	Presettlement; swamp conifer	—	Michigan	Whitney 1986
1,500	Old-growth hemlock–northern hardwoods	Disturbance that destroys >60% of canopy	U.P. Michigan	Frelich and Lorimer 1991
5,600	Old-growth hemlock–northern hardwoods	Tornadoes >75 mph winds	U.P. Michigan	Frelich and Lorimer 1991
300	Old-growth hemlock–northern hardwoods	Disturbance that destroys 30–50% of canopy	U.P. Michigan	Frelich and Lorimer 1991
6,031	Old-growth hemlock–northern hardwoods	Tornadoes >75 mph winds	N. Wisconsin	Frelich and Lorimer 1991
1,210	Presettlement; hemlock–white pine–northern hardwoods	Catastrophic wind >2.5 acres	N. Wisconsin	Canham and Loucks 1984
1,238	Presettlement; hemlock–white pine–northern hardwoods	Catastrophic wind >25 acres	N. Wisconsin	Canham and Loucks 1984
1,470	Presettlement; hemlock–white pine–northern hardwoods	Catastrophic wind >250 acres	N. Wisconsin	Canham and Loucks 1984
2,903	Presettlement; hemlock–white pine–northern hardwoods	Catastrophic wind >2,500 acres	N. Wisconsin	Canham and Loucks 1984
540	Old-growth hemlock–northern hardwood	—	N.W. Pennsylvania	Whitney 1994
1,000–2,000	Hemlock–white pine–northern hardwoods	—	N.W. Pennsylvania	Whitney 1990
9,000	Northern hardwoods	—	Central New York	Marks et al. 1992
2,000	Hemlock–northern hardwoods	—	W. New York	Seischab 1990
1,150	Northern hardwoods–spruce–fir	—	N. Maine	Lorimer 1977

[a]Unless otherwise noted, disturbance type is catastrophic winds that cause stand regeneration.

quent burning by Indians promoted nut- and berry-producing trees and shrubs and bountiful grasslands. At the landscape level, anthropogenic fire produced a mosaic of fire-dependent ecosystems. European settlers noted extensive eastern prairies, oak openings, meadows, orchards, barrens, and oak and pine forests and savannas.

Europeans imposed a new and unique suite of disturbances that changed the nature and distribution of the ecosystems they encountered in North America. Conversion of forest, prairie, and wetland to agriculture, along with commercial logging, forest grazing, and woods burning, changed the landscape in a way and at a rate unparalleled in recent history. Following the exploitation phase of European settlement, forests reclaimed abandoned farmland, invaded former prairies, and regenerated lands that had been cutover and repeatedly burned. On these lands, as a consequence of fire prevention and suppression programs that abruptly removed fire as a primary forest disturbance, oaks dominated the regeneration. We are just now beginning to witness the long-term effects of these major shifts in disturbance regimes and to recognize the ecological role of fire in the maintenance of many of our forest ecosystems.

Modern oak forests are a product of this disturbance history, because trees are relatively long-lived and processes such as succession are long-term. Previous disturbances set successional pathways that led to oak dominance because oaks are well adapted to high disturbance and stress environments where fire, drought, grazing, and logging occur periodically. In turn, current disturbance regimes are shaping the nature of future forests. Disturbance regimes characterized by less frequent, intense, or extensive burning, and less harvesting and grazing promote competitors of oak, especially on productive sites where environmental conditions are less limiting to tree growth.

In the past, wind and fire acted together to favor oak. Today, wind promotes succession to oak's competitors in the absence of fire. Succession following small gap disturbances favors shade-tolerant species, and catastrophic windthrow promotes fast-growing intolerant species at the expense of oaks.

Knowledge of historical disturbance regimes is important, because modern oak forests originated at a time when fire, logging, and other human disturbances were more common than they are today. A study of the fire ecology and disturbance history of oak and its associates can provide an understanding of the widespread occurrence of oak, and establish a basis for management strategies to restore and sustain oak ecosystems.

Chapter 5

The Ecological Basis
for Oak Silviculture
in Eastern North America

DANIEL DEY

Quercus is a dominant genus throughout North America and has been for the past 10,000 years or more (see Chapters 2 and 3). Oak distribution has shifted in response to changes in climate, disturbance regime, and human population and culture. Oak dominance has increased throughout the Holocene period. However, a recent successional trend is the replacement of oak-dominated ecosystems throughout the range of oak.

The inability of oak reproduction to compete with either large shade-tolerant advance reproduction or aggressive pioneer species is the fundamental cause of problems in oak regeneration and sustainability (Lorimer 1993). Oak regeneration problems and reductions in oak stocking are most likely on higher-quality mesic sites (site index > 60 feet, base age 50). Oaks appear to be successionally most stable on xeric sites, under current disturbance regimes, which are typified by frequent small-scale disturbances that cause isolated mortality to overstory trees and the absence of fire (Johnson 1993a). However, increased competition from shade-tolerant trees and shrubs threatens oak regeneration potential even on these drier sites.

Oaks are adapted to environments characterized by disturbance and stress. The primary factor leading to the successional displacement of oak in eastern North America has been a change in the historic disturbance regime that has altered the competitive relationship between oak and its associates. The widespread distribution and dominance of oak is a result of a long history of frequent fire, which peaked shortly after the invasion and population of America by Europeans (Abrams 1992). Since the 1930s, fire suppression has nearly eliminated wildfire as a forest disturbance. This drastic reduction in fire frequency is the most often cited cause of the recent oak regeneration problem, especially on high-quality sites (Little 1974, Van Lear 1991, Lorimer 1993).

It appears that, under current disturbance regimes and without human intervention through management, oak dominance will be increasingly confined to the less productive sites. The more productive sites will succeed to other species, with a possible loss of the species diversity and mast production that are important to so many wildlife species (see Chapters 14–17). Sustainable management of oak dominated ecosystems is predicated on an understanding of oak silvics, regeneration ecology, and response to disturbance. This chapter describes some of the more important ecological requirements of oaks, their responses to disturbances, and their competitive relationships with co-occurring tree species.

Light Relations

Inadequate light often limits oak regeneration and recruitment into the overstory (Lorimer 1993). Oak is much less shade tolerant than many of its competitors (see Table 5.1). Acorn germination and initial seedling development are not limited by light levels, because the seed is relatively large and supplies the bulk of the carbohydrates for growth until seed reserves are exhausted.

Growth and survival of oak in shaded microsites depend upon photosynthetic CO_2 fixation exceeding the respiratory requirements of seedlings, that is, net photosynthesis > 0 (Figure 5.1). The minimum light level required by oak seedlings to produce enough carbohydrate to meet their respiration needs (i.e., the light compensation point, where net photosynthesis $= 0$) is low, about 2% to 5% of full sunlight for northern red oak (Gottschalk 1987, Hanson et al. 1987). With higher levels of photosynthetically active radiation, net photosynthesis increases to a saturation point at which it remains relatively stable with further increases in light intensity.

Although survival of oak seedlings at low light levels may be possible, sufficient carbohydrate to support the production of new tissue requires

Table 5.1

Shade tolerance in oak species and common competitors

Oak species	Shade tolerance	Competing species	Shade tolerance
Bear	Very intolerant	American beech	Very tolerant
Black	Intolerant–intermediate	American elm	Intermediate
Blackjack	Intolerant	Black cherry	Intolerant
Bur	Intermediate	Black gum	Tolerant
Cherrybark	Intolerant	Black locust	Intolerant
Chestnut	Intermediate	Box elder	Tolerant
Chinkapin	Intolerant–intermediate	Cucumber tree	Intermediate
Laurel	Tolerant	Eastern cottonwood	Very intolerant
Live	Intermediate	Eastern hop hornbeam	Tolerant
Northern pin	Intolerant	Flowering dogwood	Very tolerant
Northern red	Intermediate	Green ash	Tolerant
Nuttall	Intolerant–intermediate	Hackberry	Intermediate
Overcup	Intolerant–intermediate	Pin cherry	Very intolerant
Pin	Intolerant	Quaking aspen	Very intolerant
Post	Intolerant	Red maple	Tolerant
Scarlet	Very intolerant	Silver maple	Tolerant
Shingle	Intolerant	Sourwood	Tolerant
Shumard	Intolerant	Southern magnolia	Tolerant
Southern red	Intermediate	Striped maple	Very tolerant
Swamp chestnut	Intolerant–intermediate	Sugar maple	Very tolerant
Swamp white	Intermediate	Sweet birch	Intolerant
Turkey	Intolerant	Sweetgum	Intolerant
Water	Intolerant	Water tupelo	Very tolerant
White	Intermediate	White birch	Intolerant
Willow	Intolerant	Yellow-poplar	Intolerant

Sources: Adapted from Burns and Honkala 1990, Smith 1993.

greater light. Light saturation of photosynthesis in oak seedlings (e.g., northern red oak and cherrybark oak) occurs at 30% to 50% of full sunlight (Teskey and Shrestha 1985, McGraw et al. 1990, Ashton and Berlyn 1994). Growth in height and diameter is near maximum at light intensities approaching 50% to 70% of full sunlight (Hodges and Gardiner 1993, Gottschalk 1994). In contrast, shade-tolerant species (e.g., red maple and beech) show maximum net photosynthesis at light intensities as low as 5% to 10% of full sunlight, and they have better whole-plant carbon balances in low light than less tolerant species such as the oaks (Bazzaz 1979, Kozlowski et al. 1991). Shade-intolerant species require full sunlight for light saturation of net photosynthesis, which promotes maximum growth rates that exceed those in oak.

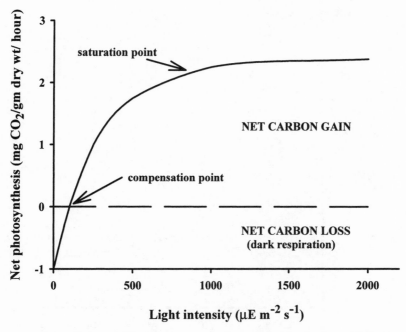

FIGURE 5.1.   Relationship between rate of net photosynthesis and light intensity.

Response of Understory Light to Overstory Manipulation

Oak reproduction rarely persists long or grows much in the understory of mature hardwood forests, because the light intensity is usually less than the light compensation point for oaks (Canham et al. 1990). Dense overstory cover and subcanopies of trees and shrubs intercept most of the sunlight.

The amount of light after harvesting does not increase in linear proportion with the amount of canopy removed (Figure 5.2). Shelterwood harvests in northern hardwoods need to remove more than 50% of the basal area to increase light intensities at the forest floor to 35% to 50% of full sunlight (Marquis 1988, Dey and Parker 1996). In group selection harvests, light levels rise at an increasing rate as the size of canopy gap increases until a threshold gap-size is reached. Light increases in the center of gaps as the ratio of gap diameter (d) to the average height of the

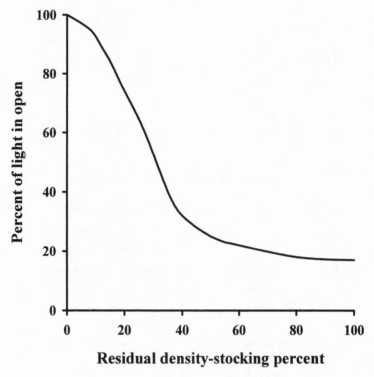

FIGURE 5.2.   The effect of residual overstory stocking on light intensity 8 inches above the ground in the central hardwood forests. The absolute values may change for forests at different latitudes, but the general relationships remain the same. (Adapted from Sander 1979.)

adjacent stand (h) increases, leveling off when d/h approaches 2.0 (Minckler et al. 1973). In openings of any size, light increases from the edge of the gap to the center. Whether shelterwood or group selection methods are used to increase light for oak reproduction, understory strata of shade-tolerant species must be controlled for oak reproduction to benefit fully from reductions in overstory density.

Light and Regeneration Methods

Regeneration methods from single-tree selection to clearcutting can be used to create canopy gaps ranging from the smallest of openings (i.e., a single-tree gap) to the largest of disturbances (several acres or more), which have much the same effect as natural disturbances on forest structure. The single-tree selection method produces understory light levels that are similar to unmanaged, mature hardwood stands. Shade-intolerant and intermediate species can germinate under heavy shade, but understory light levels are often insufficient for their long-term survival. Shade-tolerant species such as sugar maple and beech can survive in low light (e.g., < 5% of full sunlight) for decades, and respond to release following reduction in overstory density (Poulson and Platt 1989). In this manner, they can grow into the overstory through repeated cycles of suppression and release (Canham 1985). Therefore, shade-tolerant species have the competitive advantage over oaks in stands managed by the single-tree method or that experience other small-scale canopy gap disturbances (Jenkins and Parker 1998).

Group selection harvesting has been used to increase understory light levels to encourage regeneration of species of intermediate shade tolerance. However, these openings are often dominated by large shade-tolerant advance reproduction (Canham 1988, 1989, H. C. Smith 1981) or by yellow-poplar (Weigel and Parker 1997, Jenkins and Parker 1998), because competing vegetation is not controlled and oak advance reproduction is lacking before harvest. Oak advance reproduction grows well in gaps if it receives 20% to 50% of full sunlight, which occurs in gaps with a d/h ratio equal to 1.0 (Marquis 1965). Large stems of shade-tolerant species may need to be felled during harvest to permit adequate light at the forest floor. Once reproduction is established, larger openings are better for oaks, especially if future harvesting will be delayed. In larger openings, postharvest treatments to control competing vegetation may be necessary to maintain oak dominance, especially when yellow-poplar is present (Brose and Van Lear 1998).

The shelterwood method is often recommended for promoting oak regeneration. Factors common to the successful use of this method for regenerating oak include the presence of oak advance reproduction before harvest, control of competing woody vegetation, and reduction of overstory density to moderate levels. Shelterwoods can be used to produce a wide range of environments for regeneration. The overstory may need to be reduced to 40%–60% stocking and any shade-tolerant midstory canopy removed to provide sufficient light for oak reproduction (Schlesinger et al. 1993, Lorimer et al. 1994). Higher overstory densities can be maintained without adversely affecting the growth of oak seedlings when understory competition is controlled. On higher-quality sites, control of understory competing vegetation is particularly important, as is the maintenance of higher residual shelterwood density to help control oak competitors (Loftis 1990a). However, residual overstory densities with more than 70% of crown cover or > 60 square feet per acre basal area limit oak seedling growth and survival (Larsen et al. 1997).

MOISTURE RELATIONS

The oaks as a group are quite tolerant of drought, primarily because they have large root systems, leaf morphological characteristics that reduce transpiration, and the ability to maintain gas exchange and net photosynthesis to comparatively low levels of leaf water (Abrams 1990, Pallardy and Rhoads 1993). The development of a strong taproot system in oaks provides them access to moisture from deep soil layers, a source less available to their more shallow-rooted competitors. The oaks are better adapted to xeric environments than many of their common mesophytic competitors. Although northern red oak is one of the least drought tolerant of the upland oaks of eastern North America (Seidel 1972, Kleiner et al. 1992), it is still more adapted to drought than many species, such as aspen, white birch, dogwood, black cherry, and maple (Bahari et al. 1985, Martin et al. 1987, Abrams 1990). Among the oaks, those species found on xeric uplands (e.g., post oak and blackjack oak) are more drought tolerant than the more mesic species (e.g., northern red oak) (Reich and Hinckley 1980, 1989).

Despite their adaptations to drought, oaks are still subject to injury from water stress. Drought can cause declines in leaf gas exchange, dysfunction of their xylem water transport system, decreases in shoot and root growth of seedlings, and increases in the risk of mortality (Kozlowski et al. 1991, Tyree and Cochard 1996). Under water stress, oak seedlings exhibit lower leaf area and new root production, delayed bud break, reduced shoot elongation, and increased shoot dieback, and they produce less xylem tissue and fewer and smaller vessels.

Moisture and Regeneration Methods

Surface soil moisture is important to seed germination and early seedling establishment. Adequate surface moisture is needed until seedling roots can obtain moisture from the lower soil zones. High evaporation rates in recently clearcut and large group selection openings dry the upper few inches of the soil and cause high mortality in young seedlings. A shelterwood overstory produces more favorable surface soil moisture conditions through increased canopy throughfall of precipitation and reduced overstory transpiration (Minckler et al. 1973, Crunkilton et al. 1992, Breda et al. 1995). Advance reproduction and stump sprouts are better able than seedlings to access moisture in the lower soil horizons, because they have well-established root systems. Compared to seedlings, they have better water relations under drought conditions and are better adapted to survive the high summer temperatures and vapor pressure deficits common in clearcut environments. Moisture throughout the soil profile, however, is generally lower for longer periods in fully-stocked, mature forests than in larger clearcut or group selection openings (Bormann and Likens 1979, Kramer and Boyer 1995).

REGENERATION

Potential for Regeneration

Regeneration potential refers to the ability of trees, from seedlings to mature individuals, to contribute to stand regeneration through sexual or vegetative reproduction (Johnson 1993a). Mature trees may produce seed and contribute to the regeneration potential by adding to the population of new seedlings or by increasing the storehouse of seed in the forest floor. Reproduction may also arise from dormant or adventitious buds that sprout after shoot injury or death in overstory and understory trees.

In many hardwood ecosystems, the flora present at the time of disturbance control to a significant extent the future composition and structure of the ecosystem. In other words, the capacity of a species to regenerate and dominate the growing space made available by a stand-initiating disturbance is determined by the characteristics of the parent stand (Loftis 1990a, Dey et al. 1996). For example, the amount of seed produced is strongly influenced by the basal area of the species in the upper-crown classes of the parent stand (Bjorkbom 1979). Similarly, the abundance and size of advance reproduction are related to overstory characteristics, including species composition, density of mature trees, and stand structure (Bjorkbom 1979, Johnson 1992, Larsen et al. 1997). Characteristics of the understory vegetation such as the abundance, size and composition of advance reproduction significantly influence species dominance after stand-initiating disturbances (Sander 1971, Loftis 1990a, Dey et al. 1996).

The regeneration potential of a stand, therefore, is a measurable and predictable attribute (Dey 1993). For oaks and other hardwoods, it can be determined from predisturbance characteristics, including the advance reproduction and the overstory. To quantify the regeneration potential of a stand, the various sources of reproduction must be considered. Oaks and other hardwoods regenerate as new seedlings, seedling-sprouts, and stump or basal sprouts from overstory trees.

Reproduction by Seed

Most oak seedling establishment occurs in years of good acorn production (Lorimer 1993). Seed production is highly variable among oak species, between individual trees, over the years, and from one location to the next. For all oak species, some trees are consistently good producers and others are consistently poor producers. Ability to produce acorns is most often attributed to the genetic capability of the tree. However, other factors, such as weather, insects, soil fertility, stand density, diseases, and wildlife, are also important in determining the size and frequency of acorn crops. In the long term, tree characteristics such as size, crown area, crown class and age, and genetics are probably more important than environmental factors in determining acorn production (Beck 1993). In general, oaks have large seed crops at 2- to 10-year intervals.

Acorns must maintain relatively high seed-moisture content to remain viable. Seed viability drops rapidly when moisture content falls below 30% to 50% for species in the white oak group and 20% to 30% for those in the red oak group (Korstian 1927). On the ground, dessication can cause rapid deterioration of acorns. Much of the seed that escapes predation by insects, birds, and mammals fails to produce seedlings because of low seed-moisture content. Thus, acorns do not remain in the forest floor seedbank for more than a year. Acorns that are buried 2.54 cm to 5.08 cm in the soil and those in contact with mineral soil beneath the leaf litter show the best germination. The presence of a litter layer, provided that it is not too thick (i.e., > 5 cm), protects acorns from desiccation and extreme fluctuations in temperature. Thick litter layers present a physical barrier to the radicle's reaching mineral soil, and acorns mixed with or on top of leaf litter are more likely to suffer desiccation than those in contact with mineral soil.

In areas of persistent snow cover, acorns stand a better chance of surviving the winter, because conditions under the snow cover are ideal for stratification and storage of acorns. In regions lacking snow cover, acorns are exposed to sun, wind, and often to warm temperatures throughout the winter months. White oaks, which germinate in the fall, are less susceptible to desiccation than red oaks, but either seed or germinants can dry out over winter and early spring if they are not protected by a covering of litter and soil.

The use of fire to reduce litter depth usually is not necessary for oak regeneration except on high-quality sites (Stringer and Taylor 1999). Fires that burn after seed dispersal cause high mortality in acorns that are on the forest floor or mixed in the upper litter layers (Auchmoody and Smith 1993). Acorns buried under moist litter layers or in mineral soil are better protected from the heat of surface fires. Oak seedlings germinating from acorns that survive burning may benefit from increased availability of nutrients, reduced competition, and higher light intensity. However, fire can also enhance the germination and establishment of oak competitors, such as sweetgum, pin cherry, yellow-poplar, and *Rubus* (Shearin et al. 1972, Little 1974, Van Lear and Waldrop 1988, Kruger and Reich 1997a,b).

Acorns provide food for hundreds of animal species, and in the process of storing acorns, a few birds and mammals become important in seed dispersal (see Chapter 12). In years of low to moderate acorn production, wildlife and insects can consume or damage the entire seed crop, thus reducing or eliminating seedling establishment for that year (Beck 1993). Periodic surface fires may benefit oak regeneration, because fire temporarily destroys the habitat of small mammals and insects and may reduce populations of acorn insects (Wright 1986, Galford et al. 1988). However, wildlife that store, or "cache," acorns in the soil to provide a winter food source can improve the regeneration of oak by their "planting" of acorns. Burial of acorns in the soil or beneath the litter layer protects the seed from desiccation, from further predation or damage by other animals, and from disturbances such as fire (Barnett 1977, Auchmoody et al. 1994).

Reproduction by Seedlings

Shoot growth of oak seedlings is relatively slow, because oak seedlings possess moderate leaf photosynthetic capacity, relatively thick leaves, and preferential carbon allocation to the roots (Kolb and Steiner 1990, Walters et al. 1993). Suboptimal environmental conditions also slow shoot activity and trigger an allocation of current photosynthate to the roots.

This carbohydrate both supports continued root growth and is stored in the taproot for future use in shoot growth.

For most oak species, slow height growth relative to that of competing vegetation is the most often cited cause of oak regeneration failure, especially on high-quality sites (Lorimer 1993). In the open, young oak seedlings are at a competitive disadvantage when growing with large advance reproduction of shade-tolerant species (e.g., maples) and shade-intolerant reproduction such as yellow-poplar. In mesic and hydric ecosystems, oak species such as northern red oak and water oak can be regenerated successfully from seed if there is an abundant acorn crop and a low to moderate level of competition at the time of overstory removal (Johnson and Jacobs 1981, Loewenstein and Golden 1995). More frequently, intense competition and a lack of acorns or advance reproduction result in oak regeneration failures on these sites.

The preferential maintenance of root growth over shoot growth is an important ecological adaptation that enables oaks to dominate on xeric sites and to persist in high-disturbance environments. A competitive rate of growth for oak reproduction depends on the development of seedlings with a large, physiologically vigorous root system and high root-to-shoot ratio (Johnson 1993a). The number of sprouts, shoot growth, and probability of multiple shoot flushes increase with increasing root mass or its correlate, basal diameter of the stem (Johnson 1979, Dey et al. 1996, Dey and Parker 1997). A large root system provides carbohydrates, water, and nutrients in amounts required to produce new shoots capable of rapid growth after release from overstory cover.

Reproduction by Seedling Sprouts

Seedling sprouts arise from vegetative propagation of seedlings that experience shoot dieback. Shoot dieback releases dormant buds from apical growth hormones, and sprouts begin to grow from buds located under the bark along the stem and clustered near the root collar (Kozlowski et al. 1991). Oaks are better adapted than many of their competitors to disturbances or environmental stresses that cause shoot dieback; they can repeatedly produce new sprouts from their large supply of dormant buds located at the root collar, which is often beneath the soil surface, where buds are protected from fire and herbivores. Oaks that suffer serious injury to their shoots compartmentalize the damaged tissue and do not allocate resources toward the recovery of photosynthesis in that tissue. Instead, they reallocate resources to the production of new sprouts when environmental conditions become more favorable or after the disturbance-induced shoot dieback subsides. Through repeated shoot dieback and sprouting events, oaks can build a large root system, provided there is adequate light.

A well-developed root system results in a high shoot-growth potential in oak seedling sprouts, which makes them more competitive than true oak seedlings (i.e., seedlings that have not experienced shoot dieback and sprouting). Greater net photosynthesis in seedling sprouts results from a higher root area–to–leaf area ratio (Kruger and Reich 1993a,b) and, for oak seedling sprouts developing after a fire, improved leaf nitrogen content (Reich et al. 1990). Therefore, successful oak regeneration is dependent upon there being an adequate number of large advance reproduction (primarily seedling sprouts that are present before overstory removal).

Advance Reproduction on Xeric Sites. Oak advance reproduction can accumulate over successive acorn crops on xeric sites throughout eastern North America, even in the absence of disturbances such as fire (Johnson 1993b). Oak accumulator ecosystems are characterized by a relative abundance of large advance reproduction but not necessarily high numbers of seedlings. The probability of having an abundance of large oak advance reproduction increases as overstory density decreases to ≤ 58% stocking (Larsen et al. 1997). Sufficient light reaches the oak advance reproduction when site conditions (e.g., shallow soils, stony soils, fragipans, or hot and dry exposures) limit the structural complexity of the overstory and understory, and thus the intensity of competition (Johnson 1993a,b). Oaks can persist in the open understories on xeric sites despite recurrent shoot dieback. Individual root systems of oak advance reproduction are capable of becoming large and reaching ages of 50 years or more (Merz and Boyce 1956, Tryon and Powell 1984).

Advance Reproduction on Mesic and Hydric Sites. In the absence of fire or similar disturbances, mesic sites (coves, lower slopes, and northeast aspects) and hydric sites (bottomland) do not accumulate oak advance reproduction (Larsen and Johnson 1998). Pulses or waves of oak advance reproduction, as many as 50,000 to 150,000 per acre, may result from a single good acorn crop on mesic and hydric sites, but the large seedlings do not develop, because oak survival is low in heavy shade (Johnson 1975, Loftis 1990b). Hence, there are frequent and prolonged periods with little or no oak advance reproduction. Those seedlings that do persist are small and have low regeneration potential.

Accumulation of oak advance reproduction on mesic and hydric sites requires recurrent disturbances, which historically had been the role of fire before European settlement (see Chapters 3 and 4). Fire increases light at the forest floor by decreasing the density and size of woody species in the understory and by reducing overstory density. In the absence of periodic fires, mesic and hydric sites develop dense overstory canopies and complex vertical structure.

Oak-dominated stands on mesic or hydric sites do not usually regenerate to oak in the absence of disturbances that limit competing vegetation in both the understory and overstory (Johnson 1993b). However, occasionally high densities of small oak advance reproduction or a bumper crop of acorns at the time of overstory removal have led to successful oak regeneration on these sites. Normally, small oak seedlings have little chance to recruit into the overstory, when regenerating in openings in mesic sites, because they cannot compete with species such as cottonwood, green ash, sweetgum, or yellow-poplar. If, however, there are 100,000 or more oak seedlings per acre, then, by stochastic probability, some of them will recruit into the overstory following a regeneration harvest.

Reproduction by Stump Sprouts

Stump sprouts are stems of reproduction that arise from overstory trees (stems ≥ 5.08 cm dbh [diameter at breast height]) cut in a timber harvest or topkilled by fire. The separation between advance reproduction and stump sprouts by stem diameter is arbitrary (Roach and Gingrich 1968). Stump sprouts are physiologically identical to seedling sprouts, and both classes of reproduction have the potential to produce basal sprouts when the parent stem is cut or suffers shoot dieback. Stump sprouts that originate from buds at or near the ground are more likely to survive and produce a quality stem (Kelty 1988, Johnson 1994b).

Stump sprouts are the fastest growing form of oak reproduction. When growing in the open, oak stump sprouts have high probabilities of capturing growing space and maintaining dominance in the overstory. Initially, they are capable of producing four or more flushes of shoot growth per year, even in drought conditions (Johnson 1979, Cobb et al. 1985). This growth advantage is due, in part, to their having a large root system that can deliver sufficient water, nutrients, and other metabolites to the shoot, reducing water stress effects on growth.

The capacity for stump sprouting varies among species (see Figure 5.3) but also depends on a number of environmental and physiological

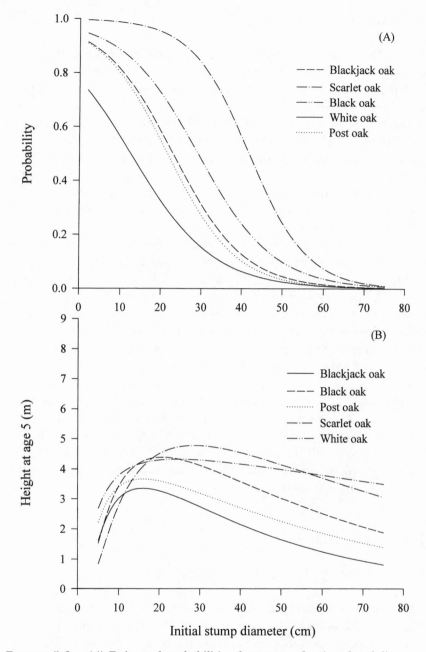

FIGURE 5.3. (A) Estimated probabilities that a tree of a given basal diameter will produce a sprout that survives to age 5 after clearcutting. (B) Estimated 5th-year heights of oak stump sprouts growing in clearcut openings in
relation to basal diameter of the parent stem. (Both adapted from Dey et al.
1996.)

factors (Solomon and Blum 1967, MacDonald and Powell 1985). The production of sprouts is governed primarily by the availability of root carbohydrate reserves, necessary to support the growth of these new shoots. Trees weakened by defoliation, oak decline, and other stresses have reduced root reserves and exhibit decreased sprouting (Powell and Tryon 1979, Wilson and Kelty 1994). Following the flush of shoot growth in the spring, root carbohydrate reserves are low, and trees have less ability to sprout vigorously if they are cut at this time. Cutting upland oaks during the dormant winter months results in higher densities of more vigorous sprouts per clump compared to trees cut during the growing season (Wendel 1975, Kays et al. 1985).

Stump sprouting frequency and growth of sprouts increase with the size of the parent tree, up to a threshold beyond which sprouting capacity declines (Figure 5.3) (Wendel 1975, Johnson 1977, 1979, Dey et al. 1996). Most oak species have high sprouting potential (nearly 100%) for stems with diameters in the range from 5.08 cm to 20.32 cm dbh. Increasingly larger trees produce fewer sprouts, and oaks larger than 50.8 cm in dbh sprout infrequently. Sprouting potential and growth also decrease in older trees (Oliver and Larson 1990, Kozlowski et al. 1991). The decrease in stump sprouting potential with increasing age and size may be due to the inability of dormant buds to penetrate the thicker bark on older trees. Although they are the most aggressive form of oak reproduction, stump sprouts should not be relied upon alone to sustain oaks in the future stand, because the stocking of sprout-producing trees is usually not high enough to maintain stand composition, unless stands are very young when harvested.

FIRE RELATIONS

Plant Characteristics and Response to Fire

The death of vascular tissue in plants occurs when they are exposed to $\geq 60°C$ for 60 seconds or longer (Hare 1965). However, longer exposure to less severe temperatures may also cause the death of tissue. The mortality of trees following fire varies by species, size of tree, bark thickness, physiological activity, stage of plant development, and tree vigor. Mortality is also influenced by fire season, fire intensity and type, and fire frequency.

Bark thickness is a major determinant of fire resistance regardless of tree species (Hengst and Dawson 1994). Thickness is the primary characteristic affecting the bark's ability to insulate cambial tissue and dormant buds from the heat of a fire. Small differences in bark thickness produce large differences in fire resistance, because the duration of heat required to kill tree cambium is proportional to the square of the bark thickness (Hare 1965).

Tree stem diameter has been correlated to survival after burning because it is directly related to bark thickness and tree height, and hence to a tree's ability to resist heat injury to the cambium or to the crown (Loomis 1973, Regelbrugge and Smith 1994). Low to moderately intense surface fires can topkill most hardwood stems that are < 10.16 cm in diameter (Waldrop and Lloyd 1991). Mortality is less for hardwood stems that are ≥ 10.16 cm, although they may suffer topkill or develop advanced decay in the lower bole.

Oaks are generally less susceptible to injury or mortality from repeated burnings than most of their competitors (see Table 5.2), because of relatively thicker bark. Although young and small oak stems (< 10.16 cm dbh) are just as likely as their competitors to suffer shoot topkill from a single fire, they are better adapted to frequent burning because of their ability to repeatedly produce sprouts long after their competitors have perished (Waldrop et al. 1987).

Fire Season

The probability of fire mortality is higher from burns during the growing season than from those during the dormant season. In the summer, lethal temperatures are reached more rapidly because the ambient temperature is higher (Whelan 1995). In addition, summer fires usually burn with greater intensity than winter fires do and therefore can kill larger diameter trees (≥ 12.7 cm dbh), especially in drought years (Waldrop and Lloyd 1991). High physiological activity and cell hydration in the summer predisposes plants to the heat of fire and increases the probability of mortality (DeBano et al. 1998). Growing season fires reduce a tree's ability to sprout following topkill because root carbohydrate reserves are relatively low after leaf expansion and shoot growth. Death of the shoot in the summer lowers root carbohydrate levels due to a loss of leaf area available for photosynthesis, a shortened season of photosynthesis, and reduced translocation of photosynthates from the leaves. Thus, sprouting capacity and sprout vigor are greater after dormant season fires than following growing season burns.

Table 5. 2

Common hardwood and conifer species in eastern North America, by vulnerability to long-term, repeated burning

Very sensitive	Sensitive	Intermediate	Resistant	Very resistant
Balsam fir	American holly	Cherrybark oak	Aspen (mature)	Bear oak
Eastern red cedar	Aspen (small)	Nuttall oak	Blackgum	Blackjack oak
Hemlock (young)	Basswood	Overcup oak	Black oak	Bluejack oak
Northern white cedar	Beech	Pin oak	Bur oak	Dwarf chinkapin oak
Red pine (young)	Bigtooth aspen	Scarlet oak	Chestnut oak	Longleaf pine
Virginia pine	Black cherry	Southern red oak	Chinkapin oak	Post oak
White pine (young)	Black walnut	Swamp chestnut oak	Cottonwood (mature)	Slash pine
White spruce	Cottonwood (young)	Swamp white oak	Hickory	Turkey oak
Yellow-poplar (repro)	Dogwood		Northern pin oak	
	Elm		Northern red oak	
	Ironwood		Red pine (mature)	
	Laurel oak		Shortleaf pine	
	Magnolia		White oak	
	Red maple		White pine (mature)	
	Sassafras		Yellow-poplar (mature)	
	Silver maple			
	Striped maple			
	Sugar maple			
	Sweet birch			
	Sweetgum			
	Sycamore			
	Water oak			
	White ash			
	Willow oak			

Sources: Gleason 1913, Nelson et al. 1933, Garren 1943, Beilmann and Brenner 1951, Curtis 1959, Heinselman 1973, Wright and Bailey 1982, Grimm 1984, Harmon 1984, Lorimer 1985, Simard et al. 1986, Burns and Honkala 1990, Reich et al. 1990, Boyer 1990, 1993, Nowacki and Abrams 1992, Abrams 1992, Lorimer 1993, Whitney 1994, Hengst and Dawson 1994, Orwig and Abrams 1994, Huddle and Pallardy 1996, Kruger and Reich 1997a,b, Hicks 1998, Brose and Van Lear 1998.

Fire Intensity

Fire intensity is a primary determining factor in the damage or mortality suffered by trees. The probability of crown scorch and cambial and root injury increases with increasing fire intensity. High-intensity fires can topkill much of the mature overstory trees in hardwood and pine-hardwood stands, but in young to middle-aged stands complete mortality from high-intensity fires is low (e.g., 3% to 8%) (Barden and Woods 1976, Regelbrugge and Smith 1994), because many of the overstory trees have a high sprouting potential. Most overstory hardwoods and conifers are able to tolerate low intensity fires, because of their high crowns and thick bark (Wendel and Smith 1986, Henning and Dickmann 1996).

Dormant or adventitious buds and other reproductive structures buried in mineral soil (below 2.54 cm to 5.08 cm deep) are protected from most surface fires, because soil is a poor conductor of heat, especially when it is dry (DeBano et al. 1998). Oak advance reproduction seedlings, with their root collar buds beneath the soil surface are less likely to suffer fire-induced mortality than species that carry their reproductive structures above ground. Survival after fire of oak advance reproduction is a function of root mass, carbohydrate reserves, and number of adventitious buds that are buried in mineral soil, for given weather conditions and fire behavior

Fire Frequency

Fire usually increases the density of trees in the smaller size classes, by causing shoot dieback and formation of sprout clumps and by promoting the establishment of new seedlings (Waldrop and Lloyd 1991). These stems are more susceptible to mortality or shoot dieback in subsequent fires, because death and injury is size dependent. Thus, frequent fires keep trees in the seedling or stump sprout state, preventing them from increasing their fire resistance by growing larger and developing thicker bark.

Fire frequency affects the accumulation and structure of fuels, which influences fire intensity and tree mortality. Increases in fuel loads during fire-free periods result in higher fire intensities and higher probabilities of tree mortality in future fires. In the absence of fire, growth of understory vegetation increases the vertical structure of fuel in the stand and hence the chance of stand-replacing crown fires, especially in drought years (Heinselman 1973). In contrast, long-term annual burning results in lower fire intensity by reducing the volume of fuel and simplifying the fuel structure.

A single fire set to improve conditions for oak regeneration after an extended fire-free period (e.g., > 20 years) often produces disappointing results, because competing trees have grown large enough to be fairly fire resistant (Johnson 1974, Wendel and Smith 1986, Van Lear and Waldrop 1988). In fact, a single fire frequently increases the density of competing vegetation, by causing stump sprouting and promoting seedling establishment. Repeated fires are necessary to improve oak regeneration in areas where fire has been suppressed for decades, especially on mesic sites.

CONCLUSION

Oak species dominate many of the forest types in eastern North America across a diversity of climate, soil, topography, and hydrology. Oaks are well adapted to persist in stressful and high-disturbance environments where drought, fire, and herbivores are prevalent. Oaks are more tolerant of drought and better adapted to surviving repeated fires than many of their competitors. However, they are only moderately able to unable to prosper in heavy shade. Historically, periodic fire has helped to control oak's competitors, allowing oaks to dominate even on high-quality sites. In the past, frequent fires created a more open stand structure than exists today, which provided adequate light for the development of large oak advance reproduction. Other human disturbances, such as logging, woods grazing, and agriculture, have contributed to oak's dominance across the landscape today. Oak seedling sprouts, with large root systems, were able to grow rapidly once agricultural lands were abandoned, and woods burning and grazing were stopped: within 40 years, in the absence of these disturbances, oak woodlands, barrens, and savannas became closed-canopy forests.

Prevailing disturbance regimes discriminate against recruiting oak on all but the driest site types. Declines in oak dominance are largely the result of man's suppression of fire for the past 50 to 100 years. Without frequent fire, forest stocking has increased, dense understories of shade-tolerant trees and shrubs have invaded oak forests, and oak advance reproduction has failed to develop in the low-light conditions. Overstory mortality due to natural disturbances such as windthrow, insects, and disease usually leads to displacement of oak by its competitors in many forest ecosystems.

Even-aged silvicultural systems (e.g., clearcut and shelterwood) and catastrophic natural disturbances favor shade-intolerant and intermediately tolerant species, while uneven-aged systems (e.g., single-tree and group selection) favor shade-tolerant and intermediately tolerant species. Oak regeneration and recruitment have been most successful where large oak advance reproduction is able to accumulate in the understory and competing vegetation is controlled. On more productive sites it may be necessary to control understory competition and reduce overstory shade to permit growth of oak advance reproduction.

Oak silviculture is inherently complex because successful regeneration is dependent upon having adequate large oak advance reproduction before completion of the regeneration harvest. The process of developing adequate oak advance reproduction normally requires time and a series of planned disturbances to the overstory and understory. Underplanting shelterwoods with oak seedlings can reduce the time required to get adequate advance reproduction established, although this method may be costly. Historically, fire was key to maintaining oak dominance, and realization of this fact has led to increased efforts to develop prescribed burning methods for oak regeneration (see Chapter 18). Successful oak regeneration follows from silvicultural practices that increase light to oak advance reproduction by reducing overstory density while controlling competing vegetation.

Unfortunately, the silvicultural practices that are most useful for perpetuating oak, such as shelterwood harvesting and prescribed burning, can be socially unacceptable or too time-consuming and expensive to interest private landowners. This is particularly significant because most of the oak resource is owned by nonindustrial private landowners. It is a challenge to develop innovative silvicultural systems for oaks that are socially acceptable and affordable. Forest managers need to understand the regeneration and stand replacement process and the critical need to incorporate measures of stand regeneration potential into the management process. Planning for oak regeneration should begin well in advance of the regeneration harvest, and assessments of stand regeneration potential can be used to develop silvicultural prescriptions that are more likely to regenerate oak.

Chapter 6

Native Diseases and Insects That Impact Oaks

STEVEN W. OAK

In the past, diseases and insects were thought of as impediments to normal forest development. Outbreaks of them were the targets of aggressive suppression efforts, when economic threats to fiber or solid wood production were perceived. More recently, there has been increasing recognition that they are integral parts of functioning forest ecosystems and have beneficial effects (U.S. Forest Service 1988a). While non-native pathogens and insects are still marked for quick and forceful eradication, alternative responses to infestations or epidemics of natives are filtered through a more complex combination of human values that includes recreation, aesthetics, biodiversity, and wildlife habitat in addition to more finely tuned economic thresholds. Increasingly, these considerations are leading to decisions to accept the consequences and take no action.

It is important to clarify several terms at this point, to avoid confusion in the terminology used to describe tree disease. *Disease* can be defined in many ways, but it is perhaps best understood as the interaction between a suscept (or host) and some agent, resulting in a sustained impairment of function, structure, or form of the suscept (Tainter and Baker 1996). Diseases caused by biotic agents (fungi, bacteria, viruses, etc.) are termed infectious, and the causes are referred to as pathogens. Those caused by abiotic agents (chemical or physical environmental factors) are termed noninfectious. Visible expressions of the disease interaction in the suscept (such as branch dieback, stem swellings, foliage yellowing or decay) are called symptoms, while macroscopic physical manifestations of the pathogen (such as mushrooms or other reproductive structures) are signs. Because disease is not synonymous with cause, there are no native or non-native diseases, only native or non-native pathogens.

The effects of non-native pathogens and insects are similar in kind to those produced by natives, but the magnitude is more extreme. This is due to the lack of co-evolved resistance mechanisms in the host and the absence of the parasites and predators that serve to regulate populations of the non-native pathogens and insects in their ecosystems of origin. *Cryphonectria parasitica* (Murrill) Barr (the cause of chestnut blight) and the European gypsy moth (*Lymantria dispar* L.) are examples that have had profound effects on oak forest ecosystems in North America. They continue to exert change themselves, and they influence the effects of native agents. This chapter will focus on the ecosystem roles of native insects and diseases with native causes; the interactions among insects, diseases, and past disturbance; and the ways that specific native insects and pathogens can alter wildlife habitat in oak forests. Chestnut blight will be discussed briefly as an interacting factor affecting native agents, while European gypsy moth is the topic of the next chapter in this book.

ECOSYSTEM ROLES OF NATIVE PATHOGENS AND INSECTS

Haack and Byler (1993) have summarized the roles of native pathogens and insects in forest ecosystems. One role is a part of basic ecosystem process, that is, recycling carbon and other nutrients through decay. Others involve beneficial relationships with plants through mycorrhyzal symbiosis and pollination. Still others involve more complex interactions with wildlife and habitat. Pathogens and insects can serve as food sources for vertebrates, invertebrates, and microorganisms and create specific habitat components, such as dead standing and down trees, that provide decayed wood for nesting cavities or forage. They also regulate populations of woody and herbaceous vegetation directly or indirectly through forest succession.

The action of pathogens and insects in successional change is through growth loss and mortality of host trees. These agents can directly regulate host population size and genetic composition and restrict host distribution at various spatial scales. Community diversity may be diminished or enhanced, relative competitive advantage altered, and the availability of food and cover for animals changed by the creation of canopy gaps ranging in size from single trees to large groups or even landscapes (Gilbert and Hubbell 1996).

Changes in community structure and composition during epidemics or outbreaks are a function of the distribution of susceptible individuals on the landscape. For this discussion, community structure refers to the physical arrangement of canopy gaps, down trees, snags, live trees, and other associated vegetation. Composition refers to the distribution of plant species (emphasizing trees) and genetic diversity within species. Both elements can be considered at various spatial scales. There is a common perception that native pathogens and insects prey on the weaker members of the population, culling the less fit and thereby increasing the overall vigor of the population. If the population is normally distributed with respect to vigor, and the low-vigor individuals are randomly distributed on the landscape, then the impacts are minimal and may be barely noticeable. However, these conditions do not always prevail. Large segments of the landscape may be occupied by a single susceptible host species or species group, and/or outbreak conditions may overwhelm even the most vigorous individuals. Interactions are complex and changes in composition and structure highly variable, based on existing community structures, host susceptibility, and the virulence or aggressiveness of the agent. Ultimately, wildlife habitat is altered, for better or worse, as a result of changes in forest composition and structure triggered by mortality or growth loss in host trees.

INTERACTIONS OF DISEASES, INSECTS, AND PAST DISTURBANCE

The tendency of forest managers is to simplify when assessing the impact of diseases and insects in forest ecosystems and to consider them in compartments separated from each other in space, time, and effect. In fact, multiple agents often interact over long time spans to shape forest composition and structure at any given point in time. One mechanism by which they effect change is predisposition, or weakening the host in a way that makes later invasion or attack by another agent more likely. Factors in the physical environment (e.g., soil depth or texture, fertility, moisture availability) or pathogens and insects themselves may act as predisposing agents for other pathogens or insects. Another mechanism is to cause a breach of host defense barriers by wounding the tree so that another agent can more effectively colonize. Usually insects act as wounding agents, enabling attack by pathogens that cannot otherwise overcome host defenses. Insect *vectors* differ from simple wounding agents in that they not only provide ingress to the host but also actively transport and transmit a pathogen to the wound site in such a way that infection and disease result.

The impacts of past disturbance on the present composition and structure of forests is often overlooked as an interacting factor in attempts to understand the effects of pathogens and insects. In many cases, the present composition and structure of North American oak forests bear little resemblance to those under which hosts, native pathogens, and insects co-evolved. Most of the changes have been triggered during the past 150 years, as a result of the increase in the human population, and they reached a peak around 1900. The forces that drove these changes at the turn of the century included wide-scale logging to supply needed fuel, building materials, and other wood-derived products; importation, for commerce or by immigrating people, of non-native trees and other plants and thereby, inadvertently, their pathogens and insects; conversion of forested land to marginal agricultural land, followed by abandonment and reversion to forest as agricultural production became more efficient; and widespread wildfire (both natural and human-caused), followed by aggressive fire suppression.

The lasting effects of these forces on oak forests are illustrated by the case of chestnut blight, which resulted in some of the greatest changes in a forest ecosystem caused by a non-native insect or pathogen ever recorded. The disease was first detected in American chestnut (*Castanea dentata* [Marsh.] Borkh.) in the Bronx, New York, in 1904 (Merkel 1905). Judging by the status of the epidemic when first detected, the pathogen (*C. parasitica*) was undoubtedly introduced several years earlier. Aided by additional introductions and/or spread on plant materials moved well outside of New York (Gravatt 1925), the pathogen spread very rapidly in the population of susceptible native trees. By 1940 it had killed 50–99% of the American chestnuts throughout nearly all of its botanical range (Clapper and Gravatt 1943).

The chestnut blight pathogen had only minor direct effects on oaks. Nonlethal infections occurred on *Quercus alba* L., *Q. coccinea* Muenchh., *Q. stellata* Wangenh., and *Q. virginiana* Mill. (Sinclair et al. 1987). However, the indirect effects were profound. American chestnut was the most important hardwood tree species in eastern forests, composing 82.5 to 100% of hardwood forest in some localities (Ashe 1911, Braun 1950). Estimates of American chestnut composition at large landscape scales ranged from 25 to 50 percent in the heart of the native range in the Southern Appalachian Mountains (Ashe 1911, Buttrick 1925). As chestnuts died, newly available growing space was quickly occupied by other species already positioned in the mid- and understory by earlier disturbances, such as repeated fire and logging. Chestnut replacement was variable, but typically oak species (*Q. prinus* L., *Q. rubra* L., and *Q. velutina* Lam., in particular) increased (Korstian and Stickel 1927, Stephenson 1986). On sites where previous fire had not been an important disturbance agent, thin-barked species such as red maple (*Acer rubrum* L.), black gum (*Nyssa sylvatica* Marsh.), and yellow-poplar (*Liriodendron tulipifera* L.) succeeded along with the oaks (Arends and McCormick 1987, Lorimer 1980). These changes occurred over a very short span of time on millions of acres from New England to north Georgia and west through the botanical range of chestnut to the Mississippi Valley. With the advent of state-federal cooperative fire control programs, public land acquisition, and lower rates of harvest compared with those experienced near the turn of the century and before, these new and historically unprecedented oak forests have aged relatively free of disturbance. Age structure and species composition are simplified and less diverse than those of the forest that existed prior to these large-scale human impacts.

Oaks in these forests may fill some of the ecological niches once occupied by American chestnut, but their substitution has been incomplete at best. We can expect that diseases like oak decline and insects like the fall cankerworm (*Alsophila pometaria* Harris) affect these forests in different ways than they did in the past and that consequent wildlife habitat impacts will be less predictable. These conditions and others will be discussed later in this chapter.

WILDLIFE HABITAT EFFECTS OF NATIVE PATHOGENS AND INSECTS IN OAK FORESTS

The effects of pathogens and insects on the plant community are ecologically neutral. What constitutes "severe" or otherwise unacceptable damage is subject to the screen of human values and the resource that is at risk. To the production forester, the death of overstory oaks from oak decline may be a severe economic loss. The same mortality may be viewed by a wildlife biologist as positive, because it adds dead standing or down trees to habitat structure and diversity, or negative, if hard mast production potential is reduced in the long term. The wildlife biologist's interpretation varies depending on the particular habitat needs of different species of interest.

For this discussion, wildlife habitat effects are arbitrarily classified as alterations of forest vegetation composition, structure, or reduction of hard mast production potential, and oak forests are defined as those where oaks form a major component of the overstory. Oaks in lower-canopy positions may constitute significant numbers of trees in a stand or landscape, but their contributions to structural habitat features are not unique to *Quercus* spp. Further, they make little contribution to hard mast production potential, because they are relatively intolerant of shade.

Increases in oak composition resulting from pathogens or insects result only where oaks would be released after the elimination of competing species or where preexisting sources of oak reproduction could compete in the new environment. This might occur where pines in a mixed stand were eliminated or reduced by an insect or disease specific to pines, like the southern pine beetle or annosum root rot. Detailed examination of this dynamic is beyond the scope of this chapter.

Oak composition is reduced by insects and diseases that kill oaks and allow associated species to flourish in the new growing space. When species composition changes as a result of the loss of oaks, the effects of the loss are primarily in the reduction of acorns that would otherwise supply wildlife food and a source of new seedlings for restocking the stand. Composition change may be for a relatively short time, as when oaks are lost from the overstory while still retaining oak regeneration capacity. This regeneration capacity can be in the form of sproutable root systems or reproduction in lower canopy positions that can compete with associated species under the new environmental conditions. With time, they grow into the overstory to constitute a species composition comparable to the earlier forest. Alternatively, an oak component may be lost or altered for a relatively long period if insufficient sources of competitive oak reproduction remain after the loss of larger overstory trees. Competing species then take the newly available growing space, excluding new oak reproduction that is incapable of competing in the new forest. The consequences of a shift in species composition towards fewer or a less diverse mixture of oaks depend on the habitat needs of particular wildlife species and the spatial extent of that shift. Loss of oaks is inconsequential to species that do not depend on acorns for food, regardless of the scale of the loss. For mobile species that do depend on acorns, local changes would be unnoticeable, because the wildlife can easily move to unaffected locations. The effects of landscape-scale changes would be most severe if there were sparse distribution or absence of desirable habitat and/or long distances to refugia.

Structural changes caused by diseases and insects could include a decrease in overstory density, which in turn would produce a growth response in understory vegetation due to increased light to the forest floor; an increase in the number, size, and distribution of canopy gaps; and an increase in the dead standing trees (snags) and down wood. Taken together, compositional and structural changes resulting from the loss of oaks to diseases and insects translate into habitat changes such as a reduction in hard mast quantity and quality (thereby affecting food supply and oak regeneration opportunities), an increase in canopy openings, short- or long-term shifts in species composition affecting oak abundance and diversity, reduced canopy density, and increases in denning sites and down woody debris.

The abundance and diversity of oaks in North America are matched by the number of diseases and insects that affect them. Most diseases and insects alter multiple habitat components to some degree at a local or stand level. For example, when oaks are killed by a girdling canker rot fungus, not only is habitat structure for forage or denning created, but acorn production potential is reduced locally, and the oak composition of the stand is smaller, at least in the short run. The effects on acorn production potential and oak composition in this case are relatively minor, because the incidence of lethal canker rots in oak stands is typically low. Surrounding oaks released from competitive pressures might even increase mast production, if they are given time to increase crown volume and improve carbohydrate status. The primary habitat effect in this case is on structure, for even the addition of a few snags per acre can be significant where this attribute is lacking. There are, however, a few conditions in which multiple habitat effects are significant, primarily because they occur across a large, landscape scale. Specific diseases (Table 6.1) and insects (Table 6.2) will be grouped and discussed according to their primary habitat effects or significant multiple effects.

Table 6.1
Pathogens of oaks native to North America causing substantial wildlife habitat effects

Disease	Symptom	Scientific name	Habitat effects	Scale
Acorn rot	Acorn discoloration and decay	*Fusarium solani*	Acorn production	Local/stand
Botryodiplodia/ botyrosphaeria cankers	Twig dieback	*Botryodiplodia gallae, Botyrosphaeria quercum* and others	Acorn production	Local/stand
Canker rots	Stem decay	*Inonotus, Poria, Irpex* spp.	Structure, food	Local/stand
Drippy nut	Wet ooze from damaged acorns	*Erwinia quercina*	Acorn production	Local/stand
Leafy mistletoe	Parasitic plant in crown, dieback	*Phoradendron serotinum* and others	Acorn production, food	Local/stand
Oak anthracnose	Foliage and twig dieback	*Discula quercina*	Acorn production	Local/stand
Oak decline	Crown dieback, tree death	Multiple interacting factors including drought, physiologic maturity, defoliation, root disease (*Armillaria* spp.), hypoxylon dieback (*Hypoxylon atropunctatum*), twig cankers, two-lined chestnut borer (*Agrilus bilineatus*)	Structure, composition, acorn production	Landscape
Oak wilt	Crown dieback, tree death	*Ceratocystis fagacearum*	Structure, composition	Landscape
Stem and butt decay	Decay	*Inonotus, Polyporus, Hericium, Pleurotus* spp.	Structure, food	Local/stand
Stem cankers	Discrete dead areas on stem	*Nectria, Strumella, Endothia* spp.	Structure	Local/stand

Table 6.2
Insect pests of oaks native to North America causing substantial wildlife habitat effects

Insect	Plant parts affected	Scientific name	Habitat effects	Habitat scale
Acorn moth	Acorns	*Blastobasis glandulella*	Acorn production	Local/stand
Acorn weevils	Acorns	Many *Curculio* spp.	Acorn production	Stand/landscape
Filbertworm	Acorns	*Cydia latiferreana*	Acorn production	Local/stand
Gall wasps	Acorns	*Callyrhytis operator, C. fructosa*	Acorn production	Local/stand
Spring defoliators (skeletonizer; leaf roller, leaf tier, elm spanworm, spring and fall cankerworms, California oakworm)	Foliage, acorns, roots	*Bucculatrix ainsliella, Archips semiferanus, Croesia semipurpurana, Alisophila pometaria, Palaecrita vernata, Ennomus subsignarius, Phryganidia californica,* others	Acorn production, food	Landscape
Wood borers (carpenter worm, red oak borer, white oak borer)	Stem	*Prionoxystus robiniae, Enaphalodes rufulus, Goes tigrinus,* others	Structure, food	Local/stand

DISEASES AFFECTING MULTIPLE HABITAT COMPONENTS AT LANDSCAPE SCALES

Oak wilt and oak decline are two diseases affecting forest composition, structure, and acorn production at landscape scales. The broad scale of these diseases is largely due to the alteration of native oak forest by human activities or to influences that yielded new oak forests with different characteristics that made them more prone to these diseases.

Oak Wilt

Oak wilt is a systemic vascular wilt disease caused by the fungus *Ceratocystis fagacearum* Bretz. Trees in forest stands become infected when the fungus is vectored by contaminated adults of one of several species of sap-feeding or oak bark beetles (Nitidulidae and Scolytidae, respectively). Sap-feeding beetles are attracted to fresh wounds on healthy oaks in spring and deposit spores, while contaminated bark beetles emerge from oak wilt–killed trees and introduce the fungus to wounds they create in small twigs. Newly formed vessel elements become infected and fungus colonies soon develop, producing spores that are carried in the water column. Xylem vessels eventually become clogged with spores, mycelium, and fungus-produced compounds, some of which are phytotoxic. Infections progress into branches, twigs, and even roots where they can persist for many years. Localized spread occurs through functional root grafts between infected and healthy trees. Root grafting occurs most frequently among trees of the same species but does not occur with equal regularity in all locations. Interspecific grafts are not unheard of (Gibbs and French 1980).

Leaves on infected limbs develop a water-soaked or bronzed appearance, droop, and abscise in mid–growing season. Species in the red oak group are more susceptible to this disease and may die within weeks of the first symptom expression, especially in cool, moist northern areas. Species in the white oak group are somewhat more resistant, and symptoms progress more slowly in them. Some trees display a remission or even recover completely. In hotter, drier southern areas, progressive dieback symptoms persist in red oaks for several years before trees succumb, mimicking the symptomatology among white oak species in the north. After infected trees die, the fungus grows saprophytically from the current growth ring deeper into the sapwood, where it can persist in stems and larger branches for 1–3 years. The disease cycle is completed when beetle vectors become contaminated by spores produced by the fungus beneath the bark of infected trees. Sap-feeding beetles become contaminated after being attracted to the sweet odor emanating from spore-bearing fungus mats, while bark beetles become contaminated as brood emerge from infected trees. Fungus fruiting is more common on red oaks than white oaks but is inhibited on all hosts by hot, dry conditions.

Control of oak wilt is a two-pronged effort. Localized spread is limited by efforts to disrupt common root systems by trenching around a buffer surrounding existing infection centers. Minimizing long distance spread is accomplished by timely salvage and sanitation of infected trees to prevent feeding, breeding, and dispersal of the insect vectors.

The pattern of oak wilt in diverse eastern hardwood forests is one of small, scattered, slowly expanding infection centers consisting of one to a few trees (Gibbs and French 1980). However, where forest composition is more uniform and covers large areas, extremely large infection centers can quickly develop. Alterations of native forests by humans have resulted in the latter form of oak wilt expression. In Wisconsin, coppice forests of northern pin oak (*Q. ellipsoidalis* E.J. Hill) developed after logging and fire constricted the diversity of the historic forest (Gibbs and French 1980). Similarly, the concurrent introduction of domestic livestock and fire suppression in the Texas Hill Country has resulted in expansion of the red oak–live oak–juniper forest into former grasslands and an increase in the area of clones of root-suckering live oaks (*Q. fusiformis* Small and *Q. virginiana*) with common root systems (Reisfield 1995). Due to these changes in composition, structure, and distribution, the new Wisconsin and Texas oak forests are subject to larger infection centers, wider distribution, and higher incidence of oak mortality after infection by the oak wilt pathogen than were the original forests. In the case of the pin oak coppice forest, oaks, along with their hard mast production potential, are lost for the long term as nonhost species mixtures fill the voids created by the disease. A more diverse structure is created as dead standing and down trees are added. The ultimate habitat consequences of oak wilt in this setting may be positive, however, as new, more diverse species mixtures are created.

In the case of Texas oak wilt, the consequences to wildlife habitat of at least two endangered species of neotropical migratory birds are more far-reaching. Habitat for the golden-cheeked warbler (*Dendroica chrysoparia*) is at risk from both oak wilt and the efforts instituted to control the disease, while disease-induced structural change could be positive but compositional change negative for the black-capped vireo (*Vireo atricapillus*).

Both birds winter in Central America but nest in the oak-juniper woodlands of central Texas. The golden-cheeked warbler is most successful in large tracts (50 ha and larger); it requires both the oak for insect foraging and the juniper for nesting materials. Wahl and Diamond (1990) identified habitat loss through fragmentation as the primary threat to the bird's survival. Oak wilt infection centers add to the fragmentation that is caused by urbanization and other land use pressures, by creating openings and destroying oak sprout regeneration potential for the long term. Control of oak wilt infection centers in otherwise suitable habitat would have the same effect, because openings would be created by the trenching, salvage, and sanitation procedures. Control actions would preserve oak sprout regeneration potential in the surrounding landscape by reducing the probability of new infection centers.

By contrast, the black-capped vireo is most successful in somewhat disturbed oak-juniper woodlands, which have an open, patchy, two-layered structure. The lower shrub layer is used for nesting habitat while the oak component in the upper layer provides insect foraging (Texas Parks and Wildlife Department 1988). Oak wilt infection centers would create the desirable patchy aspect in the short run. However, Texas oak wilt tends to persist, spread, and kill oaks until nearly all suitable hosts in an area are killed. The long-term loss of oak sprout regeneration potential from unchecked oak wilt would be detrimental to black-capped vireo habitat.

Oak Decline

Oak decline is a disease of complex etiology affecting physiologically mature trees. It involves interactions between long-term predisposing stress, such as that caused by climate or site productivity; short-term inciting stress, such as that caused by drought or spring insect defoliation; and contributing organisms of secondary action, such as armillaria root disease (caused by *Armillaria mellea* [Vahl. Ex Fr.] and perhaps other *Armillaria* spp.) and the two-lined chestnut borer (*Agrilus bilineatus* Weber) (Manion 1991). The temporal sequence of these three groups of factors is important in the ultimate expression of oak decline.

Predisposing factors such as climate and site productivity determine the onset of physiologic maturity of the host. This is the point in a tree's life history when critical levels of physiologic processes such as water transport efficiency, translocation efficiency, and the balance between photosynthesis and respiration are reached (Hyink and Zedaker 1987). Inciting stress factors such as extended drought or spring defoliation by insects or late frost alter carbohydrate chemistry in physiologically mature trees to the extent that photosynthesis shuts down and starch stored in roots is converted to soluble sugars to support metabolism. This change in carbohydrate chemistry stimulates *A. mellea,* a saprophyte ubiquitous in oak forests, to colonize even more root mass, which further compromises water relations (Wargo and Houston 1974). Twigs and branches die back progressively over a period of years in an effort to accommodate an impaired root system. This dieback typically progresses from the top and outside of the crown downward and inward and is often associated with weakly pathogenic canker fungi. In drier parts of the southeastern United States, *Hypoxylon atropunctatum* (Schw. ex Fr.) Cke. is an important interacting canker fungus (Starkey et al. 1989). The two-lined chestnut borer is attracted to stressed oaks and, together with root disease, kills them (Wargo 1977). While some decline-associated mortality shows little prior dieback evidence, most trees exhibit evidence that can be dated back 2–5 years. Radial growth increment reveals differences between healthy and decline-afflicted oaks of the same species and age class that date decades earlier (Tainter et al. 1990).

Red oak species are more prone to mortality from decline than white oak species, and black oak are considered most susceptible (Starkey et al. 1989). The pattern of oak decline on the landscape varies with initial stand species composition, stand age structure, decline severity, mortality incidence, and the duration of decline before inciting stress is eased. Patches of mortality can range from a few trees in stands with diverse species composition and age structure, to several hundred acres in stands with a more uniform composition of physiologically mature red oaks defoliated repeatedly by gypsy moth.

Periodic episodes of oak decline have been reported in virtually the entire range of eastern oaks since the early 1900s (Long 1914, Balch 1927). Widespread incidence during the mid-1980s in the southeastern United States reflects the coincidence of physiologic maturity of oak cohorts that developed after chestnut blight and fire control on a regional scale (Mueller-Dombois et al. 1983) and extended regional drought coupled with repeated gypsy moth defoliation, in Virginia, or native insect defoliators, elsewhere (Starkey et al. 1989). Overall, 9.9% of upland oak forest acreage in 12 southeastern states (3.9 mill. a.) are affected by oak decline with the largest portion (2.5 mill. a.) in Virginia, North Carolina, and Tennessee (Hoffard et al. 1995).

Habitat impacts of oak decline were interpreted by Oak et al. (1988) to include structural changes such as creation of small to large canopy openings, reduced canopy density, short-term stimulation of understory species, potential increases in cover type diversity, and increased denning and cavity nesting sites. Long-term shifts in species composition can occur where competitive oak reproduction is absent or in short supply. Overall, decline sites have lower oak composition, and the remaining oak component is less diverse as the more-susceptible red oak species dwindle relative to white oaks. Effects of oak decline on mast production potential can be significant. For a moderately affected stand in Virginia, production was estimated, from standard forest inventory and crown health measures, to be 41% below what the same stand would have produced if not infected. Losses were projected to increase to 58% within 5 years. These projected reductions will persist for a long time, because residual oaks are themselves prone to decline, and competitive oak reproduction to replace dead overstory oaks is lacking.

The incidence, severity, and potentially long duration of effects of oak decline over large landscapes in the southeastern United States make it one of the most serious forest disease problems for habitat managers in the region. Risk rating systems have been developed (Oak et al. 1996, Oak and Croll 1995) to guide managers in identifying vulnerable stands where silvicultural methods can be employed to preserve and enhance an oak component. Key stand and site factors for risk rating include oak basal area, soil texture and depth, slope gradient, stand age, and site index.

DISEASES, INSECTS, AND PARASITIC PLANTS AFFECTING ACORN PRODUCTION POTENTIAL

This group of agents produces more or less direct effects on acorn production potential, by affecting reproductive structures or acorns themselves. Exceptions are leafy mistletoes (*Phoradendron* spp.) and spring defoliating insects, which have more indirect effects.

Foliage and Twig Pathogens

Some foliage and twig pathogens have the potential, under epidemic conditions, to affect wildlife habitat by reducing acorn production potential. In white oaks, this effect may last for only one crop if the outbreak lasts one year and only the current year's flowers are affected. In red oaks, however, the same one-year outbreak may affect two crops of acorns, by destroying year-old female flowers pollinated the previous year as well as the current-year flower crop. The most notable disease of this type is oak anthracnose, caused by the fungus *Discula quercina* (West) v. Arx (and perhaps additional, closely related fungi). The most damaging situation occurs when frequent rains coincide with moderate temperatures in early spring. These conditions are conducive to spore production by the pathogen and subsequent infection of succulent, susceptible foliage (Sinclair et al. 1987). The fungus eventually grows down the leaf petiole and into the twig, causing dieback and loss of reproductive structures. Suitable conditions for infection may occur later in the growing season, but in that case the disease normally stops with leaf spotting or blotching and has no effect on acorn production potential. Twig dieback caused by oak anthracnose has been recorded for many oak species, but white oak is considered most susceptible to this form of the disease. I have observed fungus outbreaks in a northern red oak seed orchard, where it caused foliage blight and twig dieback. Some individuals showed symptoms in most years, exhibiting putative genetic susceptibility. Symptoms may be common in localities and across large landscapes in epidemic years, but the varied phenology of different oak species found in eastern mixed hardwood forests probably buffers the effects of anthracnose at landscape scales. Weather conditions conducive to epidemics may occur in consecutive growing seasons, but they seldom persist for longer periods.

Twig canker fungi that do not infect foliage can have similar effects on acorn production potential. *Botryosphaeria* spp. (including *B. rhodina* [Berk. & Curt.] von Arx, *B. dothidea* [Moug. ex Fr.] Ces. & de Not., *B. obtusa* [Schw.] Shoemaker, and *B. quercum* [Sch.: Fr.] Saccardo) and *Botryodiplodia gallae* (Schw.) Petrak & Sydow. are all opportunistic pathogens that can cause twig or branch dieback on oaks (Sinclair et al. 1987). Unable to infect healthy, vigorous trees, they require drought stress or wounds created by freezing, insects, or other mechanical means to overcome host defenses. They may be the proximal causes of twig and branch dieback observed in oak decline. These fungi are ubiquitous in forest ecosystems, but environmental conditions that predispose trees to infection rarely persist for more than a few consecutive growing seasons. Except in cases where these fungi are involved in oak decline, the habitat impact on acorn production would be local and relatively ephemeral.

Insect Pests of Acorns

Many insects are well-documented destroyers of acorn crops. Several *Curculio* weevil species are the best known and have infested up to 90% of the acorns in some collections (Gibson 1982). Virtually all oak species are attacked by one or more of 22 different weevils recorded by C. E. Williams (1989). Adults themselves cause little damage, but they lay eggs in late summer in niches they create beneath the shell. The developing larvae consume large quantities of the nut tissue in a few weeks. Oak embryos in infested acorns may escape damage and germinate, but they will grow more slowly than those in acorns free of weevils. Due to the large number of species attacking acorns and their ubiquitous distribution, acorn weevils most likely have impacts on light to moderate acorn crops at the landscape scale. In heavy mast years, enough acorns apparently escape infestation to both establish seedlings on the forest floor and supply the needs of wildlife.

Feeding by larvae of the filbertworm (*Cydia latiferreana* Walsingham) is usually lethal to infested acorns, but infestation levels of this insect are much lower than for acorn weevils (Gibson 1982). Nevertheless, they have been responsible for a high percentage of the losses in poor seed years (Drooz 1985). Adult moths lay eggs on leaves near acorn clusters in midsummer, and newly hatched larvae bore directly through the shell.

Two species of gall wasps *Callyrhytis operator* (O.S.) and *C. fructuosa* (Weld.) also can kill intact acorns, but they are typically found at a lower incidence than either *Curculio* spp. or the filbertworm (Gibson 1982). Acorns infested by *C. fructuosa* look normal on the outside but may contain as many as two dozen larvae individually encased in small stony galls. *C. operator*, on the other hand, reveals its presence by inducing a gall on the side of the acorn shell. Infested acorns abscise prematurely. It is unlikely that gall wasps have impacts on acorn crops beyond the local level, in most years.

Some agents increase losses by invading acorns that are sprouted, cracked, or previously infested. Secondary invaders include the acorn moth *Blastobasis glandulella* (Riley) and several species of *Conotrachelus* weevils (Drooz 1985, Galford 1986), putatively pathogenic fungi including *Fusarium solani* (Mart.) Appel. & Wollenw. Emend. Snyd & Hans.

[S&H, M&C] and *F. oxysporum* Schlecht. emend. Snyd. & Hans. [S&H, M&C] (Vozzo 1984; author's data), and the bacterium *Erwinia quercina* sp. n. Drippy nut disease in California live oak and interior live oak (*Q. agrifolia* Née and *Q. wislizenii* A DC.) acorns results when *E. quercina* infects the oviposition wounds of gall wasps (Hildebrand and Schroth 1967). The disease is named for the wet ooze emanating from the wound site, which is a product of anaerobic fermentation of tree sap and acorn contents.

Leafy Mistletoes

Leafy mistletoes (*Phoradendron* spp.) are chlorophyllous plants that can indirectly reduce acorn production potential by parasitizing young branches that subtend flower-bearing shoots. Four native species occur on oaks in North America: *P. serotinum* in the East, and *P. tomentosum, P. villosum,* and *P. coryea* in the West and Southwest (Sinclair et al. 1987). The seeds provide food for birds, which deposit them on susceptible branches. Heavy populations can build up in favored host trees as birds consume seeds and reinfest branches while roosting. Well-developed mistletoe plants induce water stress, which can cause abortion of flowers or young acorns and, eventually, branch dieback. Infestations can be locally heavy but seldom extend to the stand level scale.

Spring Defoliators

Oak defoliators are common at all times of the growing season, but spring defoliators have the most important habitat impacts. Mid- and late-season defoliators may be just as noticeable, but these insects remove foliage at a time of year when new shoot and radial growth has been at least partially supplied and carbohydrate storage is underway. Spring defoliators, on the other hand, remove foliage just after enormous expense of stored carbohydrate. If defoliation is sufficiently severe, host trees are stimulated to refoliate, further drawing down carbohydrate reserves. This compromises root health and a reallocation of carbohydrate away from flower and seed production. Flower and young acorn abortion can result (Gottschalk 1989). This source of loss is added to the direct consumption of flowers by these insects. Outbreaks in successive years, and/or compounding stress, such as that caused by drought, set the stage for oak decline, already discussed.

Many Lepidoptera species are counted among spring defoliators. The most injurious in North America are the elm spanworm, oak leaf tier, oak leaf roller, and oak skeletonizer (*Ennomos subsignarius* Hubn., *Croesia semipurpurana* Kearf., *Archips semiferanus* Wlk., and *Bucculatrix ainsliella* Murt., respectively, all mostly eastern species); the California oakworm (*Phryganidia californica* Pack. which feeds on many western oak species inside California); and the fall and spring cankerworms (*Alsophila pometaria* Harr. and *Paleacrita vernata* Peck, respectively, both transcontinental in distribution) (Johnson and Lyon 1976). In all cases, the larval stage is the damaging one. Life cycles vary somewhat, but egg hatch is synchronized with early leaf expansion. Early instars cause shotholing or partial defoliation injury, while later instars eat the largest portion of leaf biomass before pupating.

Populations of native spring defoliators are usually kept in check by a suite of native predators, parasites, and diseases, but outbreaks can occur when populations of these agents lag behind. Outbreaks usually do not last for more than a few consecutive growing seasons, but more persistent outbreaks have occurred. An outbreak of elm spanworm occurred between 1878 and 1881 which at its peak covered up to 1.5 million acres in western North Carolina, northern Georgia, and eastern Tennessee. A similar area was affected again between 1955 and 1963 (Drooz 1980). Multiple defoliators, including the oak leaf roller and oak leaf tier, were responsible for an outbreak that at its peak covered over 600,000 acres in Pennsylvania between 1958 and 1965 (Drooz 1985, Nichols 1968) and again in 1970–72 (Johnson and Lyon 1976). Because of outbreak dynamics, these insects can by themselves reduce acorn production potential at a landscape scale, but this impact is usually confined to the years of defoliation and one or two additional years or until carbohydrate metabolism returns to normal.

In addition to predisposing oak to decline and affecting flower and seed production, native defoliators influence habitat by serving as a food source for a variety of arthropods, small mammals, and birds. These agents, along with epidemics of viral or fungal diseases, are largely responsible for the collapse of outbreak populations.

Diseases and Insects Primarily Affecting Forest Structure

This group of agents alters habitat by creating specific structural attributes such as dead standing trees, snags, or down woody debris useful as den or nesting sites, foraging habitat, and perch or roosting points. Included are decay fungi, stem cankers, and wood-boring insects. They are very common in oak forests but seldom have serious effects beyond the local or stand scale, as they usually affect single trees or small groups of trees.

Wood Decay Fungi

Most decay fungi are in the class Basidiomycetes and fall into one of two general categories, stem and butt decay or canker rot. Stem and butt decay fungi invade wounds made by insects, fire, storms, or mechanical means and decay a core of heartwood compartmentalized to varying degrees by wood anatomy and active tree defenses (Shigo 1979). There are many genera and species of stem and butt decay fungi that act on oaks; some of the most common are *Hericium erinaceus, Inonotus sulphureus, Polyporus lucidus, P. fissilis,* and *Pleurotus sapidus.* The fruiting structures of some of these are edible and serve as food for vertebrates and invertebrates alike.

Canker rot fungi usually begin decaying heartwood after invading through dead and dying branch stubs, but, unlike other decay fungi, they can invade and kill sapwood and vascular cambium. By overcoming compartmentalization mechanisms of host trees in this way, an open and ever-enlarging canker face is produced. Common canker rot fungi include *Inonotus andersonii,* a transcontinental organism particularly important as a mortality agent on oaks in the southwestern United States (Sinclair et al. 1987); *I. hispidus,* common on many southern oak species; *Phellinus spiculosa;* and *Irpex mollis.*

Decay can cause tree mortality in severe cases where butt rot involves a large part of the root system, but more typically, affected trees fall over or break at points of structural weakness in high winds or after being loaded with ice and snow while still alive.

Wood-Boring Insects

Most of the wood borers affecting oak are beetles among the so-called roundheaded and flatheaded borers (Coleoptera: Cerambycidae and Buprestidae, respectively), named for the general shape of the damaging larval stage. Some of the most common are the white oak borer (*Goes tigrinus* De Geer), the red oak borer (*Enaphalodes rufulus* Haldeman), and the two-lined chestnut borer (*Agrilus bilineatus* Weber). The carpenterworm (*Prionoxystus robiniae* Peck) is among the most common of oak wood borers across North America, but it is a moth larva, rather than a beetle.

The life cycles of these insects are usually long (2–5 years from egg hatch to adult emergence). In some species, ovipositing females find suitable hosts by sensing host-produced volatile chemicals (Dunbar and Stephens 1976), sap from around stem wounds (Soloman 1995), and/or sweet exudates from sporulating cankers such as those produced by the chestnut blight fungus (Tainter and Baker 1996). After egg hatch, larvae bore through the bark and into phloem, sapwood, and heartwood, passing through several instars before pupating. Adults then emerge from infested trees to mate and repeat the process. Repeated infestations of the same tree over years results in brood trees containing dozens of larvae at various stages of development.

Wood borers usually do not cause tree mortality, but at least one species, the two-lined chestnut borer, is a significant contributor to oak decline mortality. More commonly, they provide infection courts for decay fungi. Repeated infestation breaches some compartmentalization responses and thus contributes to an ever-increasing decay column. In addition to providing this habitat component, wood borer larvae are a food source and prime objective of wood excavating woodpeckers. While they are extremely common in oak forests, wood borers seldom affect habitat beyond the local or stand level.

CONCLUSION

Native pathogens and insects have many ecosystem roles, including serving as food sources for some wildlife. Their effects are ecologically neutral and take on significance only in terms of human values. Therefore, the same effect might be seen as positive when considering one value set or wildlife species but negative for another set or species. Past disturbance has shaped today's oak forests, which in many cases bear little resemblance to oak forests of the past. Consequently, disease and insect attacks sometimes yield habitat effects unlike those experienced in the past.

Pathogen and insect effects can be classified as influencing tree species composition, forest structure, and/or acorn production at various spatial scales from local to landscape or even regional. Among diseases, oak wilt and oak decline produce effects on multiple habitat components at landscape or regional scales and have the potential to cause long-term loss of oak. Among insects, spring defoliators and acorn weevils can reduce acorn production over large landscapes.

Chapter 7

Gypsy Moths and Forest Dynamics

JOSEPH S. ELKINTON, WILLIAM M. HEALY,

ANDREW M. LIEBHOLD, AND JOHN P. BUONACCORSI

GYPSY MOTH POPULATION DYNAMICS

The gypsy moth (*Lymantria dispar*) is a major defoliator of oak forests throughout the Northern Hemisphere. In most regions it remains at low densities in most years, but it occasionally erupts into outbreak phase. In 1868 it was accidentally introduced into North America near Boston, Massachusetts. Because the female does not fly, the gypsy moth has taken 130 years to spread from Massachusetts to the current leading edge of the infested region in Wisconsin, Illinois, West Virginia, and North Carolina. In North America, gypsy moth outbreaks occur at irregular intervals but approximately every ten years (Elkinton and Liebhold 1990, Williams et al. 1992, Williams and Liebhold 1995). Gypsy moths in the former Yugoslavia appear to follow a regular cycle of outbreaks (Montgomery and Wallner 1988, Turchin 1990) that is apparently driven by a numerical response of Tachinid parasitoids (Sisojevic 1975). In North America, most of these same parasitoids have been introduced and established, but they typically cause much lower mortality than in Europe and there is scant evidence for regular cycles of gypsy moth outbreaks (Elkinton and Liebhold 1990, Williams and Liebhold 1995, Liebhold et al. in press).

Outbreaks, however, are frequently synchronized over large regions (Liebhold and Elkinton 1989a, Williams and Liebhold 1995). The causes of such outbreaks and their regional synchrony have recently been linked to acorn crops (Elkinton et al. 1996, Jones et al. 1998). Yearly fluctuations in acorn crops strongly influence the density of white-footed mice (*Peromyscus leucopus*), the dominant predator of gypsy moths at low density. Earlier research had suggested that changes in density of regular populations of gypsy moths were determined by survival during late instars (Campbell 1967). Experimental studies involving mouse exclosures (Bess et al. 1947) and mouse removal (Campbell and Sloan 1977) showed that predation by *P. leucopus* was the largest source of mortality of late instars, and the recent findings of Elkinton et al. (1996) and Jones et al. (1998) confirm this earlier work.

Other studies have elucidated the effects of pathogens on gypsy moth population dynamics. High-density populations are typically decimated by nucleopolyhedrosus virus (LdMNPV), a disease that has been associated with gypsy moths in North America since the early 1900s (Doane 1970, 1976, Glaser 1915). Epizootics from viral diseases are common among forest defoliators (Anderson and May 1980, 1981). Field studies (Woods and Elkinton 1987) and mathematical models (Dwyer and Elkinton 1993) confirm that mortality from LdMNPV reaches epizootic proportions near the end of the larval stage and only in high-density populations. In contrast, generalist predators such as white-footed mice have virtually no effect on outbreak populations of the gypsy moth, because mouse population densities are constrained by various factors and do not respond numerically to changes in gypsy moth population density (Elkinton et al. 1989, Elkinton and Liebhold 1990). Mouse population densities typically remain below 125 mice per hectare. At these densities the mice can easily consume most of the gypsy moths in low-density populations, which are typically below 100 egg masses per ha. Outbreak densities of gypsy moths can exceed 10,000 egg masses per ha (approximately 3,000,000 larvae), and at these densities 125 mice per ha would consume less than 1% of the population even if they ate almost nothing but gypsy moths. Consequently, the mortality factors governing the dynamics of the gypsy moth are completely different in high- and low-density populations (Campbell 1975). This pattern is typical of many insects that exhibit outbreak dynamics (Southwood and Comins 1976).

Adding to the complexity of the gypsy moth system is *Entomophaga maimaiga*, a fungal pathogen of gypsy moths that appeared unexpectedly in the northeastern United States in 1989 (Andreadis and Weseloh 1990, Hajek et al. 1990). This agent had been known for many years as an important source of mortality on the gypsy moth in Japan but had not been previously reported in North America. There had been an intentional introduction of *E. maimaiga* in 1911 near Boston (Speare and Colley 1912), but there is no evidence that it was ever established. Although we will probably never know for sure, it appears likely that there was an accidental introduction of a new strain of this disease from the Far East sometime in the 1980s (Hajek et al. 1995). In subsequent years, *E. maimaiga* spread south and west (Elkinton et al. 1991), aided in part by intentional introductions of soil laden with fungal resting spores to Virginia (Hajek et al. 1996) and Michigan (Smitley et al. 1995). By 1996 it had become established throughout North America wherever gypsy moths occurred. In many locations and years *E. maimaiga* has continued to cause high levels of gypsy moth mortality, including in low-density populations. It has evidently prevented several incipient gypsy moth outbreaks. Like most fungal pathogens it depends on moist conditions (Shimazu and Soper 1986). It is too early to predict the long-term effects of this disease on the dynamics of North American gypsy moth populations or its relative importance in preventing outbreaks compared to white-footed mice.

EFFECTS OF ACORNS ON MICE

Tree seeds, and acorns in particular, are an important part of the diet of *Peromyscus*, forming the bulk of the winter diet (Hamilton 1941, Batzli 1977). Increases in the abundance of *Peromyscus* have been associated with large acorn crops (Hansen and Batzli 1978), and declines have been associated with mast failures (Hansen and Batzli 1979). The mechanisms for population increase following good acorn crops include winter breeding, increased overwinter survival, and earlier onset of breeding in the spring (Hansen and Batzli 1978). Winter breeding is uncommon, but it has been observed in eastern populations during years of good acorn crops (Wolff 1986). Similar relationships have been reported in English oak woodlands for *Apodemus sylvaticus*, whose ecological role is similar to *P. leucopus* (Watts 1969, Flowerdew 1972). In most years, the abundance of *Peromyscus* in the spring is positively correlated with seed production during the previous autumn (Gashwiler 1979, Kaufman et al. 1995), but good mast years are not necessarily followed by abundance of mice in spring (Kaufman et al. 1995). Wolff (1996) showed that densities of deer mice in Virginia were positively correlated with acorn crops.

The results of supplemental feeding experiments have depended on the type of food and the season of feeding (Vessey 1987). Experimentally fed white-footed mice living in hardwood forest woodlots in Illinois showed increased densities, earlier breeding in the spring, and heavier body weights than unfed populations (Hansen and Batzli 1978). Mast was an important food for these populations, and a decline in the control population coincided with a mast failure. Even though feeding increased population density, Hansen and Batzli (1978) believed that female behavior set upper limits on summer densities. White-footed mouse populations in British Columbia also exhibited increased density and earlier breeding when given supplemental food (Fordham 1971). Similar results have been obtained with deer mice (*P. maniculatus*) (Taitt 1981), although in one population, feeding with oats had no effect and feeding with sunflower seeds led to a two- to threefold increase in density (Gilbert and Krebs 1981). Deer mouse populations in western coniferous forests have shown greater overwintering survival, earlier spring breeding, and marked increases in density in years after good seed crops (Gashwiler 1979, Halvorson 1982).

GYPSY MOTHS, MICE, AND ACORNS

Recent studies have solidified the links between gypsy moths, white-footed mice, and acorn production (Elkinton et al. 1996, Jones et al. 1998). Gypsy moth and mouse populations, and acorn crops were monitored over a 10-year period in mature oak forests in central Massachusetts. Observations were made in eight stands scattered over a 20,000 ha forest; the mean distance between stands was 8 km. Changes in population densities of white-footed mice and gypsy moths and in acorn production were all partially synchronized across the study region (Figure 7.1). Increases in gypsy moth density were associated with declines in density of the white-footed mouse. Furthermore, changes in density of mouse populations were positively correlated with the density of acorn crops from the previous fall. A novel bootstrap regression method was used to demonstrate these effects statistically (Elkinton et al. 1996).

These interactions have been confirmed experimentally at the stand level in southern New York. Removing mice from grids during June and July led to significantly greater survival of female gypsy moth pupae. Similarly, adding acorns during autumn in a poor mast year enhanced mouse survival and reproduction and resulted in significantly increased mouse densities the following spring and summer (Jones et al. 1998). Although these relationships are clear at the stand and local levels, intriguing questions remain about the synchrony of acorn crops and their

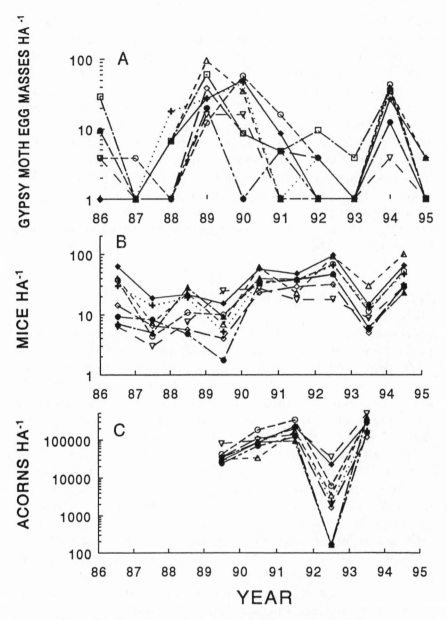

FIGURE 7.1. Yearly estimates from each of eight stands of oak in western Massachusetts: (A) gypsy moth egg masses per ha as determined prior to hatch in each year, (B) white-footed mice per ha, and (C) sound acorns per ha from each stand beginning in 1989. (Reprinted by permission from Elkinton et al. 1996.)

effects on wildlife at regional scales, especially in eastern mixed-oak forests (see Chapter 9).

EFFECTS OF GYPSY MOTHS ON ACORN PRODUCTION

The interactions described above have one missing piece. Defoliation by gypsy moths has a major impact on acorn production. In oak forests in Pennsylvania, Maryland, and Virginia, acorn production was reduced by 50% to 100% during years of defoliation, and acorn production was essentially eliminated in years of heavy defoliation (Figure 7.2, Gottschalk 1989, Drake 1991). The short-term reduction in acorn production is caused by the elimination of carbohydrate reserves in defoliated trees. Moderate acorn production may occur during the first year of defoliation. Recovery of acorn production following severe defoliation requires at least two growing seasons for members of the red oak group.

Mixed-oak stands from 42 to 58 years of age in central Pennsylvania experienced five successive years of gypsy moth defoliation. Defoliation was moderate in 1980, nearly complete in 1981 and 1982, light in 1983, and very light in 1984. Acorn production was moderate in 1980, but absent for the four consecutive years, 1981 through 1984. Moderate acorn production occurred in 1985 and 1986, but crops failed in 1987 and 1988 when stands again experienced light to moderate defoliation (Figure 7.2, Drake 1991).

Gypsy moth outbreaks may extend periods of poor acorn production well beyond what would normally be experienced by wildlife populations. In the absence of defoliation by gypsy moths, complete acorn crop failures are uncommon in mixed-oak stands, and they rarely occur in successive years (Chapter 10; Healy et al. 1999). Even low levels of defoliation by invertebrates can significantly reduce acorn production. English oak (*Q. robur*) trees from which all invertebrates were excluded with the application of insecticides consistently produced 2.5 to 4.5 more acorns than unprotected trees, which lost only 8–12% of their leaf area (Crowley 1985). This effect could account for poor acorn production observed during periods of light to moderate defoliation during gypsy moth outbreaks (Drake 1991, Figure 7.2).

In addition to reducing the acorn production of individual trees, severe defoliation also causes significant mortality of overstory oaks, beginning about two years after defoliation (Herrick and Gansner 1987).

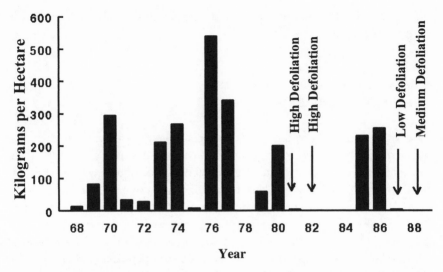

FIGURE 7.2.   Influence of gypsy moth on acorn crops in a mixed-oak stand in Huntington County, Pennsylvania. (Redrawn from Gottschalk 1989; data from the Pennsylvania Game Commission.)

Thus, acorn production potential is often reduced at the stand level for a decade or more following defoliation, due to a reduction in the density and basal area of oaks.

The other determinants of acorn production are complex, involving both genetics and weather-related factors (Olson and Boyce 1971). Each oak species tends to produce large acorn crops at certain intervals, but weather effects are superimposed on these inherent tendencies and explain much of the yearly variation in acorn crop size (Sork et al. 1993). Over much of the northeastern United States, northern red oak (*Quercus rubra*) is by far the most abundant oak species (Brooks et al. 1993). Good and poor acorn crops have been reported for relatively large areas (Downs and McQuilkin 1944, Goodrum et al. 1971, Christisen and Kearby 1984, Wentworth et al. 1992). Good mast years result from exceptional production from one species, or coincident fair to good production among several species. Poor production by one species is often offset by good production from another (Beck 1977, Christisen and Kearby 1984). Although complete mast failures are considered rare, synchronous poor production among species has been reported frequently within stands (Burns et al. 1954, Beck 1977, Rogers et al. 1990, Sork et al. 1993) and occasionally over large areas (Uhlig and Wilson 1952,

Wentworth et al. 1992). Weather effects, such as those of late spring frosts on acorn crops (Goodrum et al. 1971), and the synchrony of such crops over regions of at least 1,000 km^2 (Christisen and Kearby 1984) may explain the synchronous fluctuation of gypsy moths and white-footed mice evident in our data. These findings may thus explain the regional synchrony of gypsy moth outbreaks reported in previous studies (Williams and Liebhold 1995, Liebhold and Elkinton 1989a, Liebhold and Mc-Manus 1991) and the regional onset of gypsy moth outbreaks.

GYPSY MOTH DYNAMICS
FOLLOWING AN OUTBREAK

Gypsy moth dynamics following an outbreak remains one of the least understood aspects of this population system. As we have described, gypsy moth outbreaks virtually eliminate acorn production for several years. Outbreak populations of gypsy moths are not significantly affected by mouse predation. Instead, mortality during this phase is dominated by mortality from pathogens, especially LdMNPV. Outbreak populations typically experience epizootics of this pathogen that cause populations to decline dramatically to very low densities. Most such populations then remain at low density for periods that can extend over several decades (Campbell 1967, Elkinton and Liebhold 1990). Since mouse densities presumably also decline to very low levels during and immediately following outbreaks, because of low acorn production, one might ask why gypsy moth populations do not rebound into outbreak phase immediately following a decline?

For one thing, populations sometimes do rebound into outbreak phase. Several data sets show that on a regional scale gypsy moth outbreaks often persist for periods of up to a decade. This pattern was evident in the well-known Melrose Highlands data, which showed that outbreak conditions persisted across much of New England from 1911 to 1922. Bess (1961) reported a persistent outbreak near Freetown, Massachusetts. Campbell (1967, 1973) analyzed the Melrose Highlands data and concluded that while *regional* outbreaks would often persist for a decade, individual populations on the scale of the forest stand would usually rise and collapse over a period of 2–3 years. Campbell (1973) showed that when regional densities were high, 76% of low-density populations that had decreased the previous year increased the following year. Campbell attributed these patterns to dispersal of larvae from nearby stands of high density, but the absence of acorns following defoliation is another possible cause of this phenomenon. We have personally observed several stands on Cape Cod, Massachusetts, and in Virginia where densities rebounded within a year or two after populations declined from outbreak densities. On the other hand, it is certainly the case that many if not most gypsy moth populations remain at low density following a decline from outbreak densities, so additional explanations are needed.

One mechanism proposed by Doane (1970, 1976) is that LdMNPV levels remain high in the environment and cause substantial mortality of gypsy moths for a year or two following a gypsy moth population collapse. Woods et al. (1991) collected data on mortality from LdMNPV during and following gypsy moth outbreaks at several populations on Cape Cod. This study provided only limited support for Doane's idea. The main predictor of mortality from LdMNPV was larval density. Mortality from LdMNPV among late instars in the years following a population collapse were typically quite low, even if there were high rates of mortality among neonates, indicative of high rates of environmental contamination with LdMNPV. Modeling studies by Dwyer and Elkinton (1993) have supported these observations, so the factors constraining gypsy moth population growth immediately after an outbreak remain poorly understood.

EFFECTS OF GYPSY MOTH OUTBREAKS ON FOREST STANDS

Defoliation by gypsy moths has a number of effects on stand dynamics, including reduced acorn production, overstory mortality, altered regeneration potential, and changes in successional pathways. These effects occur at time scales ranging from annual to the natural rotation of the stand. The amount of change depends on the severity and frequency of defoliation. Susceptibility of forest stands to defoliation by gypsy moths is largely a function of the proportion of preferred tree species in stands. Gypsy moth larvae are known to feed on hundreds of tree species, but some species are preferred over others (Liebhold et al. 1995). Gypsy moth outbreaks generally are restricted to stands with > 20% of their basal area in preferred species. The most common preferred species are oaks. In addition to the direct effects of defoliation on individual oak trees, selective feeding by gypsy moths on hosts also alters competitive relationships among tree species within the stand. The net effect of the gypsy moth often is to reduce both the absolute and relative abundance of oak and to accelerate successional changes that lead to the replacement of oaks by other species.

Overstory Mortality

Defoliation reduces photosynthetic capacity, and heavy defoliation will cause trees to refoliate, severely depleting root starch reserves. Mortality may result directly from defoliation, or weakened trees may succumb to stress induced by drought or secondary attack by fungi or insects. Mortality rates increase sharply within two or more years of defoliation. In most regions, oaks are defoliated more frequently and suffer more mortality than other deciduous species. The amount of mortality increases with increasing frequency and intensity of defoliation. Mortality following defoliation has been variable among physiographic regions. Median basal-area loss (1979–84) amounted to 7% in the Pocono region and 9% in central Pennsylvania (Herrick and Gansner 1987), while 30% of the oak basal area died following defoliation on the Appalachian Plateau in southwestern Pennsylvania (Fosbroke and Hicks 1989).

Averages tell only part of the story, because the distribution of defoliation and mortality is extremely variable across the landscape and among oak species. The extensive mixed-oak forests in the ridge and valley province of central Pennsylvania experienced two gypsy moth outbreaks between 1978 and 1990. During this period, only 22% of the stands were defoliated heavily for more than one year. Basal area loss ranged from 3% to 90%; 76% of the plots had low mortality (< 15% of basal area/acre), 7% had moderate mortality (15–30% basal area/acre), and 17% of the plots had mortality exceeding 30% of the basal area per acre (Feicht et al. 1993). Most of this mortality occurred in oaks, but rates of defoliation and mortality varied among oak species (Herrick and Gansner 1987). Thus, at the landscape scale mortality may be considered minor, but catastrophic mortality may result from defoliation in some locations (Gansner et al. 1993, Feicht et al. 1993).

Defoliation by the gypsy moth sometimes causes mortality resembling a thinning that selectively removes oak. Most defoliated stands return to a fully-stocked condition over a period of 10–20 years, as growth from surviving trees compensates for mortality (Feicht et al. 1993, Gansner et al. 1993). At the landscape scale, oak growing stock volume has continued to increase as defoliated poletimber- and sawtimber-sized mixed-oak forests have matured. The volume increase in oak is primarily from growth of large trees; oak growing stock volume in trees less than 10 cm in diameter at breast height has actually decreased (Gansner et al. 1993).

Most of the increase in volume growth in these stands, however, has come from other hardwoods, primarily red maple, sweet birch, and black gum, which are growing at better rates than oaks. Thus, defoliation has caused a decrease in the relative density and dominance of oaks (Feicht et al. 1993).

Regeneration

Defoliation by gypsy moths has been consistently detrimental to oak regeneration (Allen and Bowersox 1989, Hix et al. 1991, Muzika and Twery 1995). Oak seedlings decline in abundance while seedlings of other species increase. Defoliation allows additional light, nutrients, and moisture to reach the forest floor, and species present in the understory at the time of defoliation respond to the altered conditions. The degree of regeneration response depends on both the initial undergrowth conditions and the extent of defoliation and subsequent mortality.

Oak regeneration abundance declines over time, because of seedling defoliation, increased competition from other species, and perhaps decreased acorn production. In the ridge and valley and Allegheny Mountain regions of Pennsylvania, red maple and sweet birch dominated the regeneration in mixed-oak stands following defoliation. Red maple accounted for 90% of the commercial tree regeneration in the Allegheny Mountain region, so there is potential for dramatic shifts in species composition (Allen and Bowersox 1989). On productive mixed-hardwood sites on the Appalachian Plateau of north central West Virginia, black cherry and yellow-poplar were important, in addition to sweet birch and red maple, and black cherry may dominate some future stands (Muzika and Twery 1995). None of the defoliated stands studied in these regions contained enough oak regeneration to assure the presence of oak in future stands. Defoliation by gypsy moths exacerbates oak regeneration problems on both xeric and mesic sites and accelerates the successional trend toward replacement of oaks by other species.

THE GYPSY MOTH
AND ECOSYSTEM MANAGEMENT

The gypsy moth represents a vexing problem for ecosystem management. We would like to eliminate this exotic species, but eradication from North America is no longer an option. In addition to the gypsy moth itself, numerous exotic parasitoids and invertebrate predators have been introduced to North America to control gypsy moths. Some of these generalized parasitoids have the potential to cause reductions in populations of native Lepidoptera (Boettner et al. in press). The overall effect of these introductions is unknown.

Gypsy moths have not eliminated oaks from the landscape, even in the forests of southern New England where they have been present throughout the twentieth century. In fact, overstory mortality has been relatively minor at the landscape scale, although both the absolute and the relative density of oaks have been reduced by successive defoliations. Clearly, gypsy moths contribute to the oak regeneration problem, intensifying the effects of fire exclusion, and excessive deer browsing. Mixed-oak stands in central Pennsylvania have undergone extensive changes in structure and species composition following defoliation (Allen and Bowersox 1989), and deer browsing has apparently contributed to these changes (Stromayer and Warren 1997). Oaks will form an insignificant component of future stands in these areas. Substantial input of energy will be required to reverse these trends and restore oak ecosystems.

At the stand level, gypsy moth outbreaks make it difficult to control stand structure and meet management objectives. For example, mixed-oak stands in Pennsylvania that had been thinned to enhance acorn production subsequently experienced heavy defoliation due to gypsy moths (Drake 1991). Defoliation during the experiment caused significant mortality in crop trees and reduced oak stocking below goals, thus confounding the effect of the treatment. Gypsy moth–induced mortality causes a problem whenever crop trees are to be retained for extended periods, and the problems are compounded when defoliation occurs repeatedly during the rotation. Defoliation can be particularly disruptive to the shelterwood method of regeneration, which depends on good acorn crops and close control of stand structure for a decade or more (Wolf 1988).

Silvicultural options are available for mitigating the effects of gypsy moths at the stand and landscape levels (Gottschalk 1993), but for wildlife the options are likely to be viewed as "lose-lose" situations. Treatments to protect threatened stands include regenerating the stand or thinning to reduce the abundance of oaks and preferred food species. Gypsy moth defoliation reduces the sprouting potential of oak stands, so postdefoliation treatments often facilitate the conversion of stands to other species. Direct control of gypsy moths with insecticides is an option for stands with high value for conservation, such as relict old growth, or when considerable effort has been expended to obtain regeneration or other desired attributes.

Acorns lie at the base of a complex food web upon which many species depend. The introduced gypsy moth is now an integral part of that web, with strong links to both white-footed mice and forest vegetation dynamics, and it has the potential to affect oak forests throughout eastern North America. The gypsy moth may eventually be integrated into this system to become another innocuous invertebrate herbivore, but in the meantime managers will have to mitigate the disruptive effects of this introduced species.

Acknowledgments

We are grateful to the many students and technicians who assisted us with our field work and data processing. We are grateful to the USDA NRI Competitive Grants Program (Grants Nos. 85-CRCR-1-1814, 8937250-4684) and to the MacIntyre-Stennis MS-51 project from the Massachusetts Agricultural Experiment Station for funding this research.

Chapter 8

Dynamics of Old-Growth Oak Forests in the Eastern United States

LEE E. FRELICH AND PETER B. REICH

Definitions of old-growth forest fall into three basic groups: (1) forests that have never been logged or had other severe disturbance by humans, (2) forests that have reached a state where self-replacement can occur without major disturbance, and all traces of a postdisturbance, even-aged (same-aged) cohort have disappeared—commonly referred to as the shifting-mosaic steady state (Bormann and Likens 1979, Oliver and Larson 1990), and (3) forests in which the trees are relatively large and old for the species and site.

It is often difficult to tell whether older oak forests meet the "no human disturbance" criterion for the first definition, and that definition also includes young forests that originated after natural disturbance. The second definition is demographically based and often preferred by forest ecologists. Only certain forest types, such as shade-tolerant species on good soils, attain the shifting-mosaic steady state at some spatial scale. Among the oaks, however, only a few species, such as northern red oak, are capable of reproducing in treefall gaps 150 m² in size (Lorimer 1983). The remaining species are intolerant of shade and cannot reproduce in treefall gaps, making the shifting-mosaic of treefall gaps an impractical criterion for judging old-growth status in the case of oak forest in general. These flaws in the first two definitions dictate the use of the third definition here. The third definition is also the one most commonly used by agencies that manage forests, because it can be used with inventory databases, which commonly track stand age, site quality, and tree diameter. The thresholds for just how large and/or old the canopy-dominant oaks must be to qualify as old-growth can be adjusted to meet management goals. Some examples of minimum threshold ages suggested for existing oak forests in the United States include 90 years in Missouri (Meyer 1986), 120 years in Minnesota (Rusterholz 1991), and 150 years for central deciduous forests (Parker 1989).

PRESETTLEMENT VERSUS MODERN OLD-GROWTH OAK FOREST

Most of the discussion in this chapter pertains to old-growth oak as it exists today, primarily in the eastern United States from Minnesota and Missouri east to the Atlantic Ocean. Presettlement oak forests were shaped by forces that were, in part, different from those in effect over the last century. In presettlement times, oak forests probably had a complex fire regime, with frequent surface fires (5- to 40-year recurrence) and infrequent canopy-killing fires (100- to 500-year recurrence, Abrams 1992). Fires are less frequent now, allowing most forest types to maintain themselves on more-xeric sites than they did in presettlement times (see Figure 8.1). Prescribed fires usually cannot duplicate natural fire, because they are conducted on days when the fire weather is not severe, to reduce the possibility of escape. Prescribed fires are mostly cool season fires of low intensity. Natural fires would tend to be more severe and could occur over a wider variety of times during the growing season.

Native Americans probably had a large influence on the presettlement distribution and dynamics of oak forests, by lighting fires in certain oak stands for purposes of vegetation and wildlife management (Denevan 1992). Some studies have concluded that fires set by Native Americans had only scattered, local influence on oak forests in the eastern United States (Russell 1983). However, we are aware of no studies of vegetation dynamics which take into account the fact that Native American populations were low at the time of European settlement, having been dramatically reduced between the time of first substantial contact around 1500 A.D. and the time of extensive settlement 1700–1900 (Denevan 1992). Thus, the forests had two to four centuries to adjust to the lower Native American population, which may have been accompanied by reduced frequency and areal extent of burning, before the major wave of European settlement.

The passenger pigeon (*Ectopistes migratorius*) must have been a major force in the presettlement dynamics of oak forests. Their roosts covered areas of up to 10,000 ha, and within these zones, most of the vegetation was killed because of over fertilization by pigeon dung, which often ac-

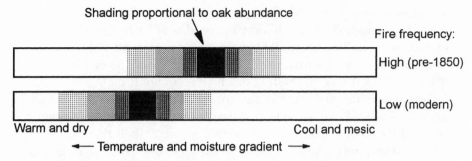

FIGURE 8.1.   Comparison of oak abundance along a temperature and moisture gradient ca. 1850 and 1990. Sites at the warm/dry and cool/mesic ends of the gradient (unshaded) are occupied by grasslands and sugar maple–hemlock forests, respectively. Oaks occupy the middle part of the gradient, shifting maximum abundance in response to changes in fire frequency.

cumulated to depths of several centimeters (Schorger 1955). It is unknown how often they returned to roost at the same site, so a rotation period for this type of stand-killing disturbance cannot be calculated. However, because only weedy herbaceous species that thrive on exceptionally high nitrogen levels could live at these roosts for several years after a roosting event, the original oak forests of the central and midwestern United States must have had numerous large openings. There were also nesting sites, which were much larger in area than the roosts—up to 200,000 ha—but less heavily populated than roosts. The nesting sites must also have altered the ecosystem dynamics by fertilization. Fertilization at these sites could have been enough to cause increased tree growth without the mortality that occurred in the roosts, and oak savannas on sandy, nutrient-poor sites could have been converted to forest; some of these stands may be in existence today because large pigeon populations were still present in the 1870s (Schorger 1955). The birds also ate acorns and must have been important in acorn dispersal in addition to fertilizing the oak forests with their droppings. We are not aware of any studies on the historical effects of the passenger pigeon on oak forest dynamics. Because the precise effects on oak forests of the passenger pigeon, Native Americans, and presettlement fire regimes remain unknown, we do not know how well current old-growth oak forests represent those that existed prior to European settlement.

The extent of old-growth oak forest in the eastern United States has declined steeply over the last century. For example, in the Lake States of

Minnesota, Wisconsin, and Michigan, an estimated 2.8 million ha of un-logged oak forest (8.5% of all forest) existed as of 1850, and about 55% of that, or 1.5 million ha, would have been old growth ($\geq$ 120 years old) under the natural disturbance regime (Frelich 1995). As of 1995, only 900 ha (0.02% of that in 1850) of oak forest had not been logged in the Lake States, although there were also approximately 140,000 ha of second-growth oak forest that was at least 120 years old. If one is willing to count these older second-growth stands as "old growth," then there is about 9.0% as much old-growth oak forest in the lake states today as there was in 1850 (Frelich 1995).

A survey of forests east of a line from Texas to Minnesota, although less detailed than Frelich's, shows that the statistics for the Lake States are representative of the rest of the eastern United States (Davis 1996). Mesic to dry-mesic old-growth oak forests are rare. Parker (1989) states that 0.07% of the original mesic old-growth oak forest still exists in the central hardwood region (Illinois, Indiana, Ohio, Kentucky). The largest remaining stands of mesic oak forest are in and around the Great Smoky Mountains National Park, in eastern Kentucky and western North Carolina (Davis 1996). Most of the remaining old-growth oak forest in the eastern United States is in the form of oak barrens on xeric and infertile sites that were not logged due to small tree size or lack of suitability of the land for agriculture. Because of fire exclusion, many of these barrens have grown up into forest over the last century. Significant stands > 400 ha in size occur on xeric ridge tops in Missouri, Arkansas, Illinois, and Indiana, throughout the Appalachians, and on sand plains in central Wisconsin and Minnesota (Reich and Hinckley 1980, Stahle and Chaney 1994, Davis 1996).

ORIGIN AND STRUCTURE OF EXISTING OLD-GROWTH OAK FOREST

Old Even-Aged Stands

Old even-aged oak stands are ones that contain long-lived oak species of the same age that have survived a long time without canopy-killing disturbance and have not yet succeeded to shade-tolerant species. Examples include northern red oak (*Quercus rubra*) and white oak (*Quercus alba*) forests in the midwestern and northeastern United States. Some old even-aged oak stands are succeeding to shade-tolerant species such as red maple (*Acer rubrum*), sugar maple (*Acer saccharum*), beech (*Fagus grandifolia*), and hemlock (*Tsuga canadensis*) (Curtis 1959, Abrams and Downs 1990, Abrams 1998), while young even-aged stands are being created by major disturbances, mostly logging in modern times. Thus, there is a constant change in the pool of old oak stands in this category, and no one stand stays in this category for more than several decades.

Extensive tracts of white pine forest across New England and the Lake States in the United States had an understory of northern red oak (Lorimer 1984, Crow 1988, Abrams et al. 1995). When the white pine was cut, the northern red oak was released and left to dominate the succeeding forest. Because most of these pine stands were cut during the 1800s, many of the stands surviving today are old even-aged oak 100 to 150 years old.

Old Multi-Aged Stands

Old multi-aged stands are those with a history of numerous partial disturbances, usually surface fires and windstorms, that kill small proportions of the canopy over two or more centuries. Oak stands in this category may be very old—up to several centuries—with respect to time since last canopy-killing disturbance, but they have sustained disturbances severe enough to prevent invasion and replacement by shade-tolerant species but not so severe as to kill most of the adult oaks. This type of oak forest is most likely to occur on dry-mesic sites where invasion by shade-tolerant hardwoods is not vigorous. A representative example is the multi-aged forest of chestnut oak (*Quercus prinus*) with smaller amounts of white oak and black oak on ridge tops in central Pennsylvania, where canopy disturbances caused by fire, wind, and harvesting occurred every 21 years, on average, during the presettlement and settlement eras (prior to 1775 and 1775–1900, respectively) and every 31 years in recent times (Nowacki and Abrams 1992, 1997). On more-mesic sites in southwestern Wisconsin and upper Michigan, northern red oak and sugar maple coexist in multi-aged stands where harvesting or spot fires have created sizable canopy gaps over time (Lorimer 1983, Frelich and Lorimer 1991).

Oaks on Severe Sites

Oaks on severe sites include stands on sandy or rocky sites, especially ridges, where the density of trees supported by the site is low. On such sites recruitment can continue even for intolerant oak species. Many such sites are classified as "woodlands" or "barrens," which are between savannas and true forest in degree of canopy closure. Shade-tolerant species cannot invade, because of dry or nutrient-poor conditions. Examples include stands of yellow oak (*Quercus muhlenbergii*) on limestone ledges, blackjack oak (*Quercus marilandica*) on sandstone ledges, and black oak/bur oak (*Quercus velutina/Quercus macrocarpa*) on extremely xeric sites (Peet and Loucks 1977, Reich and Hinckley 1980, Abrams 1992).

Grown-up Savannas

"Grown-up" savannas are areas that were once kept open by frequent fires but have now become forest. The potential range of closed-canopy forest, as determined by adequate rainfall, was not filled in presettlement times because some areas burned so frequently (1-year to 5-year return intervals) that oaks gave way to savannas or grasslands. These savannas occurred at the interface of the eastern deciduous forest and prairie biomes, from Texas and Oklahoma northward through Missouri, Iowa, Wisconsin, and Minnesota, with a notable eastward extension known as the "prairie peninsula" covering much of Illinois and parts of Indiana, Michigan, and Ohio (Nuzzo 1986). Converting most of a landscape to farms, highways, cities, and resorts is enough to break the flow of fires across the landscape, resulting in fire exclusion even without direct fire suppression. In addition, grass fires are low in intensity and, unlike many forest fires, can actually be suppressed. Such a reduction in fire has allowed former savannas to grow into forests. Savannas in the northern Midwest were dominated by black oak, northern pin oak (*Quercus ellipsoidalis*), bur oak, and white oak. Diversity of savanna-dwelling oak species increases towards the south, where *Quercus prinoides, Quercus stellata, Quercus palustris, Quercus marilandica, Quercus imbricaria,* and *Quercus falcata* are found on oak savannas in Missouri (Nuzzo 1986). Savannas on sites with deep silty to loamy soil are quick to convert to forest once fires are stopped, and virtually all such sites have grown up into mesic oak-maple forests. Savannas on excessively well-drained sandy soils are not as readily invaded, but even they can eventually grow up into forests of northern pin oak, black oak, and bur oak, frequently with red maple as an understory component that could eventually be a codominant in the canopy with oaks (Nuzzo 1986, Abrams 1992, Peterson and Reich 2001).

OLD-GROWTH OAK ON TODAY'S
FORESTED LANDSCAPES

Oak Forests

Forests of oak—those with little or no shade-tolerant hardwoods—on the modern landscape mostly originated from grown-up savannas in the Midwest and occur on sites with widely varying soil quality. They may also occur on dry-mesic to dry sites throughout the northeastern United States, where it was (and in some cases still is) difficult for shade-tolerant tree species to invade (Curtis 1959, Crow 1988, Abrams 1992, Faber-Langendoen and Davis 1995, Peterson and Reich 2001). Forests that were closed-canopy oak at the time of settlement have mostly become mixed maple–oak forests today (see below).

Oak forests on poor, sandy soils have historically had stand-killing fires at relatively short intervals of 100 years or less (Peterson 1998). On such landscapes, stands are generally even-aged, and only a few stands survive long enough to reach any of the commonly used threshold ages necessary for old-growth status (90–150 years). This is true even for sites so nutrient poor that succession to shade-tolerant tree species cannot occur.

A complex relationship among fire frequency, ability of oak to resprout after top-killing by fires, ability of adult oaks to survive fire, and competition with grasses determines whether oaks can maintain a forest canopy—rather than the more open savanna—on sites with poor, sandy soils (Davis et al. 1998, Peterson and Reich 2001). Survival of individual tree stems after surface fires is positively related to bark thickness, which is in turn positively related to tree diameter, and a function of species—oaks in general having relatively thick bark, bur oak having the thickest bark among oaks in northeastern North America (Harmon 1984, Peterson and Reich 2001). Oaks also have the ability to form seedling sprouts (commonly called grubs) with extensive underground root systems that can resprout year after year when top killed by fires. Thus, oaks can hold on to occupancy of a site through a period of very high fire frequency and then quickly grow into a forest canopy if a few decades go by without fire. Once a new canopy has formed, retrogression to savanna or grassland is difficult to accomplish with fire alone. For example, at Cedar Creek Natural History Area in Minnesota, experimental burning nearly every year for 35 years failed to rid a savanna of its bur oak trees. Conversion from forest to savanna or grassland can be facilitated by drought, which causes significant mortality of oaks on poor sites, and by windstorms in combination with frequent fire (Faber-Langendoen and Davis 1995, Peterson 1998).

The length of time between stand-killing fires is relatively long on some landscapes with slightly better soil conditions, and there may or may not be a lot of old-growth oak stands in such cases, depending on the length of time it takes for shade-tolerant species to replace the oaks. Note that windstorms and surface fires affect succession in opposite ways: windstorms can accelerate succession in old-growth oak forests by rapidly killing off large oaks and releasing the understory shade-tolerant species (Abrams 1992, Abrams and Nowacki 1992), whereas surface fires kill seedlings of invading shade-tolerant species (Reich et al. 1990, Kruger and Reich 1997a). Oak seedlings on the forest floor have a better ability than those of sugar maple to rebound after fire, because of their high root starch concentration and below-ground dormant buds (Kruger and Reich 1997b). Thus, fires promote continuing dominance by oaks, and if surface fires occur every 10 to 20 years, advanced regeneration of shade-tolerant species like sugar maple can be set back, while new cohorts of oak are established, leading to long-term maintenance of old multi-aged oak forest. Severe fires will set it back to young even-aged forest, and total lack of fire will allow succession to other species. Long-term maintenance of old-growth oak as a multi-aged forest therefore depends on the precise timing in the disturbance regime. Because the pre-settlement disturbance regime no longer exists and the modern regime varies with time, succession in oak forests is exceedingly difficult to predict. However, without prescribed fire it is likely that most such forests will succeed to maple (Reich et al. 1990, Abrams 1998).

Mixed Maple-Oak Forests

Mixed stands of mesic-forest species such as sugar maple, beech, hemlock, and oaks were noted by early land surveyors prior to European settlement, and remnants of these still occur today throughout the northern hardwood region of eastern North America. The history of fire in a given stand over the past few centuries is the major variable explaining the exact mixture of species present. With no fire over several tree generations, only a very small population of oaks—on the order of 1–2% of all canopy trees—can be maintained by chance recruitment in treefall gaps. It is possible to have a 50:50 (or any other proportion) oak:maple mix if the fire regime is just right. For example, if oak is the primary regeneration after severe, stand-killing fire, and holds dominance for 150 years before shifting to maple dominance, and all stands burned at age 300 years, then half of the landscape would be dominated by oak and half by maple. On a real landscape, intervals between fires vary from region to region, from one century to the next; and even within an area with a homogeneous fire regime, not all stands burn at the mean stand age. Two interesting questions arise. Why would old-growth maple forests (commonly referred to as "asbestos forests") burn, allowing oaks to retake dominance from maple? Assuming maple forests will burn, what frequencies of fire would be conducive to maintenance of mixed forests of fire-dependent oaks and non-fire-dependent maples?

Although stands dominated by sugar maple and other mesophytic species are not very flammable, severe fires can occur in windfall or logging slash. These fires can cover areas as large as the slash, which can be up to 10,000 ha for windfalls and much larger for logged areas like those created by European settlers in the Midwest and Northeast between 1880 and 1920. White pine, northern red oak, and paper birch are common forest dominants after such fires (Hough and Forbes 1943, Frelich and Lorimer 1991). The historic natural rotation periods for stand-killing fires that allowed oaks to take over dominance from shade-tolerant hardwoods probably was a few thousand years at the northern edge of oak's range in eastern North America (Frelich and Lorimer 1991). Near the southern edge of the northern hardwoods, in the mesic maple-oak woods extending from Iowa and southern Minnesota to southern New England, severe storms have a much shorter recurrence interval, as short as 250 years; and this forest belt also experiences more hot weather, making fires in the resulting slash more likely (L. E. Frelich, 1999 unpublished data). Near the prairie-forest border, frequent prairie or savanna fires were likely to burn into any mesic hardwood stands that had substantial windfall slash. Therefore, it is reasonable to assume that, under the historic disturbance regime, even northern hardwoods could burn frequently enough to allow a large component of oaks within many stands or to allow a significant portion of the landscape to be dominated by oak stands.

Studies of witness trees from nineteenth-century land surveys show that the locations of savanna, oak forest, and maple forest depended on the locations of fire breaks (Grimm 1984, Leitner et al. 1991). Oak savanna was on the southwest side of major fire breaks, such as rivers or major ridges, whereas mixed maple–oak forest occurred on the northeast side. Thus, rivers and ridges running in a northwest-southeast direction were most likely to allow existence of old-growth forest, because of the relatively low frequency of severe fire on the northeast side. In large patches of forest, such as the 7,500 km² Big Woods of southeastern Minnesota, oak was most abundant within 10–20 km of the forest edge, then gradually declined in abundance towards the interior of this closed-canopy forest patch. This distribution seems to have been a direct response to fire frequency: the farther into the forest, the fewer fires had occurred and the more dominance shifted from oak to sugar maple (Grimm 1984, Leitner et al. 1991).

Several studies have shown that intervals between fires ranging from 100 years to 500 years will allow for the persistence of mixed oak–maple stands (Hough and Forbes 1943, Oliver 1978, Will-Wolf 1991, Abrams 1992). Many forest landscapes in eastern North America with mesic soils fall within this range of severe fire frequency that supports mixed forests: southeastern Minnesota (the "Big Woods") (Grimm 1984), southern Wisconsin (Curtis 1959, Kline and Cottam 1979, Leitner et al. 1991), "mixed mesophytic" forests of Ohio and Pennsylvania (Runkle 1982, Abrams and Downs 1990), and mixed woodlands of New England (Oliver 1978, Lorimer 1984).

Human activities have modified natural disturbance regimes in various ways. Many former maple-beech-hemlock forests were converted to oak after logging that was followed by slash fires; white pine stands were converted to oak by selective removal of the pine; and fire suppression in forests that were oak at time of settlement has allowed succession to shade-tolerant hardwoods (Lorimer 1984, Abrams 1992). Thus, human disturbance in the past century has allowed shifts toward more or less oak in mixed oak–maple forests, depending on the history and type of human disturbance in a given area.

Oak as a Minor Component of Old-Growth Mesic Forests

Areas so well protected from fire by mesic soils, wet climate, and topographic fire breaks that fires occur at intervals of more than 500 years have oak as a minor component of the landscape, having only a few stands, groves, or even individual oak trees (Hough and Forbes 1943, Frelich and Lorimer 1991). The maintenance of widely scattered oaks within mesic forests otherwise heavily dominated by shade-tolerant species is important for the diversity of the forest, and the mechanism for establishment of these oaks is a mystery. One theory is that spot fires or low-intensity surface fires somehow favor establishment of these isolated oaks. Small groves of northern red oak in the Porcupine Mountains Wilderness State Park, Michigan, are usually associated with fire scars on the surrounding hemlocks and maples, indicating that a small lightning-caused spot fire occurred prior to establishment of the grove (L. E. Frelich, 1999, personal observation). Gaps caused by the death of a few canopy trees in such spot fires probably attract seed dispersers, such as blue jays (*Cyanocitta cristata*), which can move large numbers of acorns up to 4–5 km (Bossema 1979, Darley-Hill and Johnson 1981). We are not aware of any studies, however, that determine whether jays preferentially disperse acorns to gaps within the forest (rather than non-gaps) or to fire-caused gaps as opposed to windthrow gaps. Although it is reasonable to suppose that germinating acorns that landed within a wind-caused gap would lose out in competition with sugar maple, ironwood, and other shade-tolerant regeneration (Lorimer et al. 1994), this dense advanced regeneration is often killed in a gap caused by fire (Reich et al. 1990, Kruger and Reich 1997a). In most of the midwestern and northeastern United States, rock outcrops are scattered across the landscape, and these often have stands of oaks that can provide a permanent refuge from competition with shade-tolerant species and a seed source for oaks within large tracts of mesic forest.

Old-Growth Oak in a Fragmented Landscape

Most old-growth oak forest in the midwestern United States today is in the form of small remnants surrounded by agricultural fields, highways, or housing developments. The integrity of these examples of old growth is threatened by multiple forces, including grazing and browsing by white-tailed deer (*Odocoileus virginianus*), invasion by exotic plants and animals, severe windstorms, fire exclusion, and further development. The following sections give a brief synopsis of these threats to the integrity of oak forests. The role of diseases, such as oak wilt (*Ceratocystis fagacearum*), is covered in Chapter 6.

White-Tailed Deer Grazing and Browsing

Deer can change the plant community within old-growth oak forests by summer grazing of herbaceous plants and winter browsing of woody species, including tree seedlings. Browsing of woody stems during the dormant season is an important selective force to determine which species of tree seedlings are successful in old-growth forests. The effect of deer on successional trajectories depends on which species among those present is highest on the deer preference list in a given forest. White-tailed deer in eastern North America prefer certain conifers, such as white cedar (*Thuja occidentalis*) and eastern hemlock, within mixed hardwood–conifer forests, but they prefer oaks within hardwood forests (Frelich and Lorimer 1985, Strole and Anderson 1992). Thus, deer may well help cause succession from stands dominated by hemlock to a mixture of hardwoods, including oaks, in one region, while speeding up succession from oak to other hardwoods, such as maples, in other regions. Deer browsing in old-growth oak-maple forests can also prevent long-term maintenance of oaks as a component of the forest. Old-growth sugar maple forests have gaps large enough that intermediately shade-tolerant species such as white oak and northern red oak sometimes are able to grow through the gaps into the forest canopy (Lorimer 1983, Frelich and Lorimer 1991). A combination of deer browsing and light levels that are marginal for oak growth, however, may still give the advantage to maple in these gaps.

Studies have suggested that a population density of 2–5 deer per km^2 will maintain tree and herbaceous diversity in midwestern forests (Frelich and Lorimer 1985, Waller and Alverson 1997, Augustine and Frelich 1998).

Exotic Species

Two species of European buckthorn (*Rhamnus cathartica* and *R. frangula*) and Tatarian honeysuckle (*Lonicera tatarica*) invade the understory of oak forests in eastern North America and shade out the native shrubs, herbs, and tree seedlings. These berry-producing species are distributed by birds and are concentrated around the edges of oak forests. Thus, the more fragmentation of the forest, the bigger the problem. Some oak forests that originated as woodlands or savannas without dense canopy closure have been invaded throughout by these shrubs. In addition, even beneath dense canopies, buckthorn is in green leaf earlier and later than native understory plants and can accumulate substantial photosynthates during spring and fall, before canopy leafout and after canopy leaf drop, respectively (Harrington et al. 1989).

Several species of European earthworms are invading hardwood forests in eastern North America (Lee 1995). Of these, the nightcrawler (*Lumbricus terrestris*) is the most aggressive at eating the forest floor leaf litter (Alban and Berry 1994), leaving the bare soil susceptible to erosion and changing the environment in which seeds germinate. We know that European earthworms will cause species- and ecosystem-specific changes in seed bank and seed bed dynamics in old-growth forest remnants, creating the potential for major changes in successional trajectory in the future, for both the tree layer and the herb layer (Nielsen and Hole 1964, Nixon 1995). Because of the newness of investigation into earthworm effects on forests, however, we have more questions than answers.

Severe Windstorms

Old-growth oak forests can be leveled by severe thunderstorms, which produce tornadoes and / or straight line winds in excess of 200 km / hour (Canham and Loucks 1984, Frelich and Lorimer 1991). The forests are no longer old growth after the blowdown, and they may never recover to oak at all. The natural tree regeneration on the forest floor of old-growth oak forests is likely to be of shade-tolerant species. Since seedlings and saplings are not likely to be killed in a windstorm, the removal of the canopy will in many cases accelerate succession to late-successional species (Abrams and Nowacki 1992), or possibly to exotic species such as buckthorn.

Fire Exclusion

Surface fires that formerly enhanced survival of oak seedlings relative to that of shade-tolerant species like sugar maple (Reich et al. 1990, Kruger and Reich 1997a) cannot spread across a fragmented landscape. The probability of naturally occurring fire in a small isolated forest remnant is much less than in a forested landscape, where a fire starting many kilometers away may reach a given stand. Although prescribed fires may not have the same effect as natural fires, which sometimes burn hotter, they may still be useful in keeping exotic plants such as buckthorn and late-successional tree species at bay, while ensuring the establishment of young cohorts of oak that can repopulate the forests after the inevitable disturbances caused by severe windstorms.

CONCLUSIONS

Old-growth oak forests were historically, and still are, in a state of constant fluctuation of species composition. There has always been a tension between oaks and grasslands, on the one hand, and between oaks and forests dominated by shade-tolerant tree species on the other hand. The modern forest landscape experiences lower fire frequencies than forests did prior to European settlement, and as a result maple is invading former oak forests while the oak is invading former savannas and grasslands. On forested landscapes, any one old-growth oak stand will probably exist only for a few decades to a century; rarely will the circumstances of a series of disturbances be such that old multi-aged oak forest will last for hundreds of years at one site.

The forces that now shape old-growth oak forests are so different from pre-European settlement times, in terms of herbivory, activities of humans, absence of passenger pigeons, and the nature of fire regimes, that one can no longer assume that any old-growth oak remnant is truly representative of presettlement forests. The forces will change even more in the future, as multiple effects and interactions of exotic plants, insects and diseases, fragmentation, and climate change lead to outcomes that are difficult to predict. Although research on multiple effects in oak forests is limited at this time, one can postulate some reasonable scenarios. For example, small fragments of oak forest flattened by a severe windstorm, which have buckthorn around the periphery—a common occurrence in the midwestern United States—could be immediately invaded by buckthorn. Browsing of oak seedlings by white-tailed deer could ensure that oak seedlings stay too small to compete with the buckthorn. Thus, the combination of three forces—wind, deer, and exotic species—could work together and lead to oak regeneration failure and replacement of the historic natural oak forest community by something else. These factors are likely to be the most important force shaping the future of old-growth oak forests, especially in those regions where forests are isolated in an agricultural or suburban landscape. Even in the absence of invasion by exotic species, the overall effect of modern disturbances affecting oak forests (e.g., deer browsing, windstorms, fire exclusion, oak wilt) is to accelerate succession away from oak and towards shade-tolerant species. Invasion by exotic shrubs and European earthworms, and deer grazing on herbaceous plants will not necessarily threaten the old-growth canopy of oaks. However, it is apparent that these forces will change the entire forest community into something with no historical precedent.

Part II

Ecology and Patterns of Acorns

Chapter 9

The Behavioral Ecology
of Masting in Oaks

WALTER D. KOENIG AND JOHANNES M. H. KNOPS

Few genera of organisms permeate the cultural and biological fabric of our world as extensively as oaks. One reason for this is that oaks, through the production of acorns, provide a major food resource for a vast array of wildlife and (at least historically) for many native peoples. Furthermore, oaks and acorns have traditionally garnered much attention due to the considerable variability of acorn production. The most obvious source of such variability is the wide range in acorn crop size from year to year, which characterizes the concept of "mast seeding" or "masting" behavior (see below). Variation is also significant on several other levels, including from individual to individual, population to population, and species to species. We are just beginning to understand the correlates and the causes of this monumental variability.

Among the questions we address in this chapter are (1) How variable is annual acorn production? (2) Is annual acorn production bimodally distributed such that mast years and nonmast years are distinct? (3) What is the interval between mast and nonmast years? (4) What are the environmental correlates of annual acorn production? (5) How geographically synchronous is acorn production within a species? (6) Is acorn production synchronous within a community? (7) Do oaks switch resources between growth and reproduction? and (8) What are the functional consequences of masting behavior in oaks? Our goal is to summarize these key aspects of acorn production to more clearly illuminate the magnitude, scope, and causes of masting behavior by oaks.

Definition of Masting

Masting is the intermittent production of large seed crops by a population of plants (Kelly 1994). Anecdotally, crop years are often divided into those when many seeds are produced ("mast years") and other years ("nonmast years"). However, for the data sets that have been examined thus far it is rare that the distribution of annual seed crop size is strongly bimodal, and thus the distinction between mast and nonmast years is usually arbitrary (Kelly 1994, Koenig and Knops 2000). Taking this into consideration, Kelly (1994) devised three labels for masting behavior: "putative" masting, in which seed crops vary greatly from year to year, "normal" masting, in which seed crops vary greatly and there is either weak bimodality or there is switching of resources, and "strict" masting, in which seed crops vary but with strong bimodality. Most of our discussions of masting behavior in this chapter relate to these distinctions. In general, however, we use the term *masting* to refer only to the high annual variability in seed production apparently exhibited by most oak populations, without implying one or the other of the more specific types distinguished by Kelly (1994).

Research Materials and Methods

We draw on two sources of data. The first is our own work on acorn production in California (which we shall call "the Hastings data"). Central to this work is a continuing study of acorn production by five oak species at Hastings Reservation in central coastal California begun in 1980 and thus currently encompassing 19 years of data. The five species (number of individuals surveyed) are *Quercus lobata* (87), *Q. douglasii* (57), *Q. agrifolia* (63), *Q. kelloggii* (21), and *Q. chrysolepis* (21). These include species that require a single year to mature acorns ("1-year" species), *Q. lobata,* *Q. douglasii,* and *Q. agrifolia,* and species that require two years to mature acorns ("2-year" species), *Q. kelloggii* and *Q. chrysolepis.* They also represent winter deciduous species (*Q. lobata, Q. douglasii,* and *Q. kelloggii*) and evergreen species (*Q. agrifolia* and *Q. chrysolepis*).

Trees were individually tagged and acorns were visually surveyed following the methodology of Koenig, Knops, et al.(1994). In brief, we visited trees once each autumn during September to early October, just prior to acorn fall, at which time we scanned each tree's canopy and counted as many acorns as possible in 15 seconds. Counts were then added to yield acorns per 30 seconds ("N30"). N30 values were log-transformed ($\log[N30 + 1]$) prior to analysis.

The second source of data is acorn production data extracted from the literature (the "oak data set"). If data were originally categorical we ranked the categories in order of increasing crop size, giving the highest category a 10, the lowest category a 0, and making the difference between all intermediate categories equal. For example, if only three categories were used (e.g., good, fair, and poor), years when the crop was rated as good were given a 10; those in which the crop was rated as fair were given a 5; and those in which it was rated as poor were given a 0.

If values presented were interval or ratio-level data, such as the actual counts or number of acorns falling in traps, we analyzed both the original and log-transformed values. Only interval and ratio-level data sets were used in the analyses involving coefficients of variation.

In all, the oak data sets included 80 data sets from 24 separate studies involving a total of 743 years of data between 1900 and 1998 (mean = 9.3 years/data set). The majority of studies were North American (69 data sets from 16 studies), but a few were European (7 data sets from 6 studies) and Japanese (4 data sets from 2 studies). Data were recovered for 23 different species, including 12 1-year species and 11 2-year species. Most analyses were divided according to acorn maturation time. Taken together, these data allow us to investigate patterns of acorn production on a taxonomic and geographic scale much larger than would otherwise be possible.

Variability in annual acorn production was assessed by calculating the coefficient of variation across all years of data within each noncategorical data set that contained at least six years of information. Bimodality was examined by counting the frequency of years in each third of the overall range. The series was considered bimodal if the number of years in the middle third of the range was less than the number in both the upper and lower thirds. Periodicity was examined using the 14 data sets for which at least 15 years of data were available. Geographic synchrony in acorn production was determined using the "modified correlogram" technique of Koenig and Knops (1998b). In brief, this involves calculating Pearson correlation coefficients (r) for all pairwise combinations of sites for which data from at least four years were in common. Two matrices are then extracted, the first consisting of the distance between sites and the second consisting of the r-values between the acorn crops of the two sites measured over the years for which they overlapped. Sites are then divided into distance categories and the significance of the overall mean r-values tested using a randomization procedure. Results provide a picture of the absolute degree of spatial or geographic synchrony that can be expected in annual acorn production between sites a given distance apart.

RESEARCH FINDINGS

Variability in Annual Acorn Production

All studies of acorn production to date have found considerable variability from year to year. For example, using the Hastings data, the mean annual number of acorns counted in 30 seconds per tree (transformed or not) has ranged over nearly two orders of magnitude or more, with coefficients of variation (CV) on the order of 84–112% (untransformed values) or 55–87% (transformed values) (see Table 9.1). Whether acorn crops ever fail entirely is probably unanswerable, but crops can clearly

Table 9.1
Variability in mean annual and overall mean individual acorn productivity in five species

Species individuals	Annual mean			Individual mean			
	Minimum	Maximum	Coefficient of variation	Minimum	Maximum	Coefficient of variation	N
	Acorns per 30 seconds (N30)						
Q. lobata	0.69	75.69	93.4	0.00	71.84	83.7	85
Q. douglasii	0.93	70.29	108.8	0.00	65.05	91.7	56
Q. agrifolia	0.38	47.79	91.6	0.42	49.68	85.7	61
Q. chrysolepis	0.00	55.76	83.7	0.42	49.68	85.7	21
Q. kelloggii	0.00	39.29	112.0	0.47	39.58	63.6	18
	Log-transformed acorns per 30 seconds (LN30)						
Q. lobata	0.20	3.84	56.6	0.00	3.89	52.3	85
Q. douglasii	0.39	3.69	60.2	0.00	3.56	56.7	56
Q. agrifolia	0.13	2.98	68.5	0.04	3.43	53.4	61
Q. chrysolepis	0.00	3.49	54.8	0.19	3.15	48.9	21
Q. kelloggii	0.00	3.36	87.2	0.18	2.54	35.7	18

Notes: Data from Hastings Reservation measured from 1980 through 1998 (19 years) as the mean and log-transformed mean number of acorns counted in 30 seconds. Only individuals with complete data are included.

range from being very poor (only a small proportion of trees containing any acorns whatsoever) to being extremely good (nearly all trees containing a heavy crop of acorns).

Equally dramatic are differences in productivity by individual oaks. For example, in our study at Hastings Reservation, some trees have produced acorns (usually a very good crop) in all or nearly all 19 years of the study, whereas other individuals have not produced one acorn in 19 years; mean variability across individuals again ranges over nearly two orders of magnitude or more, with CVs in the range of 64–92% (untransformed values) or 36–57% (transformed values).

Within the population, however, within-year acorn production is relatively synchronous despite differences among individual trees (Koenig, Mumme, et al. 1994). Thus, even in the best of years there are trees with relatively few acorns, but for those individuals a few acorns constitute a relatively large crop; in most years, such trees do not produce any acorns at all.

The distribution of the CVs in annual acorn production for 1-year and 2-year species from the complete oak data set are summarized in Figure 9.1. Variability among the untransformed data on masting behavior is generally similar to other published compilations, including 42 studies of plants summarized by Kelly (1994), a wide variety of woody species by Herrera et al. (1998), and various northern hemisphere conifers and broad-leafed genera (including oaks) by Koenig and Knops (2000). The majority of CVs fall between 41% and 160% using the untransformed data and almost uniformly below 80% using the log-transformed data. There was no statistical difference in CVs of 1-year and 2-year species using either the untransformed or log-transformed data (Mann-Whitney U-tests, both $z < 1.0$, $P > 0.3$), suggesting that there is no clear difference between these groups in the extent to which they exhibit masting behavior. Using the untransformed data, only 5 of the 40 populations yielded CVs greater than 160%, with the maximum of 225% being achieved in a 12-year study of *Q. prinus* in the Appalachians by Beck (1977) in which 8 of the 12 years were near or total failures. Such extreme variability is the exception rather than the rule.

For both the 1-year and 2-year species there were negative correlations between latitude and CV which were significant or nearly so for the log-transformed data (untransformed data: $r_s = -0.30$, $n = 21$, $P = 0.19$ [1-year species], $r_s = -0.31$, $n = 19$, $P = 0.19$ [2-year species]); log-transformed data: $r_s = -0.48$, $n = 21$, $P = 0.03$ [1-year species]), $r_s = -0.43$, $n = 19$, $P = 0.07$ [2-year species]). However, the statistical sig-

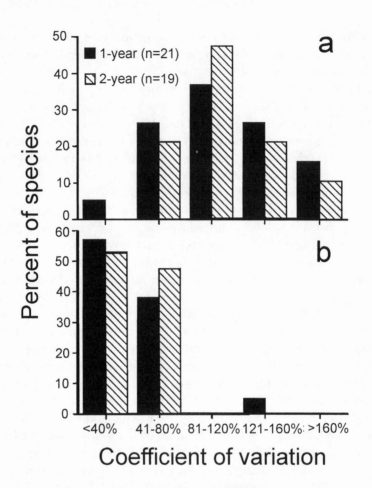

FIGURE 9.1. Distribution of the coefficients of variation in annual acorn production of oaks requiring one and two years to mature acorns; only data sets presenting interval or ratio-level data at least six years in length were included. (*a*) untransformed data (mean ± SD = 108.6 ± 49.4% [1-year species] and 110.3 ± 35.8% [2-year species]); (*b*) log-transformed data (mean ± SD = 37.8 ± 30.9% [1-year species] and 40.5 ± 23.2% [2-year species]).

nificance of these latter relationships was lost in multiple regressions controlling for the number of years of data.

Distribution of Annual Acorn Crop Size

Prior analyses of the oak data set revealed that normality could be rejected for only one of the untransformed and none of the log-transformed sets of data using Kolmogorov-Smirnov one-sample tests (Koenig and Knops 2000). Thus, it is generally not possible to reject the null hypothesis that acorn crops are normally (or log-normally) distributed. Dividing the data sets (including those that are categorical) into thirds, approximately one-third (18 of 49) of the untransformed and one-fourth (13 of 49) of the log-transformed data sets are bimodal in that the frequency of years yielding intermediate acorn crops was less than the frequency of years with larger and smaller crops. Neither of these frequencies is significantly different from the number expected by chance alone (χ^2 tests, both P > 0.2). Using the untransformed data, there was no difference between 1-year and 2-year species in the frequency of bimodality (1-year: 9 of 26 [34.6%] data sets; 2-year: 9 of 23 [39.1%] data sets). Only 11 of the 49 (22.4%) data sets contained no years of intermediate acorn crops. These 11 data sets were relatively short (mean ± SD length = 8.1 ± 2.9 years) compared to 38 data sets with at least some intermediate years (13.1 ± 6.8 years; Mann-Whitney U-test, z = 2.7, P < 0.01). Thus, failure to record intermediate crops was likely an artifact of small sample size. Identical conclusions are reached using the log-transformed data.

An alternative way to investigate bimodality in these data is to combine the standardized data sets and then compare the total number of years of data falling into the middle third of values against those in the upper and lower thirds. If we do this with the untransformed oak data sets, restricting the analysis to sites with at least six years of information, the number of years in the lower, middle, and upper thirds of the distribution are 350, 122, and 115. Using the log-transformed data sets, comparable values are 148, 164, and 275 years. Thus, neither of these combined data sets is bimodal, even with the conservative definition of bimodality used here.

Using the Hastings data, bimodality in acorn production, again defined as fewer years in the middle third of the overall range of annual acorn production values, is significant for individuals of four of the five species (*Q. lobata, Q. douglasii, Q. agrifolia,* and *Q. kelloggii;* Koenig,

Mumme, et al. 1994). However, overlap in the "tails" of the two modes is considerable for all species.

Intermast Interval

The intermast interval is the average time between good acorn crops (mast years). Although it is commonly held that most populations of oaks exhibit some sort of significantly cyclic behavior in acorn production, leading to a regular intermast interval (e.g., Schopmeyer 1974), it is by no means certain that intermast intervals are of regular length, or indeed that they exist at all, except in a few populations where masting intervals are unambiguous, such as in the alternate-bearing population of *Q. robur* studied by Crawley and Long (1995). Such clear cycles in acorn production are generally not evident for oaks, because, in the absence of a bimodal distribution or some other discontinuity in the distribution of annual seed crops, it is not possible to objectively distinguish between mast and nonmast years (Kelly 1994, Herrera et al. 1998). An alternative way to detect intermast intervals or the presence of cycles of acorn production is to perform time-series analysis. We performed such analyses on the oak data set using only sites providing at least 15 years of continuous data—still a marginal sample size for time-series analyses. Nine 1-year species and five 2-year species met this criterion.

Of the 14 data sets, 4 (29%; two 1-year and two 2-year species) exhibited no significant periodicity as tested by a Kolmogorov-Smirnov one-sample test of the periodogram values against that expected from a uniform distribution; the remaining 10 populations exhibited significant periodicity. Peak spectral density of all three 2-year species falling into this latter category corresponded to periodicities of between 4.6 and 6.3 years (mean 5.7 ± 1.0 years), whereas peak spectral density of five of the seven 1-year species corresponded to relatively short periodicities of 2.0 to 2.4 years. However, two populations of *Q. mongolica* were significantly periodic and had peak spectral densities at long periods of 10 years. Consequently, there was no significant difference between the peak spectral densities of 1-year and 2-year species (Mann-Whitney U-test, $z = 1.7$, P = 0.09). Excluding these two *Q. mongolica* populations, 1-year species cycled at significantly shorter periodicities than 2-year species (1-year: 2.1 ± 0.1 yrs, 2-year: 5.7 ± 1.0 yrs; Mann-Whitney U-test, $z = 2.9$, P < 0.005). Examples of spectral density plots for a typical 1-year species (*Q. alba*) and 2-year species (*Q. velutina*) are presented in Figure 9.2.

These results support the hypothesis that most populations of oaks ex-

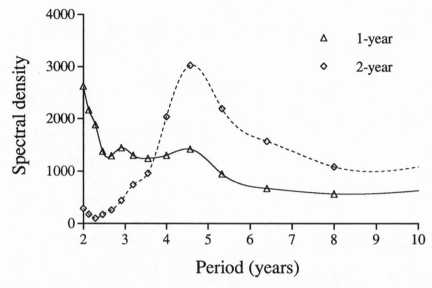

FIGURE 9.2.   Representative spectral densities of *Q. alba* (1-year species) and *Q. velutina* (2-year species) based on 32 years of data (1949–1980) from Missouri. (Reported by Christisen and Kearby [1984].)

hibit some sort of significantly cyclic behavior in acorn production. However, the fact that peak spectral densities generally were at fractional (noninteger) values indicates that cycles are often of irregular length rather than occurring at the same interval of time. They also suggest that periodicity in 1-year species is often, but not always, shorter than in 2-year species, in contrast with prior results (Sork 1993) based on the intermast intervals reported by Schopmeyer (1974).

Environmental Correlates of Acorn Production

Freezing of flowers in the spring can clearly result in poor acorn crops or apparent total failures the following autumn for 1-year species and in the subsequent year for 2-year species (Uhlig and Wilson 1952, Goodrum et al. 1971, Neilson and Wullstein 1980). Beyond this extreme, attempts to correlate annual acorn production with meterological correlates have met with mixed success (Table 9.2). The only notable generalization that emerges is the finding in five populations of 1-year species in the white oak subgenus (three populations of *Q. alba*

Table 9.2

Reported relationships between annual acorn production and weather

Species	Duration of study (years)	Relationship	Reference
1-year species			
Q. agrifolia	19	More rainfall year x − 2 and more rainfall year x = larger crop[a]	This study
Q. alba	14	Warmer April temperatures = larger crop	Sharp and Sprague 1967
Q. alba	8	Warmer spring temperatures = larger crop	Sork et al. 1993
Q. alba	6	Maximum temperature and increased days of hail during pollination = smaller crop	Cecich 1997
Q. douglasii	19	Warmer April temperatures, less rainfall year x and warmer summer temperatures = larger crop	This study
Q. lobata	19	Same as for Q. douglasii	This study
Q. robur	15	No clear significant correlation found	Crawley and Long 1995
2-year species			
Q. chrysolepis	19	More rainfall year x − 1 = larger crop	This study
Q. ilicifolia	—	Low humidity = larger crop (among individuals)	Wolgast and Stout 1977
Q. kelloggii	19	None found	This study
Q. rubra	8	Dry summer year x − 2, warmer spring temperatures, absence of spring frost year x − 1 = larger crop	Sork et al. 1993
Q. velutina	8	Warmer spring temperatures = larger crop	Sork et al. 1993
Q. velutina	6	Maximum temperature during pollination = smaller crop	Cecich 1997

Note: Data from this study update Koenig et al. 1996. Multiple relationships are in decreasing order of significance.

[a]"Year x" refers to the year of the acorn crop; rainfall is from September of year x − 1 through August of year x.

and one each of *Q. douglasii,* and *Q. lobata)* of a significant relationship between conditions during the spring flowering and fertilization period and acorn production the following autumn, the only caveat being that one of these (Cecich 1997) found a negative effect of warmer temperatures on the acorn crop rather than the positive effect found by the remaining studies. No clear significant relationship was found between any environmental variable tested and acorn production in *Q. robur,* while acorn production in the live oak *Q. agrifolia* is correlated with rainfall rather than spring weather conditions. For the white oaks, at least, this suggests that acorn crops of many species are proximately determined either by pollen dispersal or factors otherwise influencing fertilization in the spring.

For 2-year species results are more variable. No two species or studies demonstrated similar correlations, with the exception of a positive relationship once again between warm spring temperatures and the subsequent acorn crop in both *Q. rubra* and *Q. velutina* (see Table 9.2). Otherwise, each study yielded a slightly different set of correlations between weather and acorn production.

There has been considerably less investigation of the effects of weather conditions specifically during flowering on acorn production by individual trees. Among a subset of the Hastings oaks, the mean amount of solar radiation during the time period when 23 *Q. douglasii* flowered was significantly positively correlated with the size of the subsequent acorn crop after controlling for annual differences, suggesting that conditions favorable for pollination may influence not only annual acorn crop size but also variation among individuals. However, no such relationship was detected for either *Q. lobata* or *Q. agrifolia* (Koenig et al. 1996).

Geographic Synchrony in Acorn Production

Until recently, most studies of annual acorn production emphasized site-to-site variability rather than the possibility of geographic synchrony (Neilson and Wullstein 1980, Crawley and Long 1995). However, masting is a population phenomenon (Kelly 1994), and the size of the population involved determines, among other things, the extent to which predators may be affected by crop failures. Consequently, the extent of geographic synchrony in acorn production bears on the issue of the ultimate functional consequences of masting behavior.

We have approached this issue in several ways. First, comparisons of annual acorn crops for *Q. lobata, Q. douglasii,* and *Q. agrifolia* at three sites in central coastal California over a 10-year period indicate statistically significant synchrony within, and even between, these species over distances of nearly 300 kms (Koenig et al. 1996, Koenig, Knops, et al. 1999). Second, comparisons over a 16-year period between acorn production by *Q. lobata, Q. douglasii,* and *Q. kelloggii* at two sites in coastal California 320 km apart confirms a high degree of synchrony for the 1-

year *Q. lobata* and *Q. douglasii;* no significant correlation in the acorn crops between the two sites was detected for the 2-year *Q. kelloggii* (Koenig, McCullough, et al. 1999). Preliminary indications suggest that geographic synchrony is maintained by the same environmental factors that correlate with acorn productivity within a site. For example, April temperatures, which correlate strongly with annual acorn production of both *Q. lobata* and *Q. douglasii* at Hastings Reservation (see Table 9.2), exhibit high spatial autocorrelation between sites as well, more or less matching the degree of synchrony observed in the acorn crops of these two species in coastal California (Koenig et al. 1996).

We are currently expanding these studies by means of a survey of acorn production at 14 sites throughout California, encompassing 34 populations of six different species spanning a distance of nearly 1,000 km. Preliminary results based on the first five years of data (1994 through 1998) for five species, including three 1-year species (*Q. lobata, Q. douglasii,* and *Q. agrifolia*) and two 2-year species (*Q. chrysolepis* and *Q. kelloggii*), support the hypothesis that 1-year species generally show extensive geographic synchrony whereas 2-year species may not. Of the three 1-year species surveyed at multiple sites, mean (± SD) correlations between pairwise combinations of sites are all high (*Q. lobata:* 0.76 ±

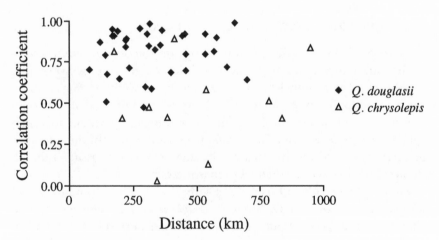

FIGURE 9.3. Pairwise Pearson correlation coefficients of mean annual acorn production between sites of two species (*Q. douglasii,* a 1-year species, and *Q. chrysolepis,* a 2-year species) measured at nine (*Q. douglasii*) and six (*Q. chrysolepis*) sites in California surveyed between 1994 and 1998. (Koenig and Knops, unpublished data.)

0.17, n = 15; *Q. douglasii:* 0.80 ± 0.14, n = 36; *Q. agrifolia:* 0.71 ± 0.29,
n = 15), whereas those for the two 2-year species are relatively low and
more variable (*Q. chrysolepis:* 0.29 ± 0.44, n = 15; *Q. kelloggii:* 0.42 ± 0.38,
n = 15). Correlation coefficients for one species in each group are presented in Figure 9.3. None of the species exhibits any significant decline
in synchrony with increasing distance between sites, based on Mantel
tests, in contrast to the pattern generally found in ecological phenomena (Koenig 1999). In fact, synchrony between sites for all species except
Q. lobata actually increases with distance, albeit not significantly. However, this is most likely an artifact of the relatively short time span (5
years) over which we currently have data.

Finally, we can analyze the larger oak database for patterns of geographic synchrony, using recently developed methods of testing for spatial autocorrelation (Koenig and Knops 1998b). Results (Figure 9.4) in-

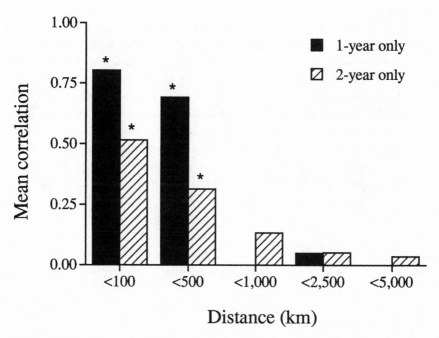

FIGURE 9.4.   Mean correlations between annual acorn crops of 1-year and
2-year oaks depending on the distance separating sites. Correlations that are
significantly greater than zero (based on randomization tests; Koenig and
Knops 1998b) for both 1- and 2-year species up to sites 500 km apart are indicated with an asterisk.

dicate significant synchrony among both 1-year and 2-year species over distances of up to 500 km and generally support the hypothesis that synchrony is greater, at least between sites up to 500 km apart, for 1-year rather than 2-year species.

Synchrony within Communities of Oaks

At least four studies have reported within-site synchrony of oak species requiring the same number of years to mature acorns but asynchrony between species requiring different numbers of years to mature acorns (Mohler 1990 [three studies cited], Koenig, Mumme, et al. 1994). Updated results from the Hastings data are summarized in Table 9.3. All four comparisons between species requiring the same number of years to mature acorns are positive and all but one is statistically significant. In contrast, all six comparisons between species requiring different numbers of years to mature acorns are negative; none, however, is statistically different from zero.

Given that their phenologies do not overlap except with the final growth of acorns (Sork et al. 1993), this pattern of synchrony within types and asynchrony between types of oaks requiring different numbers of years to mature acorns is expected if acorn production by all species is, to at least some extent, influenced by the same suite of environmental factors. For example, a freeze that kills flowers in year x will result in a crop failure of 1-year species in autumn of year x, whereas 2-year species will remain unaffected until year x + 1. Such asynchrony appears to have important consequences for several aspects of oak ecology, from reducing the frequency of total acorn crop failures within a site to facilitating co-occurrence of oak species (Mohler 1990).

Table 9.3
Spearman rank correlations between the annual acorn crops of five species

	Q. douglasii *(1)*	Q. agrifolia *(1)*	Q. chrysolepis *(2)*	Q. kelloggii *(2)*
Q. lobata (1)	0.85*	0.35	−0.26	−0.07
Q. douglasii (1)	NA	0.58*	−0.35	−0.27
Q. agrifolia (1)		NA	−0.01	−0.21
Q. chrysolepis (2)			NA	0.56*

Source: Data from Hastings Reservation, central coastal California, 1980–1998.

Note: Number of years species requires to mature acorns is shown in parentheses.
* = $P < 0.01$; other $P > 0.05$. $N = 19$ years.

The surprising feature of such interspecific asynchrony is that, by reducing the annual variability in total acorn abundance and the probability of total crop failures, it also facilitates the persistence of generalist species dependent on acorns. For example, acorn woodpeckers (*Melanerpes formicivorus*) are virtually restricted to sites containing at least two species of oaks, presumably because of the lower probability of crop failure and reduced annual variability as oak species diversity increases (Bock and Bock 1974, Koenig and Haydock 1999). Such facilitation is not predicted by the hypothesis of predator satiation, as discussed below.

Switching of Resources into Acorn Production

Two sorts of evidence have been used to test for evidence that oaks divert resources from elsewhere to produce large seed crops. The first is a negative autocorrelation between seed crops in successive years. Such negative autocorrelations have been documented within individual trees with a 1-year time lag for the 1-year species *Q. lobata* and *Q. douglasii* (Koenig, Mumme, et al. 1994) and *Q. alba* (Sork et al. 1993) and at longer (2- or 3-year) time lags for the 2-year species *Q. rubra* and *Q. velutina* (Sork et al. 1993) and *Q. kelloggii* (Koenig, Mumme, et al. 1994). Using the combined oak data set, Koenig and Knops (2000) documented a significant negative temporal autocorrelation with a 1-year time lag for 1-year species and both 2- and 3-year time lags for 2-year species.

There have unfortunately been few direct tests of the potential tradeoff between vegetative growth and reproduction in oaks. Crawley and Long (1995) indicate a failure to find such a tradeoff in *Q. robur*, while Sork and Bramble (1993) found suggestive evidence only for a tradeoff based on five years of data in *Q. velutina* and *Q. rubra*. Our own work at Hastings Reservation currently involves comparing acorn production between 1980 and 1994 with tree-ring growth for the deciduous species *Q. lobata, Q. douglasii,* and *Q. kelloggii.* Comparisons of annual acorn counts with standardized tree-ring widths within individuals yields no significant correlation between growth and reproduction during the current year for any of the species (see "within individuals" analyses in Table 9.4); in fact, if anything there is a positive correlation in *Q. kelloggii.* Among individuals in the population, there are significant inverse correlations for *Q. lobata* and *Q. douglasii,* but values are still very small. Lagged effects are similarly mixed. For *Q. lobata,* for example, comparison of acorn production with growth the prior year yields a significant

Table 9.4
Analyses of a tradeoff between growth and reproduction in three species of deciduous oaks

	Q. lobata	Q. douglasii	Q. kelloggii
No time lag			
Within individuals	40 vs. 39[a]	19 vs. 29	14 vs. 4**
Between individuals	−0.10*[b]	−0.11**	0.13
Growth in year x vs. acorns in year x + 1			
Within individuals	42 vs. 37*	28 vs. 21	7 vs. 10
Between individuals	−0.09**	0.01	−0.00
Acorns in year x vs. growth in year x + 1			
Within individuals	51 vs. 28**	25 vs. 24	4 vs. 14
Between individuals	0.06	0.08	−0.19**

Source: Data from Hastings Reservation (Knops and Koenig, unpublished data).
[a]Tree-ring width vs. acorn production by tree i. Numbers are trees exhibiting positive and negative correlations. Tests are by Wilcoxon sign-ranks tests.
[b]Pairwise comparisons of tree-ring width of tree i vs. acorn production by all other trees in sample except tree i. Numbers are mean correlation coefficients. Tests are by randomization.
* = $P < 0.05$; ** = $P < 0.01$.

inverse correlation between individuals but a positive correlation within individuals, while acorn production compared to growth the following year is significantly positive within individuals. *Q. kelloggii* demonstrates no significant relationship between acorn production and growth the prior year, but a significantly negative correlation between acorns and growth the following year, at least among individuals in the population.

These results support the possibility of resource switching in oaks, but they apparently accomplish this more by diverting resources from energy reserves than through direct tradeoffs between growth and vegetation. This contrasts with several other major genera of temperate trees for which switching of resources between growth and reproduction have been documented both within populations (Eis et al. 1965, Tappeiner 1969, Woodward et al. 1994, Norton and Kelly 1988) and between populations (Koenig and Knops 1998a).

DISCUSSION

Masting Behavior in Oaks

Variability in annual acorn production of oaks is high (Table 9.1; Figure 9.1), generally matching that of other temperate-zone trees including conifers and a range of other species summarized by Kelly (1994). Based on the general inability to reject the hypothesis that the distribution of annual acorn crop size is normally distributed (Koenig and Knops 2000), combined with a lack of bimodality in a greater number of annual oak data sets than would be expected by chance, oaks do not appear to be "strict masting" species, defined as species that produce widely varying annual seed crops that are bimodally distributed with no overlap between the tails (Kelly 1994). This is consistent with data from most other genera of temperate-zone forest trees (Kelly 1994, Koenig and Knops 2000) as well as the wide variety of woody species analyzed by Herrera et al. (1998).

Evidence for "normal masting," with significant (but not complete) bimodality or resource switching , is better. Annual acorn production is evident within individuals for some species, and, although direct trade-off between growth and reproduction within individual trees has not been demonstrated, oaks in general exhibit negative autocorrelations between seed crops in successive years or longer intervals, suggesting that resources are being diverted from reserves to produce large acorn crops. More work clearly is desirable in this area, to clarify the tradeoffs, if any, that are being made within individuals to produce the highly varying seed crops observed in oaks.

Among environmental factors that correlate with variability of acorn production by most, but not all, populations of oaks, the only variable that stands out is warm, dry weather conditions during the spring flowering period, which has been found usually to produce good acorn crops in several species, particularly 1-year species in the white oak subgenus. Otherwise, the variables involved appear to differ considerably among species and possibly even from population to population, although in general it is likely that the same environmental factors correlating with annual acorn production within a site are used as cues to synchronize acorn production over relatively large geographic areas.

The precise mechanisms by which weather influences the subsequent acorn crop appear to be complex and may involve pollen limitation (E. Knapp, personal communication), the initial size of the female flower crop, or subsequent flower survival (Sork et al. 1993). In general, poor female flower crops lead (unsurprisingly) to poor acorn crops, while good crops of female flowers may or may not result in good acorn crops (Sharp and Sprague 1967, Gysel 1958, Sork and Bramble 1993, Cecich 1997). Unfortunately, there are relatively few studies carefully examining the relationship between flowering and subsequent acorn production, no doubt because of the difficulty of dealing with the small, inconspicuous female flowers of oaks. This is clearly an area worthy of additional work.

Functional Consequences of Masting in Oaks

Annual acorn production in oaks is apparently an adaptive response to some environmental challenge rather than merely the product of resource tracking. Determining the precise nature of the environmental challenge that makes masting functionally significant has proved challenging. For oaks, two major hypotheses have proved the most promising, both corresponding to one or another economy of scale (Norton and Kelly 1988), whereby larger reproductive efforts are more efficient, thereby favoring irregularly large seed crops rather than more regular small crops.

The "wind pollination" hypothesis focuses on the potential advantages of concentrating flowering in years of high reproductive output, thereby increasing the efficiency of pollination (Smith et al. 1990, Kelly 1994). Although this hypothesis has proved to be a strong candidate in many temperate genera, it is logically more likely to be important in species that commit considerable energy to female growth prior to pollination, thereby increasing the energy benefits potentially attainable by synchronized flower production. This is not the case in oaks, which produce small, presumably cheap flowers and commit significant energy to them only after pollination.

Direct tests of the wind pollination hypothesis involve demonstrating a higher percentage of fertilization in high-flowering years. Unfortunately, few such tests have been performed in oaks. Sork (1993) found that female flowers did not correlate with male catkin biomass and that fertilization percentage was not greater in years of high pistillate or pollen production; neither of these results supports the wind pollination hypothesis. Patterns of acorn production in our population in California generally conform to predictions of the wind pollination hypothesis, but these latter tests are indirect and are also largely consistent with alternatives (Koenig, Mumme, et al. 1994). Thus, efficiencies related to wind pollination remain a viable reason for masting in oaks, but probably not to the extent that they are in other wind-pollinated taxa, such as conifers, in which a considerable investment in female reproductive structures occurs prior to pollination.

The second major hypothesis for mast fruiting is predator satiation, which suggests that variable seed production acts to decrease populations of predators in poor crop years, thereby increasing the probability that seeds produced during good years will escape predation. As pointed out by Kelly (1994), the importance of predator satiation is likely to be dependent on the functional, as well as the numerical, response of predators to large seed crops. Acorn predators include a wide diversity of both vertebrate and invertebrate species, making the identification of key predators difficult. At least three attempts to test the hypothesis for insects that prey on oaks have generated at least some support for the prediction of increased seed survival from insect predators in mast years (Sork 1993, Crawley and Long 1995, Koenig and Knops unpublished data).

Patterns of acorn production by individual oak species at Hastings are consistent with several predictions of predator satiation, but in general they fail to reject the alternative hypothesis of wind pollination (Koenig, Mumme, et al. 1994). Perhaps the most damaging feature of acorn production with respect to the predator satiation hypothesis is the lack of synchronization in masting behavior among species within a community, in particular between species requiring different numbers of years to mature acorns (Table 9.2). This results in a relatively high proportion of sites containing species that are effectively uncorrelated, or even possibly negatively correlated, in their acorn production patterns. Such asynchrony not only serves to reduce the probability of total crop failure in a site but in some cases may be key to maintaining viable populations of acorn predators (Bock and Bock 1974, Koenig and Haydock 1999).

Summary

The evolution of synchronous reproduction has important ecological consequences and thus is an important problem in behavioral ecology. Although there are monocarpic species such as bamboo for which masting is unambiguous, variable seed production by most polycarpic species, including oaks, cannot usually be objectively divided into mast and nonmast years, and thus it is necessary to carefully quantify and define the degree to which reproduction is synchronous both within and between populations. Among most species of oaks that have been studied, reproduction is more or less synchronous, but mean acorn production is log-normally, not bimodally, distributed. In most cases, acorn production can be correlated with some environmental factor, often conditions during the spring flowering period, suggesting that pollen limitation or some other factor involved with fertilization may be important in determining acorn crops at a proximate level. Correlations with environmental conditions also are apparently the proximate cause of relatively large-scale synchrony in acorn production within species over distances of hundreds of kilometers.

Few data are available to document tradeoffs between growth and reproduction in oaks, but there is good evidence for temporal autocorrelations in acorn crops within individuals, indicating that switching of resources occurs. This in turn indicates that patterns of acorn production are not simply the result of tracking of weather events but rather that masting is an evolved strategy. However, masting patterns are complex, particularly at the community level, and do not strictly conform to predictions of either the wind pollination hypothesis or the predator satiation hypothesis. Since many vertebrate predators of acorns are simultaneously major dispersal agents of acorns, this may reflect an attempt to balance the conflicting selective benefits of seed dispersal with predator satiation.

ACKNOWLEDGMENTS

We thank our colleagues Bill Carmen, Ron Mumme, and Mark Stanback for helping with the acorn surveys over the years and Bill Healy for comments on the manuscript. Financial support for our work has come from the Integrated Hardwoods Range Management Program of the University of California, Barry Garrison and the California Department of Fish and Game, and the National Science Foundation.

Chapter 10

Dynamics of Acorn Production by Five Species of Southern Appalachian Oaks

CATHRYN H. GREENBERG AND BERNARD R. PARRESOL

The management implications of fluctuations in acorn crop size underscore the need to better understand their patterns, causal factors, and predictability (both within a year and long term). Acorn yield has a demonstrable influence on the population dynamics of many wildlife species, both game (Eiler et al. 1989, Wentworth et al. 1992) and nongame (Hannon et al. 1987, Koenig and Mumme 1987, Smith and Scarlett 1987, Elkinton et al. 1996, Wolff 1996, McShea 2000). Wolff (1996) suggests that acorns function as a "keystone" resource in forest community dynamics, by influencing small mammal prey populations. Indeed, acorn crop size has a far-reaching influence on ecosystems. White-footed mouse (*Peromyscus leucopus*) populations, which are directly influenced by acorn crop size, affect gypsy moth (*Lymantria dispar*) populations (Elkinton et al. 1996) and even the prevalence of Lyme disease (Jones et al. 1998). Also, oak regeneration has been shown to increase following large acorn crops (Marquis et al. 1976), although a host of other factors influence seedling establishment and success (Loftis and McGee 1993). The ability to predict the size of future acorn crops (Sork et al. 1993, Koenig, Mumme, et al. 1994) and to estimate current-year production (e.g., Koenig, Knops, et al. 1994, Whitehead 1969, 1980, Graves 1980, Sharp 1958, Christisen and Kearby 1984) has received considerable attention by forest managers and researchers because of its importance to wildlife and forest regeneration.

This chapter examines temporal patterns of acorn production within and among five species of southern Appalachian oaks. The data encompass the first five years (1993–1997) of an ongoing, long-term study. Variability in acorn production among individual trees and characteristics of fruit production that contribute to such variation will be addressed. The correlation between both the number of acorns on fruiting trees and the proportion of trees bearing acorns with annual crop size is evaluated, and a simple method for estimating acorn crop yield is proposed (number of acorns per square meter of basal area [BA]). Using visual survey information and a BA inventory for each oak species, land managers can apply crop size estimates (acorns/m² BA) to areas within the southern Appalachians to calculate the acorn crop by species within years. Finally, an acorn yield table based on five-year average acorn production is provided. These tables can be used with BA inventories to calculate mean annual acorn production by species on an area basis.

Acorn Sampling

Acorn production by 765 individuals of five oak species was sampled throughout the southern Appalachians from 1993 to 1997 (see Greenberg 2000, for details). Study species included northern red oak (*Q. rubra*) (*N* = 148), scarlet oak (*Q. coccinea*) (*N* = 142), and black oak (*Q. velutina*) (*N* = 91) in the red oak subgenus, and chestnut oak (*Q. prinus*) (*N* = 201) and white oak (*Q. alba*) (*N* = 183) in the white oak subgenus. Study trees were scattered in small groups throughout national forests (NFs) in three states: the Cherokee NF in Tennessee, the Pisgah NF in North Carolina, and the Chatahoochee NF in north Georgia. Study sites were distributed generally from northeast to southwest following the orientation of the mountains and separated by ≤ 220 km. Sample trees were located at elevations ranging from 850 to 1,180 m above sea level and over a wide range of topographic features (e.g., aspect, slope position, and percent slope).

Trees were selected to represent a wide range of size (9–133 cm dbh [diameter at breast height]) and age classes. Most trees were mature and in dominant or codominant crown positions (a few were intermediate). One stand of scarlet oak (*N* = 20) and white oak (*N* = 18) in the Pisgah NF was established following a clearcut regeneration harvest in 1967 (when all trees taller than 1.4 m were felled) and was 26 years old at the start of the study.

Acorns were collected in circular, 0.5-m-diameter traps placed beneath the trees to obtain a representative sample of the crown. The number of traps per tree was approximately proportional to the BA (2–14 per tree; average 4.1 ± 2.2 standard deviation/tree). Crop-size estimates probably were conservative, because trap tallies did not account for acorns removed by squirrels or other arboreal consumers. Crown areas were measured with eight equally spaced radii from tree base to the canopy drip line, and area was computed as an octagon. Traps were checked at approximately two-week intervals from mid-August through the completion of acorn drop.

STATISTICAL ANALYSIS

Acorn production was calculated for each tree by multiplying the number of mature acorns collected per m^2 trap area by the crown area. All well-developed acorns were included in the analyses regardless of their condition (sound, animal- or insect-damaged). To standardize comparisons among different-sized trees and simplify for use by forest managers, the number of acorns per tree were converted to the number per m^2 basal area by dividing the total acorn production by the BA of each tree. Because of the correlation between BA and crown area, the number of acorns/m^2 BA is correlated with the number/m^2 crown. However, BA is more easily measured than crown area. This measure of acorn production can be tailored to stands (any size area) of varying oak composition and BA simply by multiplying the BA present by the number of acorns/m^2 BA for each species and summing.

The annual crop size for each species was ranked as "poor," "moderate" or "good" by comparing the mean number of acorns/m^2 BA for that year to its five-year mean (1993–1997). Good crop years were defined as ≥ the five-year mean, moderate as ≥ 60% of but < 100% of the mean, and poor as < 60% of the five-year species mean (adapted from Healy et al. 1999). Individual trees of each species were also ranked as poor, moderate, or good producers, by the same criteria.

Using analysis of variance (ANOVA), the mean number of acorns/m^2 BA of fruiting trees (excluding nonfruiting individuals) was compared among years for each species, and pairwise contrasts were performed using least squares means tests (SAS Institute 1989). The number of acorns/m^2 BA was natural-log transformed for ANOVA to reduce the correlation between the mean and variance (Sokal and Rohlf 1981). Statistical significance is reported at the $P < 0.05$ level unless otherwise stated.

Reduced major axis (RMA) regression was used to predict within-year crop size using the proportion of acorn-bearing trees in the population as the independent variable (Greenberg and Parresol 2000). The RMA technique rather than ordinary least squares regression was selected because in this case the independent variable (x), the proportion of acorn bearing trees, is a sample-based estimate subject to error. In cases where both the x and y variables are subject to error, the RMA technique of fitting lines is recommended (Ricker 1973, 1984, Rayner 1985, Leduc 1987).

ARE SOME SPECIES BETTER PRODUCERS THAN OTHERS?

Acorn production (number and mass) is an important determinant of habitat quality for many species of wildlife and is a focus for many wildlife managers. Hence, understanding the frequency, timing, and relative contribution of acorn production by each oak species composing a forest could assist managers in planning for wildlife food supplies. Acorn production differed significantly among the five species studied (see Table 10.1). On average ($\pm$SE), white oak produced the most acorns per m^2 BA (4,216 $\pm$ 3,118) and chestnut oak the fewest (1,274 $\pm$ 841). Both northern red and white oak produced significantly higher green weight and dry biomass than chestnut, black, or scarlet oak. The distinction between acorn quantities versus mass (green weight and dry, edible biomass) is important for land managers who wish to maintain a specified mast capability in forest stands (Greenberg 2000). Damage to acorns by insect larvae was not examined here, but it can be very high; Beck (1977) estimated that an average of 35% of acorns, in a range of 29–67% depending on species and year, were infested in the southern Appalachians. If insect damage makes acorns nonviable or inedible, their relative contribution to the total crop may differ.

Despite the importance of acorns for wildlife, local and regional yield tables for acorns are unavailable. Table 10.2 summarizes acorn production estimates by this and other studies (although the list is not exhaustive) for the five study species. Comparison of acorn production among studies (Table 10.2) is confounded by a number of factors. Most pub-

Table 10.1

Average acorn production, green weight and dry biomass conversion factors for five species of southern Appalachian oaks, 1993–1997

Species	N	Acorns (±SE) per m² BA	Green weight (±SE) (kg/m² BA)	Green weight conversion (kg/m² BA)	Dry biomass (kg/m² BA)	Dry biomass conversion (kg/m² BA)
Black oak	88	2,045 ± 966[a]	5.36 ± 2.53[a]	0.00262	2.43 ± 1.15[a,b]	0.00119
Northern red oak	111	2,511 ± 1,097[a,b]	17.07 ± 7.46[b]	0.00680	6.38 ± 2.79[c]	0.00254
Scarlet oak	124	2,807 ± 1,401[a,b]	8.48 ± 4.23[c]	0.00302	3.59 ± 1.79[a]	0.00128
Chestnut oak	161	1,274 ± 841[c]	10.26 ± 6.77[a]	0.00805	3.22 ± 2.13[b]	0.00253
White oak	155	4,216 ± 3,118[b]	13.32 ± 9.85[d]	0.00316	5.31 ± 3.93[d]	0.00126

Notes: Green weight and dry biomass conversion factors are based on a subsample of sound acorns drawn from all five years (1993–1997). Superscript letters following acorn numbers or weights denote means within the column that are significantly different based on ANOVA.

Table 10.2

Comparison of studies of acorn production estimates for five eastern oak species.

Species	Author	Number Acorns	Unit	N (sample size)	Duration of study	Location
Black oak						
	Beck 1977	4,218	m² BA	by plot	1962–1973	Asheville, NC
	Burns et al. 1954[a]	900	tree	?	1947–1952	Dent Co., Missouri Ozarks
		1,500	tree	5	1948–1952	Butler Co., Missouri Ozarks
	Christisen and Kearby 1984	115	tree	37	1973–1976	Missouri Ozarks
	Downs 1944 (from Beck 1977)[b]	6,327	m² BA	by plot	1936–1942	Southern Appalachians
	Greenberg (this chapter)	2,045	m² BA	88	1993–1997	Southern Appalachians
	Sork et al. 1993	1,050	tree	13	1981–1988	St. Louis Co., MO
Northern red oak						
	Beck 1977	16,409	m² BA	by plot	1962–1973	Asheville, NC
	Christisen and Kearby 1984	50	tree	15	1973–1976	Missouri Ozarks
	Downs 1944 (from Beck 1977)[b]	4,745	m² BA	by plot	1936–1942	Southern Appalachians
	Greenberg (this chapter)	2,511	m² BA	111	1993–1997	Southern Appalachians
	Healy et al. 1999	16	m² crown	120	1986–1996	Central Massachusetts
	Sork et al. 1993	444	tree	12	1981–1988	St. Louis Co., MO

Scarlet oak	Beck 1977	7,586	m² BA	by plot	1962–1973	Asheville, NC
	Burns et al. 1954[a]	500	tree	?	1947–1951	Dent Co., Missouri Ozarks
		2,400	tree	5	1948–1952	Butler Co., Missouri Ozarks
	Chritisen and Kearby 1984	38	tree	16	1973–1976	Missouri Ozarks
	Downs 1944 (from Beck 1977)[b]	11,126	m² BA	by plot	1936–1942	Southern Appalachians
	Greenberg (this chapter)	2,807	m² BA	124	1993–1997	Southern Appalachians
Chestnut oak	Beck 1977	2,582	m² BA	by plot	1962–1973	Asheville, NC
	Downs 1944 (from Beck 1977)[b]	2,582	m² BA	by plot	1936–1942	Southern Appalachians
	Goodrum et al. 1971	259	tree	?	1950–1954	Kisatchie Nat'l Forest, LA
	Greenberg (this chapter)	1,274	m² BA	161	1993–1997	Southern Appalachians
White oak	Beck 1977	10,717	m² BA	by plot	1962–1973	Asheville, NC
	Burns et al. 1954[a]	1,100	tree	?	1947–1952	Dent Co., Missouri Ozarks
		700	tree	5	1948–1952	Butler Co., Missouri Ozarks
	Christisen and Kearby 1984	112	tree	35	1973–1976	Missouri Ozarks
	Downs 1944 (from Beck 1977)[b]	5,552	m² BA	by plot	1936–1942	Southern Appalachians
	Goodrum et al. 1971	725	tree	10?	1950–1955	Kisatchie Nat'l Forest, LA
	Greenberg (this chapter)	4,216	m² BA	155	1993–1997	Southern Appalachians
	Sork et al. 1993	664	tree	15	1981–1988	St. Louis Co., MO

Note: Reported estimates were converted to number of acorns/m² BA if possible, and reported as in the original study if not.

[a]Same study used for Christisen and Korschgen 1955.

[b]Predicted estimates based on Beck's data and applying data on production by diameter class from Downs (see Beck 1977, Downs 1944).

lished studies are relatively short in duration (12 years is the maximum among those reviewed). Average production estimates differ dramatically depending upon which set of years was sampled, as well. For example, Healy et al. (1999) note that their perception of a "good" northern red oak acorn crop changed during the sixth and eighth years of their study; white oak produced a good crop only in the sixth year of another study (Sharp and Sprague 1967). Differences in geographic location, sampling strategies (individual tree versus area-based plots), sample sizes (often very small, not reported, or reported as combined N for all species studied), and the units in which averages are reported (number per unit crown area; per unit BA by plot; per tree; per ha) further confound comparisons among studies. Although many sources report productivity by diameter class, few note the sample size within diameter classes. These discrepancies highlight the need for long-term studies and for standardization in measurement and reporting methods.

Despite differences among estimates caused by these confounding factors, and despite potentially real regional variation in relative productivity within a species, it is clear that all species are capable of producing a crop that ranges from almost none to many thousands of acorns/m² BA. Estimates of average annual acorn production per unit area also vary widely among studies (Table 10.2). For example, in a hypothetical 1-ha stand composed of 0.8 m² black oak, 1.7 m² northern red oak, 0.5 m² scarlet oak, 1.0 m² chestnut oak, and 1.3 m² white oak, estimates of average annual number of acorns produced range from 51,576 acorns/ha (Beck 1977) to 29,906 acorns/ha (Beck 1977, using data from Downs 1944) to 14,064 acorns/ha (this study). Such large differences serve as a warning when comparing species; variability among years and locations could be misleading when computing average acorn production.

TEMPORAL PATTERNS IN ACORN PRODUCTION

Many studies report that, in most years, acorn production by some species compensates for the effect of crop failure by others (Downs and McQuilken 1944, Burns et al. 1954, Christisen and Korschgen 1955, Gysel 1956, Beck and Olson 1968, Goodrum et al. 1971, Beck 1977, Christisen and Kearby 1984, Beck 1993, Sork et al. 1993, Koenig, Mumme, et al. 1994). Hence, it is important to remember that, although some species may outperform others on an average basis, averages do not insure a consistent supply of acorns.

Differences between the floral biology of the two subgenera of oaks probably contribute to some differences in acorn production patterns among species. Species in the white oak group (*Leptobalanus* subgenus), including chestnut (*Quercus prinus*) and white oak (*Q. alba*), produce flowers in the spring. If they are fertilized, acorns develop by fall of the same year. Conversely, species in the red oak group (*Erythrobalanus* subgenus), including black oak (*Q. velutina*), northern red oak (*Q. rubra*), and scarlet oak (*Q. coccinea*), produce flowers in the spring but (if fertilized) do not develop acorns until the fall of the following year. Hence, the influence of weather or other external influences on acorn production might be expressed differently by species within the red oak versus white oak subgenera.

If external factors such as weather (Sork et al. 1993) influence flower fertilization or acorn development, it might be predicted that, regionally, species within subgenera should perform similarly. Indeed, northern red oak and scarlet oak of the red oak group exhibited similar temporal patterns of acorn production during the five-year study period (Figure 10.1). However, black oak differed, by having a poor crop year in 1994 (northern red oak and scarlet oak had moderate crops) and a moderate crop year in 1996 (northern red oak and scarlet oak had poor crops). White oak and chestnut oak exhibited similar temporal patterns of acorn production, although white oak outperformed chestnut oak in both 1994 and 1996 (the other years were poor crop years for both species). Crop failure occurred only once during the five-year study period for each species.

Indeed, poor acorn production by some species was offset by good or moderate production by others during most years. In some years (1993 and 1995), species of the red oak group produced acorns when those of the white oak group did not, whereas white oak and chestnut oak produced acorns when red oak species performed poorly (1996). In 1994 all species except black oak produced moderate acorn crops. In only one of the five years studied (1997) was there a complete crop failure (Greenberg and Parresol 2000). This and numerous other studies emphasize the importance of maintaining mixed oak stands that include multiple species within both the white oak and red oak subgenera, to enhance the likelihood of a constant acorn supply.

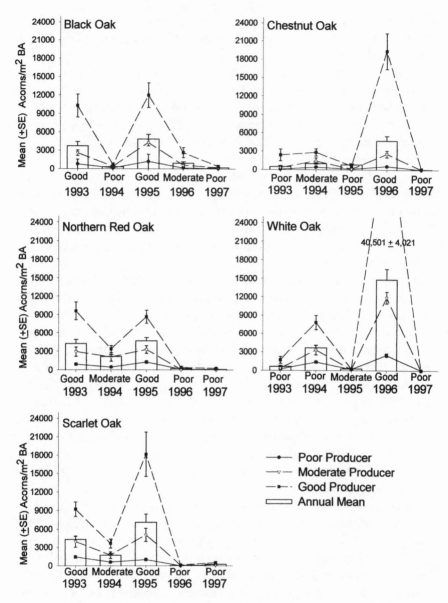

FIGURE 10.1. Annual crop size (mean ± SE number of acorns/m² BA) and relative contribution (mean ± SE number of acorns/m² BA) by good, moderate, and poor producers of five oak species 1993–1997 in the southern Appalachians. Crop-year rating is denoted for each year.

INDIVIDUAL TREE VARIATION
IN ACORN PRODUCTION

Frequency of acorn production also varies among individuals within species. A small proportion of individuals in each species never produced acorns during the study period (1993–1997). With the exception of white oak, a few individuals bore acorns every year (Greenberg 2000).

Good producers composed between 20% (chestnut oak) and 46% (northern red oak) of the sample populations (see Table 10.3). Poor producers composed over 50% of the population for every species except northern red oak. Despite their relatively low representation, good producers of all species outperformed poor and moderate producers by a wide margin of acorn production (Figure 10.1) (Greenberg 2000). Differences were most apparent during good crop years and were negligible in poor crop years. Such disparities in production performance have been reported in numerous studies (Downs and McQuilken 1944, Burns et al. 1954, Gysel 1956, Sharp and Sprague 1967, Christisen and Kearby 1984, Koenig et al. 1991, Sork et al. 1993).

Good producers were characterized by having a higher frequency of acorn-bearing years and producing more acorns/m^2 BA on fruiting trees during good or moderate crop years (Greenberg 2000). However, in any given year, good, moderate, and poor producers were represented similarly in the fruiting population. Hence, the presence of acorns during poor or moderate crop years did not distinguish good from poor producers, nor did an absence of acorns distinguish poor from good producers during good crop years (Greenberg 2000).

Acorn production potentially could be enhanced following silvicul-

Table 10.3
Proportion of poor, moderate, and good acorn producers of five oak species sampled in the southern Appalachians

		Poor	*Moderate*	*Good*
Species	*N*	*(Percentage of individuals)*		
Black oak	135	51.7	19.1	29.2
Northern red oak	111	40.4	13.5	45.9
Scarlet oak	124	53.2	12.1	34.7
Chestnut oak	162	72.2	7.4	20.4
White oak	155	54.2	12.3	33.5

tural treatments such as thinning or two-age harvesting if good producers could be identified and retained. However, three to five years (Healy et al. 1999) or more (Johnson 1994a) of monitoring individual trees for acorn production are necessary to identify good producers. Such difficulty in identifying good producers may in part explain differences in findings among studies of how thinning influences acorn production. If more good than poor producers are removed in one study and more poor than good producers are removed in another, results may differ. Results may be especially confounded when factoring in variability in acorn production among years and species (Healy 1997b).

DOES BIGGER MEAN BETTER?

Acorn production per tree is significantly positively correlated with basal area in all species (Table 10.4). This is not surprising, given the close positive relationship between crown area and BA. Acorn production increases with tree size at least in part simply because larger trees have greater crown areas for producing acorns. It is not surprising then that some studies report increasing acorn production per tree with increasing tree diameter (Goodrum et al. 1971). However, if this were the only influence of tree size on acorn production then the same volume of acorns could be produced by a few large trees or by the same area of crown distributed among several smaller-diameter trees. The key question is whether larger-diameter trees produce more acorns per unit BA (or per unit crown area) than smaller diameter trees.

Table 10.4
Correlation between basal area and mean number of acorns per tree and between basal area and crown area for five species of southern Appalachian oaks, 1993–1997

Species	BA (m^2) vs. acrorns/tree		BA (m^2) vs. crown area (m^2)	
	N	r^2	N	r^2
Black oak	88	0.2706	91	0.4957
Northern red oak	111	0.2387	148	0.5152
Scarlet oak	124	0.2051	142	0.7481
Chestnut oak	162	0.1013	201	0.7328
White oak	154	0.2677	183	0.7122

Note: All correlations are significant ($p < 0.0001$).

Alone, basal area was significantly positively correlated with the number of acorns/m^2 BA in black oak (p = 0.0003; r^2 = 0.14), northern red oak (p = 0.0581; r^2 = 0.03), and white oak (p = 0.0098; r^2 = 0.04), but not in chestnut or scarlet oak. However, size of BA explained little of the variation in acorn production among individuals (Greenberg 2000). A weak relationship between tree diameter and acorn production has been observed in numerous studies (Downs and McQuilken 1944, Burns et al. 1954, Gysel 1956, Sharp and Sprague 1967, Christisen and Kearby 1984, Koenig et al. 1991, Sork et al. 1993). Healy et al. (1999) report that thinning promoted crown and diameter growth in northern red oaks and also increased acorn production per m^2 crown. However, they note that variation among individuals and years had a much greater effect on acorn production than thinning. Given the high variability in acorn production among individual trees it is not surprising that any potential relationship between tree size and the number of acorns/m^2 BA is obscured.

However, when trees are grouped into diameter classes, some differences in acorn production among size classes are apparent (Figure 10.2). ANOVA indicated that in black oak, northern red oak, and white oak, trees ≤ 25 cm dbh produce significantly fewer acorns/m^2 BA than their larger-diameter counterparts. Acorn production appears to taper off in northern red oak and white oak trees > 76 cm (Greenberg 2000). This has been observed in other studies of acorn production (Downs and McQuilken 1944, Goodrum et al. 1971).

Differences among species in the performance of small-diameter individuals make it impossible to generalize. The fecundity of small dominant or codominant white oaks (10–25 cm dbh) and scarlet oaks (9–22 cm dbh) originating after a 1967 clearcut differed considerably. From 1993 through 1997 scarlet oak produced an average (±SE) of 4,077 ± 2,549 acorns/m^2 BA. Nearly half (45%) of the trees (*N* = 20) were good producers, and 45% were poor producers. However, white oaks (*N* = 18) produced an average of 1,535 ± 924 acorns/m^2 BA. Good producers composed only 11% of the trees, and 83% were poor producers (Greenberg 2000).

Do All Oaks Mast?

Acorn production patterns are often characterized as *masting*, a term that implies synchronous acorn production that results in boom or bust

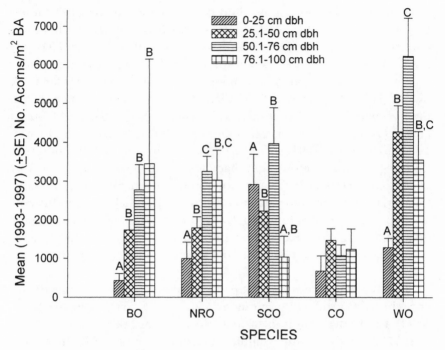

FIGURE 10.2. Mean (1993–1997) (±SE) number of acorns/m² BA produced by four diameter classes of black oak (BO) (F = 5.8; p = 0.0012), northern red oak (NRO) (F = 11.8, p = 0.0001), scarlet oak (SCO) (F = 3.85; p = 0.0013), chestnut oak (CO) (F = 1.1; p = 0.3509), and white oak (WO) (F = 6.71; p = 0.0003) in the southern Appalachians. Letters A–C denote significant differences in mean acorn production among diameter classes. Acorn data were natural log transformed for ANOVA but are presented as actual means.

crop years. Most studies of eastern oaks report large fluctuations in acorn crop size (e.g., Downs and McQuilken 1944, Burns et al. 1954, Beck 1977). However, despite the large year-to-year fluctuations exhibited by most oak species, moderate crop sizes also are common. The degree of synchronized fruiting within a population of conspecifics is not directly addressed in most studies. Koenig, Mumme, et al. (1994) found evidence of species-specific masting intervals at the individual level but not at the population level. Conversely, Sork et al. (1993) found that most northern red and white oak conspecifics (and to a lesser degree black oak) produced acorns in the same years.

Between 1993 and 1997 annual acorn production by individual trees within a population was not synchronous for any species in our study. Within a species the proportion of trees bearing acorns ranged from 70% (chestnut oak, 1996) to 90% (white oak, 1996) during the maximum crop year and from 3% (white oak) to 29% (black oak) during the poorest crop year (1996 for scarlet oak, 1997 for all other species). At least 29% of individuals of all species produced acorns every year except the poorest crop year for that species. At least one year of the five-year study was characterized by approximately one-third to two-thirds of sample trees producing acorns for each species (Greenberg and Parresol 2000). Koenig and Knops (Chapter 9) report some evidence for "normal" masting in California oaks and other temperate-zone oaks, tempered by environmental variables. Indeed, high year-to-year variability in the degree of fruiting synchrony and crop size within a population suggests that "normal" rather than "strict" masting (*sensu* Kelly 1994) better characterizes the fruiting patterns of southern Appalachian oaks.

CHARACTERISTICS OF FRUITING THAT CONTRIBUTE TO CROP SIZE

Although acorn production is not reliably synchronous among conspecifics, acorn crop size is strongly correlated with both the proportion of individuals in the population that produce acorns ($r \geq 0.87$; $p \leq 0.0123$ except chestnut oak [$p = 0.0578$]) and the number of acorns/ m^2 BA of fruiting trees ($r \geq 0.97$; $p \leq 0.0069$ using untransformed values). In addition, both the proportion of fruiting trees and the number of acorns on fruiting trees change significantly among years for all species (range of $r = 0.91 - 0.98$) and are significantly positively correlated in all species, except chestnut oak ($r = 0.68$, $p = 0.20$). In other words, good crop years are characterized both by more trees producing acorns and by more acorns per producing tree (Greenberg and Parresol 2000). Healy et al. (1999) also observed that more northern red oak trees bore acorns and there were more acorns per tree during good crop years.

A SIMPLE METHOD OF ESTIMATING ACORN CROP YIELD

Several visual survey methods are commonly used for estimating acorn yield (Koenig, Knops, et al. 1994, Whitehead 1969, 1980, Graves 1980,

Sharp 1958, Christisen and Kearby 1984). However, visual surveys are time consuming and provide only categorical estimates of acorn crop yield, which may be biased by differences among observers.

By itself, the proportion of trees bearing acorns was a significant and strong predictor of acorn crop size (mean number/m^2 BA) in any given year of this study (Table 10.5) (Greenberg and Parresol 2000). This provides an expedient tool for forest and wildlife managers or planners to quantify acorn crop sizes within years. Because the proportion of acorn-bearing trees and the number of acorns/m^2 BA of fruiting trees are correlated with one another it is inappropriate to include both in regression analysis. Because of the relative facility with which the proportion of fruiting trees can be ascertained, these equations are of greater use to forest managers than equations that use estimates of mean number of acorns/m^2 BA of fruiting trees.

Greenberg and Parresol (2000) detail a method for determining the required sample size to estimate the proportion of trees bearing acorns within a given year, and regression equations (using reduced major axis regression) for five species to estimate within-year crop size with confidence intervals. Methods are as follows:

Estimating the Proportion of Acorn-Bearing Trees to Predict Yield

The natural logarithm of acorn crop yield is estimated as

$$\hat{y} = b_0 + b_1 \hat{x} \tag{1}$$

where $\hat{y}$ is the predicted logarithm of acorn crop yield, the b's are equation coefficients (from Table 10.5), and $\hat{x}$ is an estimate of the percentage of acorn-bearing trees. To compute $\hat{x}$, it is necessary to draw a random sample of trees of size n, and count the number of successes, s, that is, of trees bearing acorns. The proportion, p, of acorn-bearing trees is unbiasedly estimated as $\hat{p} = s/n$, thus $\hat{x} = 100 \times \hat{p}$. Of course, it is desirable to estimate p within some margin of error, d, at the $(1 - \alpha)$ confidence limits. The sample size required to achieve the desired level of precision is (Zar 1984)

$$n = \frac{Z_{\alpha/2}^2 \tilde{p}\tilde{q}}{d^2} \tag{2}$$

where Z is a standard normal variate (Zar 1984 p. 483), $\tilde{p}$ is an initial guess of p (based on intuition or, preferably, a pilot survey), and $\tilde{q} = 1 -$

Table 10.5

Reduced major axis regression of the natural logarithm of acorn yield (acorns/m^2 BA) on the proportion of trees bearing acorns for five species of southern Appalachian oaks

Species	b_0	b_1	r	p-value	$\bar{x}$	S_{xx}	$\hat{\sigma}_\varepsilon^2$
Black oak	3.56472	0.055905	0.9942	0.0005	60.3	2,569.41	0.03113
Northern red oak	3.66069	0.060747	0.9861	0.0020	54.3	3,102.63	0.10617
Scarlet oak	3.51901	0.064498	0.9918	0.0009	51.8	3,952.09	0.09008
Chestnut oak	3.78155	0.064998	0.8658	0.0578	39.6	3,184.57	1.20371
White oak	2.58029	0.080842	0.9526	0.0123	47.1	5,095.65	1.05248

Note: See Greenberg and Parresol 2000.

$\hat{p}$. The use of this formula will be demonstrated in the examples following Equation 7 and in Table 10.7.

The antilogarithm of $\hat{y}$ yields the estimated crop size ($\widehat{cs}$) (number of acorns/m^2 BA) in arithmetic (untransformed) units, that is,

$$\widehat{cs} = \exp(\hat{y}) \qquad (3)$$

Confidence Intervals

Placing bounds on the predictions of acorn yield is useful, since point estimates from regression equations such as (1) are subject to error. In this case, the variance of y is a function of both residual error and the variance of X. It is given by (Madansky 1959, Kendall and Stuart 1979)

$$\hat{\sigma}_y^2 = b_1^2 \hat{\sigma}_x^2 + \hat{\sigma}_\varepsilon \qquad (4)$$

The variable X (i.e. $100\, s/n$) is based on a binomial random variable. Thus, its estimated variance is

$$100^2\, \hat{p}\, (1-\hat{p})\, /\, n \qquad (5)$$

The construction of confidence intervals on the predictions requires the standard errors of the predictions ($s[\hat{y}_i]$) and a t-value. The interval boundary points are obtained from

$$\hat{y} \pm t_{\alpha/2,\, n_r-2} \times s(\hat{y}_i) \qquad (6)$$

where n_r is the number of regression observations ($n_r = 5$). The standard errors are calculated as

$$s(\hat{y}_i) = \sqrt{\sigma_y^2 \left[\frac{1}{n_r} + \frac{(x_i - \bar{x})^2}{s_{xx}} \right]} \qquad (7)$$

Example

A manager wishes to estimate acorn production in a hypothetical forest stand composed of black oak ($0.8\ m^2$), northern red oak ($1.7\ m^2$), scarlet oak ($0.5\ m^2$), chestnut oak ($1.0\ m^2$), and white oak ($1.3\ m^2$). A preliminary walk-through indicates that about 50% of black oak, 90% of northern red oak and scarlet oak, 30% of chestnut oak, and 20% of white oak are producing acorns that year (see Table 10.6).

Beginning with black oak, equation 2 is used to determine how many trees must be surveyed to be within 3% ($d = 0.03$) of the true fruiting proportion (approximated at 50%) at the 80% confidence level ($\alpha = 0.2$, therefore $Z = 1.28$):

$$n = \frac{1.28^2 \times 0.50 \times 0.50}{0.03^2} = 455$$

The required sample size is reduced dramatically if a 5% margin of error is used, that is

$$n = \frac{1.28^2 \times 0.50 \times 0.50}{0.05^2} = 164$$

Slightly more trees should be sampled than predicted from equation 2. Since $\hat{p}$ often is not equal to the initial guess $\tilde{p}$, and if $\hat{p}$ is closer than $\tilde{p}$ to 0.5, n will be slightly larger (variance is maximized at $p = 0.5$). After surveying 174 trees for presence or absence of acorns, it is determined that $\hat{p} = 0.52$, using a 5% margin of error (for this p, the required n also is 164, so a survey of 174 is adequate). Using the coefficients from Table 10.5, for black oak the estimated yield of acorns per m^2 tree BA in logarithmic units is

$$y = 3.56472 + (0.055905 \times 52) = 6.4718$$

To place bounds on this prediction, $\hat{\sigma}_y^2$ must be calculated first, using equation 4. This yields the following result:

$$\hat{\sigma}_y^2 = 0.0559^2 \frac{100^2 \times 0.52 \times 0.48}{174} + 0.0311 = 0.0449$$

where b, and $\hat{\sigma}_\varepsilon^2$ come from Table 10.5. From equation 7 the standard error is computed as

$$s(\hat{y}_i) = \sqrt{0.0449 \left[\frac{1}{5} + \frac{(52 - 60.3)^2}{2,569.41} \right]} = 0.1009$$

where $\bar{x}$ and S_{xx} are from Table 10.5. From equation 6 the 90% confidence interval is

Table 10.6

Hypothetical forest stand of five southern Appalachian oak species illustrating use of equations 1–7 to predict within-year acorn crop size

Species	BA (m²/ stand)	Estimated % fruiting	$n = \dfrac{Z^2_{\alpha/2}\tilde{p}\tilde{q}}{d^2}$	Actual % fruiting	Acorn crop $\hat{y} = b_0 + b_1\hat{x}$	$\hat{\sigma}^2_y = b^2_1\hat{\sigma}^2_x + \hat{\sigma}_\varepsilon$
			Required sample (n)[a]			*Bounds*[b]
Black oak	0.8	50	164	52	6.4718	0.0449
Northern red oak	1.7	90	60	89	9.0672	0.0516
Scarlet oak	0.5	90	60	92	9.4528	0.0437
Chestnut oak	1.0	30	138	29	5.6665	0.0588
White oak	1.3	20	105	22	4.3588	0.0975
Total	5.3					

Note: First the required sample size for trees used to determine the proportion of fruiting trees must be estimated.

[a]Slightly more trees should be sampled than predicted from equation 2. Since $\hat{p}$ often is not equal to the initial guess $\tilde{p}$, and if $\hat{p}$ is closer than $\tilde{p}$ to 0.5, n will be slightly larger (variance is maximized at $p = 0.5$).

[b]The variance of variable x, $\hat{\sigma}^2_x$ (equation 5) and used in equation 4 to calculate bounds was calculated here using the required sample size $(n) + 10$.

$$6.4718 \pm 0.1009\,(2.353) = 6.2344 \leq \hat{y} \leq 6.7092$$

Values are converted from logarithmic to arithmetic units by applying equation 3. Black oak crop size (number of acorns/m² BA) from a 52% fruiting population is predicted to be

$$\hat{cs} = \exp(6.4718) = 646.65 \text{ acorns/m}^2 \text{ BA}$$

Applying equation 3 to the confidence limit values, the following interval is obtained:

$$510 \leq \hat{cs} \leq 819.9$$

Using the same set of equations, the minimum required number of sample trees is determined for each species, and (using the minimum required number + 10 for the n value in calculations) the crop size (number of acorns/m² BA) with confidence intervals is predicted. Crop size values, now in numbers per m² BA, must now be expanded to the

SE $s(\hat{y}_i) = \sqrt{\hat{\sigma}_y^2\left(\dfrac{1}{n} + \dfrac{(x_i - \bar{x})^2}{S_{xx}}\right)}$	90% confidence interval $\hat{y} \pm t_{\alpha/2,\, n_r - 2}$ $\times s(\hat{y}_i)$	Crop size (acorns/ m^2 BA) $\hat{c}s = exp(\hat{y})$	Acorns/stand $BA \times \hat{c}s$
0.1009	$\hat{y} \pm 0.2374$	$510 \le 647 \le 820$	$408 \le 518 \le 656$
0.1742	$\hat{y} \pm 0.4099$	$5{,}752 \le 8{,}666 \le 13{,}057$	$9{,}776 \le 14{,}732 \le 22{,}197$
0.1631	$\hat{y} \pm 0.3838$	$8{,}682 \le 12{,}744 \le 18{,}706$	$4{,}341 \le 6{,}372 \le 9{,}353$
0.1176	$\hat{y} \pm 0.2767$	$219 \le 289 \le 381$	$219 \le 289 \le 381$
0.1776	$\hat{y} \pm 0.4179$	$51 \le 78 \le 119$	$66 \le 101 \le 155$
			$22{,}012$

whole stand. The number of acorns produced in the stand is calculated by multiplying crop size ($\hat{c}s$) for each species by the BA of that species and summing (Table 10.6).

Using these equations, managers can estimate within-year acorn crop size, knowing for each species only the proportion of trees bearing acorns and the BA inventory within the survey area, by multiplying the antilog of $\hat{y}_{\text{species A}}$ by the $BA_{\text{species A}}$ (m²). To calculate total acorn production by the five oak species within an area the species values are summed. Crop yield estimates described in numbers of acorns can be converted to green weight or dry biomass (no hulls) using the conversion values presented in Table 10.1. Estimates can be applied to surveyed areas of any size within the southern Appalachian region.

In order to ensure a precise estimate of the proportion of acorn-bearing trees, it may be necessary to sample large numbers of trees per species. The sample size required depends on the proportion of fruiting trees, the margin of error, and the confidence level one is willing to accept. Moderate crop years require the highest sampling effort (164 trees with a 5% margin of error and 80% confidence level, if 50% are fruiting), while poor or good crop years require the least (as few as 59 trees if 10% or 90% of a given species are fruiting).

The relationship between acorn crop size and the proportion of acorn-bearing trees described in our study is based on data from acorn traps. It is probable that there will be some discrepancy in acorn detection (presence/absence) between visual surveys and trap data (Gysel 1956), especially in years of poor crop yield. Visual surveys, in which tree canopies are closely scrutinized, may detect the presence of very small numbers of acorns that could be missed by acorn traps. Relative to the trap data that our equations are based upon, visual surveys would probably provide an higher (although more accurate) estimate of the proportion of fruiting trees and, therefore, would inflate crop-yield estimates. Until that issue is better addressed, visual surveys of trees having very few acorns should be considered as without acorns for these calculations.

ACORN YIELD TABLES

Table 10.1 is a tool that managers in the southern Appalachians can use to determine or maintain a specified acorn production capability on an area basis. It is possible to test a variety of BA apportionment scenarios among oak species. Average acorn yield for the five study species is presented in Table 10.1. This information cannot be used to predict or even estimate within-year acorn crops, because actual production varies considerably from year to year (Figure 10.1). In addition, mean values are likely to change, perhaps even substantially, with additional years of data (see above discussion under "Are Some Species Better Producers than Others?"). However, given these words of caution, Table 10.1 can be used to estimate long-term acorn production capability. Mean acorn production by species can be calculated on an area basis by multiplying the total BA of a species within an area by the mean acorn production/m^2 BA for that species. By summing these values, we can calculate total average acorn production by the five species.

CONCLUSIONS

The ecological and land management implications of acorn crop size underscore the need to better understand acorn production patterns within and among oak species and production characteristics among individuals. For this study, the data indicated that, on average, white oak produced more acorns and biomass (along with northern red oak) than black oak, scarlet oak, or chestnut oak. However, other studies rank species differently in their production capacity. This almost certainly is in part due to differences among studies in the number of years sampled, which years were sampled, how the data are reported (per tree, per m^2 crown area, per m^2 BA by plot, etc.), and the number of trees sampled. These difficulties in comparing studies also highlight the need for long-term studies that use standardized measurement and reporting methodologies. Poor acorn production by some species in this study was offset by good or moderate production by others during most years. Hence, even if some species produce more acorns per m^2 BA, it is important for managers to retain multiple oak species within both the red oak and white oak subgenera, to enhance the likelihood of a constant acorn supply.

Individuals varied in frequency of production and quantity of acorns produced per m^2 BA. Good producers tended to produce more acorns more frequently than poor producers. Although good producers composed 20% (chestnut oak) to 46% (northern red oak) of sample populations, they contributed disproportionately to the acorn crop. However, good producers could not be identified by the presence of acorns during poor crop year nor could poor producers be identified by an absence of acorns during good crop years.

Acorn production increased with tree size, at least in part simply because larger trees have bigger crowns. If that were the primary influence of tree size on acorn production, then the same number of acorns could be produced by a few large trees or by the same area of crown distributed among several smaller-diameter trees. A more pertinent question is whether larger-diameter trees produce more acorns per unit BA than smaller-diameter trees. For the five species studied, the correlation between BA and acorn production/m^2 BA is very weak and/or nonsignificant. It becomes clear that, when grouped into diameter classes, for most species, trees ≤ 25 cm produce fewer acorns/m^2 BA than trees > 25 cm. However, the performance of small-diameter trees differs among species.

The term *masting* may not appropriately characterize the fruiting patterns of southern Appalachian oaks. Although most individuals within species in the study produced acorns in some years, or did not produce in others, one-third to two-thirds of individuals produced acorns in other years. This suggests that acorn production is not synchronous among individuals within a population.

Acorn crop size is strongly correlated with both the proportion of individuals in the population that produce acorns and the number of acorns/m^2 BA of acorn-bearing trees. Good crop years are characterized by both more trees producing acorns and more acorns/m² BA on fruiting trees. The relationship between acorn crop size and the proportion of individuals in the population that produce acorns provides an expedient tool for land managers to quantify crop size within years. Using the equations presented, land managers can predict acorn crop size for the five study species if they have an inventory of oak BA by species (within any size of area) and if they know the proportion of trees bearing acorns as estimated by simple visual surveys (presence/absence of acorns).

ACKNOWLEDGMENTS

The Southern Research Station, U.S. Forest Service, funds this study. We extend special thanks to D. E. Beck for conceiving and initiating this project and to D. L. Loftis for his continued support. Thanks are also due J. Murphy, V. Gibbs, and S. Dowsett for their invaluable efforts in launching and sustaining this study. Special thanks are also extended to U.S. Forest Service employees A. Frisbee, M. Stables, J. McGuiness, G. Miller, K. Proffitt, C. R. Lintz, W. Dalton, R. Lewis, J. Wentworth, P. Hopton, and K. Wooster for their dedication and invaluable contributions to this study by collecting acorns. T. Roof, R. Hooper, R. Brock, J. Metcalf, and J. Allen also have participated in field collections.

Chapter 11

Nutritional Value of Acorns for Wildlife

ROY L. KIRKPATRICK AND PETER J. PEKINS

Acorns are a highly digestible, high-energy, low-protein food for many species of wildlife. Nutritionally, acorns can be viewed as having three different structural parts: the meat, the shell, and the cup. Some species of wildlife, especially larger animals such as bears, eat all three parts. Deer usually eat primarily the shell and meat, and smaller animals, such as grouse, quail, squirrels, and songbirds, often eat only the meat portion. The meat is the highly digestible portion of the acorn, and the shell and cap are much higher in fiber and are less digestible. Although acorns are considered a staple food for many wildlife species, they are not nearly as reliable as a food crop for wildlife as were American chestnuts (*Castanea dentata*), because acorn crops often fail due to frost, whereas chestnut crops seldom failed, because they flowered in June (Leopold et al. 1989), well after the latest frost dates.

Martin et al. (1961) state that acorns rate at or near the top of the wildlife food list and that they are a good wildlife food staple—the "staff of life" for many wildlife species. When acorn crops fail, many wildlife species may be hard pressed for sustenance. Martin et al. (1961) listed more than 90 species of wildlife that use acorns (ranging from bears to songbirds and small mammals).

CHEMICAL COMPOSITION

The species of acorns vary considerably in fat and tannin content, therefore acorns are not a homogeneous entity nutritionally (see Tables 11.1 and 11.2). In general, acorns of the red oak group (*Erythrobalanus* sub-

Table 11.1
Total phenols, crude protein, crude fat, and metabolizable or digestible energy from acorns of three oak species

Reference	Total phenols[a]			Crude protein (%)			Crude fat (%)			Metabolizability or digestibility (%)		
	White	Chestnut	Red	White	Chestnut	Red	White	Chestnut	Red	White	Chestnut	Red
Short and Epps (1976) (ruminant DMD)	—	—	—	5.9	—	5.9	4.3	—	17.9	68.3[b] 59.8[c]	— —	66.9[b] 68.5[c]
Servello and Kirkpatrick (1988) (ruffed grouse ME)[b]	3.3	8.1	13.0	7.3	5.9	6.3	8.8	8.8	23.0	86.1	75.5	78.8
Chung-MacCoubrey et al. (1997)[b] (gray squirrel DE)	1.37	—	5.2	7.2	—	6.9	8.8	—	20.2	87.7	—	85.4

[a]Mg of gallic acid equivalents.
[b]Nylon-bag dry-matter digestibility.
[c]Estimated true dry-matter digestibility.
[d]Data are for acorn meat without shell.

Table 11.2
Nutrient composition of acorns of 11 oak species

Component	White oak group					Red oak group					
	White	Swamp chestnut	Live	Post	Sawtooth	Water	Willow	Black	Bluejack	Northern red	Southern red
Oven-dry matter	60.5	54.5	71.3	67.3	66.9	81.9	79.5	87.2	82.6	79.4	80.2
Crude protein	4.6	4.4	5.8	6.8	6.6	4.9	5.5	5.7	6.9	4.9	7.0
Crude fat	5.8	2.9	6.1	6.7	4.7	21.1	20.0	17.5	26.2	14.0	22.7
Crude fiber	18.6	22.3	14.6	18.1	12.4	17.6	18.9	16.6	15.8	26.4	19.9
Nitrogen-free exract	68.3	67.6	71.7	65.3	73.2	54.0	53.5	57.9	49.0	52.3	48.6
Ash	2.7	2.8	1.8	3.1	3.1	2.4	2.1	2.3	2.1	2.4	1.8
Calcium	0.18	0.14	0.13	0.22	0.20	0.18	0.23	0.19	0.16	0.15	0.23
Phosphorus	0.09	0.09	0.09	0.09	0.14	0.08	0.07	0.10	0.12	0.08	0.11
Silica	0.06	0.04	0.05	0.14	0.06	0.08	0.02	0.03	0.02	0.10	0.06
Neutral-detergent fiber	44.9	51.1	38.0	47.7	44.4	34.1	34.3	30.9	30.7	44.5	35.4
Acid-detergent fiber	21.8	27.8	22.2	24.5	20.0	21.3	22.7	23.2	19.6	29.3	22.8
Acid-detergent lignin	7.8	11.2	10.8	8.9	9.0	11.2	10.4	9.6	10.8	10.0	10.2
Hemicellulose	23.1	23.3	15.8	23.2	24.4	12.8	11.6	7.7	11.1	15.2	12.6
Cellulose	14.0	16.6	11.4	15.6	11.0	10.1	12.3	13.6	8.8	19.3	12.6

Source: Data from Short 1976.

Note: Oven-dry matter is expressed as a percentage of the weight of fresh acorns; the other components are given as percentages of oven-dry weight.

Table 11.3
Neutral-detergent solubles (NDS), total phenols (TP), crude protein (CP), and metabolizable energy (ME) of different types of forage obtained from crop contents of ruffed grouse (*Bonasa umbellus*)

Forage type	NDS (%)	TP (%)	CP (%)	ME (%)	ME kcals/g
Acorns without shells	89.4	11.1	8.0	78	3.55
Acorns with shells	66.6	7.9	7.0	57	2.80
Soft fruits[a]	74.9	2.3	7.2	57	2.56
Leaves of herbaceous plants	77.9	3.5	25.2	59	2.65
Leaves of woody plants	72.8	8.3	12.2	50	2.47
Buds and twigs	47.3	5.9	13.2	30	1.48

Source: Adapted from Servello and Kirkpatrick 1987.
[a]Flesh only, seeds excluded.

genus) contain higher levels of tannins and higher levels of fat than those of the white oak group (*Leucobalanus* subgenus) (Ofarcik and Burns 1971, Chung-MacCoubrey et al. 1997). The crude protein content of acorns is relatively low in comparison to other foods of wildlife (see Table 11.3) and is generally in the range of 5–8%. Crude fat (a potent source of energy) ranges from 3–9% for the white oak group and 14–23% for the red oak group. Cell solubles (also called neutral detergent solubles and calculated as 100 − neutral detergent fiber), the highly digestible portion of plant cells, compose a relatively high proportion of an acorn (50–70%). Conversely, acid detergent fiber is relatively low (20–30%).

Tannins are phenolic compounds that precipitate proteins. In vivo, they can react with the mucous lining of the gastrointestinal tract, with digestive enzymes (which are proteins), and with the proteins in foods to render them indigestible. Tannins themselves are difficult to quantify, and thus total phenols (of which tannins are a component) are often measured instead. Total phenols are much higher (300–400% higher) in the red oak subgroup than in white oaks (Table 11.1). Acorns really contain two types of tannins—condensed and hydrolyzable. Condensed tannins usually are not broken down in the intestines nor are they absorbed into the bloodstream. They form complexes with digestive enzymes, proteins of food, and proteins of the intestinal mucosa. Thus, their primary effect is to reduce protein digestibility and cause animals to be in negative nitrogen balance, that is, to lose more nitrogen than they ingest. Hydrolyzable tannins, on the other hand, are broken down (hydrolyzed) in the gastrointestinal tract into smaller phenols, many of which are absorbed into the blood and then must be detoxified and excreted, since they serve no useful purpose in the body. Hydrolyzable tannins can have detrimental effects either as a digestive inhibitor (similar to condensed tannins) or as a foreign substance that must be detoxified by the liver and excreted in the urine (Chung-MacCoubrey et al. 1997). Some wild herbivores have proline-rich proteins in their saliva that bind tannins and reduce their inhibitory effects on protein digestion and also their absorption (Robbins et al. 1991).

The higher energy value of red oak acorns (which is due to their higher fat content) is more than offset in most species of wildlife by the negative effect of high tannin levels on palatability. It was once thought that tannin levels in acorns might decline over winter due to leaching of the tannins. However, two recent studies indicate little or no decline, especially in red oak acorns (Chung-MacCoubrey 1993, Dixon et al. 1997a). Likewise, there is apparently little change in nutrient values over time. Three species of acorns monitored during six to eight months of storage in a freezer decreased slightly in crude fat and increased slightly in soluble carbohydrates. White oak acorns did not decrease as much in crude fat but did have a twofold increase in soluble carbohydrates (Clatterbuck and Bonner 1985).

PALATABILITY

Although there is some disagreement, acorns of the white oak group generally appear to be more palatable than those of the red oak group (Short 1976, Servello and Kirkpatrick 1987, Chung-MacCoubrey et al. 1997). A preference for white oak acorns over red oak acorns was measured in deer (Duvendeck 1962, Goodrum et al. 1971, Pekins and Mautz 1987). However, red oak acorns may represent > 50% of the autumn, winter, and spring diets of deer (Harlow et al. 1975, Heim 1987, Pekins and Mautz 1987) and of wild turkeys (Korschgen 1967). Animals usually eat fewer red oak acorns and thus may have lower energy intakes. This lower level of consumption is believed to be due to the higher tannin levels in red oak acorns, which gives them a bitter taste and often causes adverse physiological responses in the animal. Early American Indians and settlers apparently referred to the white oak acorns as "sweet acorns" and the red oak acorns as "bitter acorns" and often soaked the red oak acorns in water for some time in an effort to leach out the tannins.

DIGESTIBILITY

Digestibility and metabolizability of all acorn species tested are fairly high in relation to other foods, as would be expected from their high proportion of cell-soluble contents and relatively low proportion of acid detergent fiber (see Tables 11.1 and 11.3). Digestibility and metabolizability values for several species ranged from 57% to 89% (the higher values were from acorns with the shells removed). Acorns can be viewed as a highly concentrated form of food energy. Managers of domestic animals often refer to foodstuffs as being either "concentrates" (grains) or "roughages" (hay, straw, cornstalks). By such a categorization, acorns would be concentrates, whereas other, more fibrous, wildlife foods, such as grasses, stems, leaves, and woody twigs and buds, are analogous to roughages.

Nitrogen Balance

Likely because of the low protein content and high tannin content of acorns, animals who consume acorns only often exhibit negative nitrogen balances (Short 1976, Servello and Kirkpatrick 1987, Chung-Mac-Coubrey et al. 1997). Worm-infested acorns would provide consumers with a higher level of protein and it would be animal protein, but some wild animals may discriminate against acorns with worms (Dixon et al. 1997b). However, because most wild herbivores consume mixed diets, and many have physiological adaptations to recycle nitrogen when it is limited, negative nitrogen balances probably are rare during autumn, when acorns are consumed. A positive nitrogen balance was realized by wild turkeys consuming a mixed diet containing 55% red oak acorns (Decker et al. 1991).

Calcium and Phosphorus Content

Short and Epps (1976) compared calcium and phosphorus levels of several seeds and fruits of southern forests and found little difference in calcium levels among the species. However, acorns were especially low in phosphorus ranging from 0.09% to 0.14% (see Table 11.2).

ENERGY REQUIREMENTS

Animals eat to satisfy energy requirements and consume protein and minerals in relation to the ratio of these nutrients to energy. When acorns are plentiful, wildlife can consume huge quantities in a short period of time, meet their energy requirements quickly, and enter a productive state (e.g., body growth and fat deposition). Reduction of foraging time not only reduces energy expenditure but also reduces vulnerability to predation.

A comparison of the nutritional value of various forage classes consumed by ruffed grouse shows that acorns are one of the most energy rich foods consumed by this species (Table 11.3). It would take only about 12 acorns to satisfy the daily energy requirements of the ruffed grouse. Servello and Kirkpatrick (1988) found that dietary metabolizable energy (ME) levels were 1.2 times higher in ruffed grouse crops containing 63% acorns (64% ME) than in that containing only 3–4% acorns (52% ME). Although female grouse collected in a year of high acorn intake did not have significantly higher mean carcass fat levels, all 17 tested had carcass fat levels > 13.8% (15 had levels ≥ 17%), whereas 8 of 22 collected in a year with almost no acorn intake had fat levels of 3–8%.

Wild turkeys fed mixed winter diets realized higher ME from diets with red oak acorns (15–55%) than diets without acorns (Decker et al. 1991). A diet of 55% acorns had 80% ME and satisfied the predicted daily energy requirements of wild turkeys; a diet of approximately 72 acorns alone would meet the requirements. The predicted ME intake of wild turkeys during winter was substantially reduced if acorns were unavailable, particularly with prolonged snow cover and limited availability of persistent fruits.

White-tailed deer fed mixed autumn diets with variable proportions of red oak acorns (5–55%) realized highest ME (43%) and ME content/ g of intake (2.15 kcal/g) on a diet with 55% acorns (Pekins and Mautz 1988). A 40% reduction in ME/g intake occurred if acorns were eliminated from the diet. To meet their growth requirements, fawns needed 28% and 58% greater consumption of diets with medium and low acorn

Table 11.4

Proximate analysis of chestnuts and acorns

	Crude protein (%)	Crude fiber (%)	Crude fat (%)	Nitrogen-free extract (%)	Ash (%)
Chestnut (*Castanea* spp.)	6.1	2.3	3.2	86.3	2.1
Acorns (*Quercus* spp.)	4.4	18.6	4.7	70	2.5

Source: Ensminger et al. 1990.

Note: Details are not given as to species or whether shells and or cups were included.

composition, respectively. Because fat deposition is most efficient on diets with high ME (Holter and Hayes 1977), acorn-dominated diets are optimal for fat deposition that is directly related to winter survival of deer (Pekins and Mautz 1988).

Surprisingly, acorns also can play an important role in the spring energy balance of deer. Acorns represented > 50% of the spring diet in a study of deer in New Hampshire (Heim 1987), and consumption of acorns by deer increased dietary energy substantially prior to spring green-up (Kilpatrick et al. 1991). A simulated forest diet with 50% red oak acorns was similar in digestible matter and metabolizable energy intake to an agricultural diet with 40% corn. The high ME provided by acorns during spring is advantageous to the physical recovery of deer after winter and helps compensate for the exponential energy costs incurred by pregnant deer during their last trimester (Pekins et al. 1998).

A rough comparison of the proximate analysis of acorns and American chestnuts indicated that acorns are higher in fiber (poorly digestible) and lower in nitrogen-free extract (highly digestible) than chestnuts but are comparable in protein, fat, and ash levels (Table 11.4). The acorn data likely are for both meat and shell combined and indicate that chestnuts were at least somewhat more digestible than acorns.

SUMMARY

Acorns are generally an excellent wildlife energy food, ranking very high in energy content and digestibility. Presumably, acorn consumption and the relative importance of acorns to wildlife increased after the loss of the American chestnut. Acorns are low in protein, and those of the red oak group are high in tannins, which precipitate proteins and make them unavailable for digestion.

Palatability of white oak acorns is highest, but both red and white oak acorns are consumed at high levels by wildlife. Although consumption of acorns alone may cause a negative nitrogen balance, most herbivores consume mixed diets with other forages, satisfying their nitrogen and protein requirements. The highly digestible and concentrated energy value of acorns is advantageous for many wildlife species that require seasonally productive states to prepare efficiently for reproduction and face seasonal constraints of energy availability.

Chapter 12

Acorn Dispersal by Birds and Mammals

MICHAEL A. STEELE AND PETER D. SMALLWOOD

Dispersal, or movement of offspring away from the place of birth, is a critical phase in the life history of most species (van der Pijl 1972, Gotelli 1995). Among plants, this process is often accomplished by the movement of a specific life stage (e.g., seeds or fruits) by one or more agents of dispersal (van der Pijl 1972). Seed dispersal by seed-eating animals is especially important for the majority of nut-bearing trees, including the more than 500 species of oak found worldwide. Successful dispersal of the fruits of oak (acorns) by these seed consumers seems to be most closely linked to the deliberate hoarding of acorns in the ground, followed by failure of the animals to recover some of these stored reserves. Both mammalian and avian agents of acorn dispersal are known to move large numbers of the propagules in a relatively short period of time, sometimes over considerable distances, and often to sites suitable—or even optimal—for germination (Smith and Reichman 1984, Price and Jenkins 1986, Vander Wall 1990). As Henry David Thoreau observed, "so far as our noblest hardwood forests are concerned, the animals, especially squirrels and jays, are our greatest benefactors."

Here we review the importance of birds and mammals in the dispersal of oaks by discussing the empirical evidence for their impact and by reviewing a series of experimental studies by the authors on the complex relationship between oaks and their dispersal agents. We frame our discussion in the context of current literature by focusing on four questions: (1) What are the advantages of dispersal? (2) What are the primary dispersal agents of acorns? (3) How do the behavior and ecology of oak dispersal agents help to clarify the complex dispersal syndromes of oak?

(4) What are the large-scale consequences and implications of the behavioral responses of these dispersal agents?

ADVANTAGES OF ACORN DISPERSAL

For oaks, as for many angiosperms, the advantages of movement away
from parental seed sources likely include: a reduction in local competition between siblings and parent trees, a reduction in density-dependent
predation close to the parent tree, establishment at sites most suitable
for germination and survival, and an increased probability of outcross-
ing (van der Pijl 1972, Price and Jenkins 1986, Vander Wall 1990). Most
of these advantages are embodied in three competing (but not mutually
exclusive) hypotheses: the escape hypothesis (Janzen 1970, Connell
1971), the colonization hypothesis (Howe and Smallwood 1982), and the
directed dispersal hypothesis (Howe and Estabrook 1977, Howe and
Smallwood 1982). Respectively, the three models suggest that the primary advantages of dispersal are reduced mortality near parent trees,
chance colonization of some sites suitable for establishment, or dispersal directly to specific sites with precise requirements necessary for establishment (Howe and Smallwood 1982).

While assumed to be obvious, many of these benefits lack strong empirical support (Vander Wall 1990). Nevertheless a few generalizations
can be made. First, it is clear that for most tree species examined, mortality is highest near the parent tree (see reviews by Howe and Smallwood
1982, Vander Wall 1990). If the same is true for oaks, dispersal away from
the parent tree may reduce the probability of this predispersal predation. One caveat to be noted is that dispersal agents are also seed predators. Some white oak species experience greater predation than dispersal by these animal agents (see below). This may predispose these species
to establishment under parent trees (Barnett 1977, Fox 1982, Steele and
Smallwood 1994, Smallwood et al. 1998). Nevertheless, it appears that
for most species of oaks, establishment is far more likely in open sites
away from parent trees (Harrison and Werner 1984, Tripathi and Khan
1990, Crow 1992). Both germination and seedling growth rates are
closely related to specific edaphic factors, especially the nutrient content
of the soil (Sonesson 1994), and seedling survival and growth rates are
negatively correlated with overstory density (Crow 1992). Finally, it is
clear that establishment, survival, and growth of oak seedlings is often
significantly higher in the cache sites chosen by many acorn-hoarding

animals than in more open or exposed sites (Griffin 1971, Barnett 1977, Vander Wall 1990).

DISPERSAL AGENTS OF OAK

Although the observations above do not permit discrimination of the three dispersal hypotheses, they do strongly suggest that dispersal away from the parent tree increases the probability of establishment, growth, and regeneration of many oaks. If this is true for most species, then the question to follow is, How is such dispersal achieved? The contribution of birds and mammals—and especially tree squirrels and jays—to this process was first recognized and articulated in detail by Thoreau ([1860] 1993). Despite Thoreau's extensive observations and rather compelling accounts, it has taken science over a century to arrive at virtually the same conclusions.

Among the 150 or more species of birds and mammals in the United States that rely on acorn mast for food (Van Dersal 1940), only a few actually handle acorns in a way that is likely to contribute significantly to dispersal of the seeds (Table 12.1., Vander Wall 1990). Many species (e.g., deer, bear, turkey, raccoons, quail, grackles, pigeons, and crows) act solely as acorn predators, masticating and killing acorns immediately upon consumption (Goodrum et al. 1971, Barnett 1977, Johnson and Webb 1989, McShea and Schwede 1993, but see Steele et al. 1993), while many others (woodpeckers, flying squirrels, and ground-dwelling sciurids) tend to store seeds in larderhoards or in cache sites (e.g., nest cavities) that are unsuitable for germination and recruitment (Koenig and Mumme 1987, Johnson and Webb 1989). Still others disperse and accidentally drop large numbers of acorns, but due to the high probability of desiccation it is unlikely that many of these seeds survive (Barnett 1977, Johnson and Webb 1989, Steele et al. 1993). However, the accidental dropping of acorns should not be discounted entirely as a means of oak dispersal (Webb 1986, Steele et al. 1998).

It is the few species of birds and mammals that scatter-hoard acorns just below the ground surface that are most likely to contribute to the dispersal, establishment, and regeneration of the oaks (Price and Jenkins 1986, Vander Wall 1990). Scatter-hoarding is a food-hoarding strategy employed by many seed predators; it involves the rapid storage of seeds in a spatial pattern that reduces the chances of pilfering by competitors (Jenkins and Peters 1992, Jenkins et al. 1996, Wauters and

Table 12.1
Dispersal agents and type of dispersal of *Quercus*

Disperser	Scatter-hoarding	Larder-hoarding	Long-distance	Short-distance	Primary citations
Birds					
Cyanocitta cristata	X		X		Darley-Hill and Johnson 1981, Harrison and Werner 1984, Johnson and Adkisson 1986, Stapanian and Smith 1986, Johnson and Webb 1989, Scarlett and Smith 1991, Steele et al. 1993, Thoreau [1860] 1993
Garrulus glandarius	X		X		Bossema 1979, Sonesson 1994, Kollmann and Schill 1996, Monsandl and Kleinert 1998
Aphelocoma coerulescens[a]	X			X	DeGange et al. 1989, Fleck 1994
A. ultramarina	X		X		Hubbard and McPherson 1997
Corvus corone[a]	X	X	X		Waite 1985
C. frugilegus[a]	X	X	X		Waite 1985
Pica pica[a]	X	X	X		Waite 1985
Mammals					
Sciurus carolinensis	X			X	Barnett 1977, Steele and Smallwood 1994, Smallwood 1992, Steele et al. 1993, Smallwood et al. 1998, Steele et al. 1998
S. niger	X			X	Stapanian and Smith 1984
S. aureogaster[a]	X			X	Steele et al. 2001
Sciurus vulgaris[a]	X			X	Wauters and Casale 1996
Tamiasciurus hudsonicus[a]	X	X		X	Hurly and Lourie 1997
Dasyprocta punctata	X			X	Hallwachs 1994
Apodemus spp.	X				Jensen and Nielsen 1986, Miyaki and Kikuzawa 1988
Clethrionomys spp.	X				Jensen and Nielsen 1986, Miyaki and Kikuzawa 1988
Peromyscus spp.	X	X			P. D. Smallwood and M. A. Steele, unpublished data
Tamias spp.	X	X			Elliot 1974, Kawamichi 1980

[a]Effect on oak regeneration not known, or thought to be insignificant.

Casale 1996). The spacing of caches by scatter-hoarders appears to represent a compromise between increasing costs of storage at low scatter-hoarding densities and the increasing probability of cache robbery at higher densities (Stapanian and Smith 1978, 1984, Clarke and Kramer 1994, Hurly and Lourie 1997).

Regardless of causes of the behavior, scatter-hoarding results in the movement of seeds to locations that are often suitable for establishment, provided that some of these seeds escape recovery by the predator (Price and Jenkins 1986, Stapanian and Smith 1986, Howe 1989, Vander Wall 1990). Thus, it follows that successful establishment of acorns from animal caches is closely dependent on the probability of cache recovery. It is widely appreciated that both mammals (McQuade et al. 1986, Jacobs and Liman 1991, Lavenex et al. 1998) and birds (Balda and Kamil 1989, Bennett 1993, Brodbeck 1994, Clayton and Krebs 1994) that scatter-hoard are well adapted for the efficient retrieval of their stored food, capable even of remembering individual cache sites based on spatial information (Jacobs and Liman 1991, Lavenex et al. 1998), and often responsible for recovery of more than 95% of their cached acorns under some conditions (Sork 1984, Miyaki and Kikuzawa 1988, Iida 1996, Steele et al. 2001). However, it is also true that many acorns can establish in these caches, especially in years of high acorn abundance, when the premium on these stored reserves is relaxed (Silvertown 1980, Jensen and Nielsen 1986, Crawley and Long 1995, Monsandl and Kleinert 1998) or when individual scatter-hoarders die or disperse before acorns can be recovered (Smith and Reichman 1984, Vander Wall, 1990).

By dispersing and storing acorns in individual sites just below the leaf litter, many of these scatter-hoarders decrease the chances of subsequent predation, desiccation, and seedling competition, and increase the probability of germination, root establishment, and winter survival (Barret 1931, Griffin 1971, Barnett 1977, Van der Wall 1990, but see Shaw 1968b). Some acorn-hoarding species, such as eastern gray squirrels (*S. carolinensis*) or blue jays (*Cyanocitta cristata*), may even select cache sites that are coincidentally optimal for germination (Barnett 1977, Johnson and Webb 1989).

The impact of these scatter-hoarders can be further categorized by two broader ways in which they appear to influence oak establishment: long distance dispersal (often > 1 km) attributed primarily to jays, especially the blue jay, and short-distance dispersal (< 150 m), usually within forest patches by small mammals (Johnson and Webb 1989). The two processes likely exert markedly different but equally significant consequences (Johnson and Adkisson 1986, Steele and Smallwood 1994, Smallwood et al. 1998).

Birds

Jays are the primary avian agents of acorn dispersal (e.g., *Cyanocitta cristata*—Darley-Hill and Johnson 1981, Grinnell 1936, Johnson and Adkisson 1986, Johnson and Webb 1989, Johnson et al. 1997; *Aphelocoma coerulesces*—DeGange et al. 1989, Fleck 1994; *A. ultramarina*—Hubbard and McPherson 1997; *Garrulus glandarius*—Bossema 1979, Kollmann and Schill 1996, Monsandl and Kleinert 1998). However, other corvid species, including carrion crows (*Corvus corone*), rooks (*C. frugilegus*), and magpies (*Pica pica*), also may contribute to the process (Waite 1985).

Individual jays can harvest, disperse, and cache thousands of acorns several kilometers from their source (Bossema 1979, Darley-Hill and Johnson 1981, Johnson and Adkisson 1986, DeGange et al. 1989). In a single season, blue jays have been reported to cache 54% of the crop (130,000 acorns) of a single stand of willow oak acorns (*Q. phellos*, Darley-Hill and Johnson 1981), and European jays (Bossema 1979) and Florida scrub jays (DeGange et al. 1989) have been observed to cache as many as 4,600 and 6,500–8,000 acorns per individual, respectively.

In addition, there appears sufficient evidence to argue that specific caching decisions of jays enhance oak establishment. Both blue jays and European jays regularly cache acorns singly in a few centimeters of soil, almost always covering the cache (Darley-Hill and Johnson 1981, Kollmann and Schill 1996). Moreover, cache site selection is not random but concentrated in disturbed or open successional habitats such as sparsely wooded prairies, forested edges, hedgerows (Johnson and Webb 1989), pure, conifer stands (Monsandl and Kleinert 1998), mowed lawns, open grasslands (Darley-Hill and Johnson 1981, Johnson and Webb 1989, Kollmann and Schill 1996), and other habitat patches where the probability of establishment is high and the light requirements are similar to those of some oak species (Johnson and Webb 1989). Estimates of establishment from jay caches at such sites vary from 150 seedlings per hectare in forested edge habitats in Iowa (Johnson and Webb 1989) to 2000 per hectare in pure pine stands in Europe (Monsandl and Kleinert 1998).

The evidence that jays often disperse acorns to sites that favor germination, establishment, and survival suggests that the caching activity of jays may result in directed dispersal of the oaks (Johnson and Webb 1989), and that the interactions between jays and oaks may be best interpreted as a form of mutualism (Bossema 1979). However, the extent of such a relationship and its influence on oak dispersal is complicated further by the activity of acorn-feeding insects, especially weevils (*Curculio* spp.). Dixon et al. (1997a), for example, report that, like other species, blue jays consistently lose weight when maintained on a diet of only acorns. However, this weight loss is easily overcome when birds are provided with dietary supplements of acorn weevils, which may provide a critical source of protein not available in acorns (Johnson et al. 1993). Johnson et al. (1993) suggest that the presence of weevils, and possibly other insects, in the acorns may allow jays to survive on acorns during the autumn caching season. Nonetheless, jays prefer uninfested acorns (Dixon et al., 1997b).

Fleck (1994) suggests that, in the western United States, acorn weevils may in part mediate the dispersal of acorns of Gambel oak (*Q. gambelii*) by scrub jays. He observed that individual trees producing acorns with higher weevil infestations were less likely to be dispersed. He also reported that maximum dispersal occurred for those trees producing acorns with intermediate tannin levels. Those with higher tannin levels showed lower weevil infestations but were unattractive to birds, whereas those with low tannin levels exhibited much higher levels of infestation, which significantly reduced their probability of dispersal. However, experiments with Mexican jays have failed to produce any support for this tritrophic symbiosis (Hubbard and McPherson 1997),

The rather considerable distances to which jays disperse acorns (100 m to several km; Darley-Hill and Johnson 1981, Kollmann and Schill 1996) led Johnson and Webb (1989) to speculate that blue jays in North America were largely responsible for the consistent and rapid postglacial migration of *Quercus* and other fagaceous tree species at the end of the late Quartenary (Davis 1976, Delcourt and Delcourt 1987). Johnson and Webb (1989) summarize strong evidence implicating the blue jay as the only species in eastern deciduous forests of the United States that routinely disperses acorns more than a few hundred meters to cache sites suitable for germination and establishment. Johnson et al. (1997) suggest that the jay's habits of caching at the regenerating edges of forests and in recently burned areas makes it a keystone species for oak population dynamics in times of climate change and in our current, fragmented landscape. Similar effects of long-distance dispersal of red oaks (*Q. robur*) and the postglacial migration of oaks in Europe are attributed to the caching activities of the European jay (*Garrulus glandarius*—Bossema 1979, Le Corre et al. 1997, Monsandl and Kleinert 1998). Le

Corre et al. (1997) concluded that dispersal by European jays may account for oak migratory rates up to 500 meters per year.

Mammals

Mammalian agents of dispersal include tree squirrels (*Sciurus* spp.—Barnett 1977, Stapanian and Smith 1984, Steele and Smallwood 1994, Steele et al. 1996) and members of a few other genera (e.g., *Apodemus*—Jensen and Nielsen 1986, Miyaki and Kikuzawa 1988, Iida 1996, *Clethrionomys*—Jensen and Nielsen 1986, Miyaki and Kikuzawa 1988, *Peromyscus*—Vander Wall 1990, *Tamias*—Elliott 1974, Kawamichi 1980) (Table 12.1). The pine squirrels (*Tamiasciurus*) and the flying squirrels (*Glaucomys*) may also scatter-hoard and disperse acorns, but the impact of most mammals (except the tree squirrels) may be limited by their tendency to also larderhoard in tree cavities or underground dens (Vander Wall 1990, see Chapter 17 for a discussion on the evolved relationship between squirrels and oaks and pines).

In marked contrast to most jays, many rodents disperse acorns over relatively short distances, often less than 50 meters from their sources (e.g., Sork 1984, Jensen and Nielsen 1986, Steele and Smallwood 1994). Several studies, relying on a number of techniques (e.g., radioactive labels, metal tags, magnetic transponders, and a spool-and-line technique) have successfully determined the fate of acorns by following their dispersal by small mammals (Sork 1984, Jensen and Nielsen 1986, Yasuda et al. 1991, Steele and Smallwood 1994, Iida 1996). These studies indicate that mammal-dispersed acorns are scattered and deposited singly or in small larderhoards of a few acorns usually at a mean distance of < 20 m from their source (Sork 1984, Jensen and Nielsen 1986, Miyaki and Kikuzawa 1988, Iida 1996). While many of these cached acorns are recovered and destroyed before germination, modest to significant levels of establishment (up to 13%) from rodent caches do occur (Jensen and Nielsen 1986, Steele et al. 2001). From these investigations it appears that small mammals may have a significant effect on oak dispersal and regeneration at a local scale, by influencing the spatial arrangement of oaks, local patterns of gene flow, and patch succession (Steele and Smallwood 1994, Jensen and Nielsen 1986, Smallwood et al. 1998).

Other subtle effects of mammals, as well as of birds, on the dispersal of acorns include the selective caching of sound seeds by squirrels (Steele et al. 1996) and by jays (Korstian 1927, Johnson and Webb 1989, Dixon et al. 1997b), selective dispersal of larger acorns by some mammals (Hallwachs 1994), selective dispersal of smaller acorns by birds (Darley-Hill and Johnson 1981, Scarlett and Smith 1991), and the potential dispersal of partially eaten (but still viable) acorns by both birds and mammals (Steele et al. 1993, 1998).

ACORN DISPERSAL SYNDROMES

An understanding of oak dispersal further requires an appreciation of various adaptations (or preadaptations) of acorns for dispersal by birds or mammals. Dispersal syndromes are suites of characteristics of seeds and fruits that increase the probability of dissemination and establishment by particular types or assemblages of dispersal agents (Howe and Westley 1988). Examples include the starchy eliaosomes of ant-dispersed seeds and the bright, odoriferous fruits of many tropical plants that ensure dispersal by some primates and birds (Howe and Westley 1988). For many of the nut-bearing trees, such as the oaks, however, it is widely assumed that their seeds exhibit few specific adaptations for dispersal other than a high energy content, rounded form, and heavy seed coat (Howe and Westley 1988). While this is perhaps true for some species, a closer look at the characteristics of acorns as they relate to animal dispersal reveals a far more complex picture.

Evidence for the dispersal syndromes within the genus *Quercus* follows largely from the differences in acorns of the two major groups of North American oaks (Kaul 1985): the white oak group (subgenus *Leucobalanus*) and the red oak group (subgenus *Erythrobalanus*). Acorns vary in their physical and chemical properties across the genus but show considerable consistency within each of these two groups. In general, red oak species produce acorns with high lipid content (18–25% dry mass), high tannin content (6–10%), and usually require, at least in temperate regions, a period of winter dormancy before germinating. In contrast, acorns of white oak species exhibit lower concentrations of both lipid (5–10%) and tannin (< 2%) and germinate immediately after autumn seed fall (Ofcarcik and Burns 1971, Short 1976, Fox 1974).

These characteristics directly affect the behavior of acorn-consumers. The early germination of white oak, for example, poses a potential problem for any species that stores acorns for later use (Fox 1982, Pigott et al. 1991). Radicle emergence in acorns of white oaks, characterized by the development of a thick, fleshy tap root composed largely of cellulose (Korstian 1927, Fox 1974), results in the rapid transfer of energy to a form that is of little use to seed consumers (Smith and Follmer 1972, Fox 1974, and Chapter 11). This characteristic of white oak acorns is widely interpreted as an adaptation to escape predation (Barnett 1977). Likewise, the behavior of embryo excision by eastern gray squirrels, in which the animal removes the embryo with a few quick scrapes of the incisors before storing it, is viewed as a counter behavioral adaptation to prevent cache losses to germinating acorns (Fox 1974, 1982, Barnett 1977). More recent observations that this behavior is also performed by another species of tree squirrel, *Sciurus aureogaster* from central Mexico (Steele et al. 2001), suggest that the behavior may be common and widely concomitant with the distribution of white oaks across North America.

The potential effect of tannin and lipids on acorn preferences also has been a subject of considerable research (Smith and Follmer 1972, Short 1976, Lewis 1980, 1982), although conclusions from such studies have been contradictory. It is well known that tannins reduce palatability and digestibility of various plant parts, including acorns (Bate-Smith 1972, Robbins et al. 1987, Koenig 1991, Johnson et al. 1993), and Chung-MacCoubrey et al. (1997) showed that the tannins in red oak acorns have this effect on gray squirrels. However, it has also been argued that the higher lipid concentration of red oak acorns makes these seeds preferred over those of white oak (Smith and Follmer 1972). When Smallwood and Peters (1986) presented free-ranging squirrels with artificial acorn doughballs containing varying amounts of tannin and lipid, they found that tannin significantly affected feeding preferences. The gray squirrels selected acorns low in tannin, especially when levels of food abundance were high (i.e., autumn). However, consistent with the predictions of Smith and Follmer (1972), they also found that higher lipid levels attenuate the effects of tannins. They hypothesized that in the autumn, squirrels might preferentially consume acorns that exhibit early germination and lower tannin levels (i.e., those of white oaks) and store those that show the opposite characteristics (i.e., acorns of red oaks).

Tests of this hypothesis verify the predictions: squirrels and possibly other mammals consistently store viable red oak acorns more often, and disperse them farther, than they do white oak acorns (Steele and Smallwood 1994, Hadj-Chikh et al. 1996, Smallwood et al. 1998). Field experiments with metal-tagged acorns suggest that, unlike red oak acorns, white oak acorns rarely survive to germinate if discovered by small mammals (Steele et al. 2001). The ultimate cause of this behavior seems to be the differences in perishability resulting from early germination and higher rates of insect infestation found in white oak acorns (Hadj-Chikh et al. 1996). Steele et al. (1996) showed that red oak acorns are eaten, rather than stored, if they are infested with weevil larvae. Red oak acorns may gain a dispersal advantage even when consumed: Steele et al. (1993) found that at least three different vertebrate species sometimes consume only the basal (top) part of the acorn, leaving the remainder—with a viable embryo—to germinate.

A final component of the oak dispersal system involves masting: the episodic production of large acorn crops in some years and small or nonexistent crops in others (Chapters 9 and 10). Recent studies suggest that populations of seed predators and seed dispersal agents are strongly influenced by changes in acorn availability (Crawley and Long 1995, Wolff 1996, Jones et al. 1998, McShea 2000). However, it is not yet known how masting events affect the probability of seed dispersal and seedling establishment.

Results of these studies point to a striking difference in the dispersal syndromes of the red and white oaks. Collectively, they suggest that the majority of physical and chemical characteristics of red oak acorns (e.g., shape, pericarp, germination schedule, composition and distribution of various chemical compounds) represent a suite of adaptations (or preadaptations) to enhance dispersal by small mammals and birds. In contrast, it appears that most white oaks are rarely dispersed by acorn consumers and that they lack many of these adaptations (but see Monsandl and Kleinert 1998 for evidence of dispersal of *Q. robur* by European jays).

If this general characterization of red oak and white oak dispersal holds true, it follows that red oak seedlings should be widely dispersed to greater distances from parental sources, whereas seedlings of white oaks should show a clumped distribution closer to parent trees. In tests of this hypothesis, Smallwood et al. (1998) measured the distribution of seedlings and adults of two white oak and two red oak species in a single oak forest and, using the computer models of Ribbens et al. (1994), calculated the best-fit distribution of seedlings around parent trees (Smallwood et al. 1998). As predicted, the two white oak species showed truncated seedling shadows, while those of red oaks were distributed 3–6 times farther and disproportionately fewer of their seedlings were recorded near adult trees (Figure 12.1). While many postdispersal mortality factors may distort the seed shadow prior to the development of a seedling shadow (Howe 1986, Vander Wall 1990), results of Smallwood et al. (1998) are clearly consistent with the patterns of seedling dispersion predicted from this differential dispersal hypothesis.

As argued by Howe (1989), scatter- and larder-hoarding birds and mammals are likely to contribute not only to the dispersal of seeds but also, as a result of their influence on the spatial arrangement of plants, to a wide range of population and genetic attributes. If we are correct about the differential dispersal of red and white oaks, several differences in seedlings and perhaps adult trees are likely to follow. Indeed, reports of higher rates of establishment of red oak seedlings in more exposed sites (Crow 1992), greater clumping of white oak seedlings and adult trees (Abrams and Downs 1990, Berg and Hammrick 1994), higher shade tolerance in some white oaks (Crow 1992), and the ability of red oaks to withstand low soil moisture (Kolb et al. 1990, McCarthy and Dawson 1990, Kubiske and Abrams 1992) are all highly suggestive of an entire suite of characteristics that may be tied to these contrasting dispersal patterns (Figure 12.2).

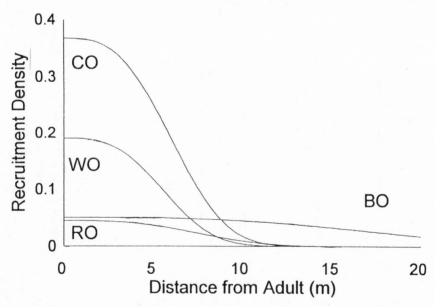

FIGURE 12.1. Estimated seed shadows for two white oak and two red oak species. (From Smallwood et al. 1998.) As predicted, the two white oak species (chestnut oak [CO], *Q. prinus,* and white oak [WO], *Q. alba*) show much shorter dispersal distances than the two red oak species (red oak [RO], *Q. rubra,* and black oak [BO], *Q. velutina*).

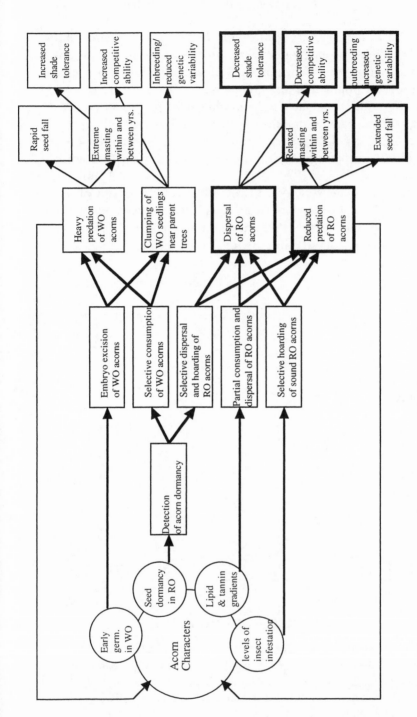

FIGURE 12.2. Conceptual overview of the relationship between acorns, mammalian acorn consumers, and the dispersal and dispersion of oaks. Shown are acorn characteristics (circles), observed and predicted behavioral responses of small mammals to these characteristics (rectangles), and the *predicted* effects of small mammals on the dispersal and dispersion of oaks (squares). Heavy directional lines represent direct ecological effects and light lines represent possible long-term evolutionary effects. Dark squares contain information pertaining to red oaks (RO) and light squares information pertaining to white oaks (WO).

ACKNOWLEDGMENTS

We thank the biology departments of Wilkes University and the University of Richmond for their support during the preparation of this manuscript and J. O. Wolff for reviewing an earlier version of the manuscript. Our research has been supported by the National Science Foundation (grants DEB-9306641 and DBI-9978807).

Chapter 13

Ecological Webs Involving Acorns and Mice

Basic Research and Its Management Implications

RICHARD S. OSTFELD

The disciplines of ecology and natural resource management have many parallels, including the use of both community and ecosystem perspectives. Decades ago, ecologists recognized a distinction between population ecology, in which the focus was on properties of single species, and community ecology, in which the focus was on interactions among species. More recently, community ecologists have expanded their focus to include interaction networks among species and the existence of both direct and indirect effects of species on one another (e.g., Wootton 1993, Holt 1984, Pimm 1991). Similarly, management of natural resources can be divided into approaches that focus on single species and those that target more inclusive entities, such as diversity or productivity. The latter is typically called ecosystem management (e.g., Grumbine 1994).

Natural resource managers have traditionally focused on single species of commercial or conservation importance. Such efforts to manage single species often fail, either because they are ineffectual or because they have unanticipated results (Pickett et al. 1997). For instance, the management of freshwater fisheries is often compromised because the fish participate in a "trophic cascade" (Carpenter et al. 1985) that complicates simplistic management approaches. Stocking of lakes with top predators can result in either suppression of primary production or in algal blooms, depending on the number of trophic levels in the lake (Carpenter and Kitchell 1988). Fisheries management has failed repeatedly because of poor information on species interactions and trophic structure in freshwater and marine ecosystems (Walters 1998). The ecosystem management approach, on the other hand, explicitly recognizes that species exist in webs of trophic and nontrophic interactions and that these interactions have strong implications for the performance of both individual species and the entire ecosystem. Moreover, this approach is required when community diversity, stability, or ecosystem processes, rather than single species, are the target of management efforts.

The oak forests of the eastern United States are being modified and degraded by human activities, and sound management approaches are needed (Healy et al. 1997). Oak forests are highly complex ecosystems

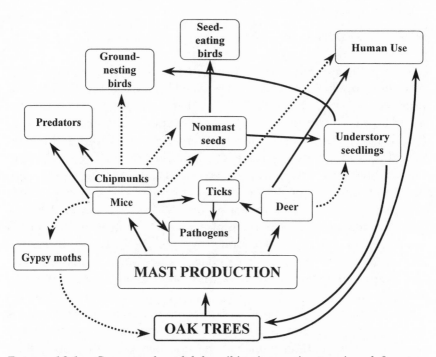

FIGURE 13.1.   Conceptual model describing interacting taxa in oak forests of the eastern United States. Arrows indicate the direction of the primary interaction between taxa, for instance, mast production affects mice, but mice do not affect mast production. Solid arrows indicate that the effect of one taxon on another is positive (e.g., more mice leads to more ticks), and dashed arrows represent a negative effect of one taxon on another (e.g., more mice leads to fewer gypsy moths). The nature and strength of most interactions, and contingencies involved in the outcomes of interactions, are described in the text. (Adapted from Ostfeld, Jones, and Wolff 1996.)

with many interacting species and processes. Some of these interactions are conspicuous, for example, the overbrowsing of tree seedlings by dense populations of white-tailed deer (Healy 1997a). But others are subtle and cryptic, for example, the interactions between nocturnal rodents, such as white-footed mice, and their predators and prey. In this chapter, I describe some recently documented interconnections among oak trees, acorn production, wildlife populations, gypsy moth dynamics, and Lyme-disease risk (see Figure 13.1). I then describe the implications of these interconnections for human use of oak forests. Finally, I advance some possible management options and anticipate their consequences in light of the complex web of interactions within these ecosystems.

MAST SEEDING BY OAKS

Masting—the episodic or periodic production of large crops of fruits or seeds by a population of plants, punctuated by one to several years of low production—is characteristic of trees that bear large seeds and is believed to have evolved as a means of escaping seed predation (e.g., Janzen 1971). During mast years, so many seeds are produced that seed predators cannot consume them all, and this leads to successful recruitment of seedlings by the masting trees. In contrast, seedling recruitment may be minimal or nonexistent following years of low seed production, because virtually all seeds are consumed (Kelly 1994, Sork et al. 1993). Seed consumers often respond to mast production by increasing population density; therefore, selection may favor trees that produce large seed crops after several years have elapsed since the previous mast year, resulting in high seed production when populations of seed predators are at a nadir.

For the past few decades, plant ecologists and evolutionary biologists have been interested in both the proximate and the ultimate causes of masting (Chapter 9). Proximately, masting may occur because pollen production or pollen dispersal is limiting in some years but not others (Lalonde and Roitberg 1992). Alternatively, seed crops may be initiated only after sufficient carbon has been stored by trees to allow them to allocate stored resources to reproduction rather than to growth and maintenance (Lalonde and Roitberg 1992). Because by definition masting behavior involves simultaneous seed production by many or most members of a plant population, plant biologists are interested in discovering the environmental cues that trigger synchronous seed production within a population. Whether individual plants respond to local or regional environmental cues, such as cumulative degree-days or seasonal rainfall, or to chemical signals produced locally or regionally by conspecifics is unknown. Plant ecologists appear to be in general agreement that the main ultimate cause of masting is the satiation of seed predators, to produce seedling recruitment (Janzen 1971, Kelly 1994, Silvertown 1980).

Until recent years, much less attention had been devoted to understanding the consequences of masting than its causes. However, the consequences of masting to the structure and function of forest ecosystems can be profound (Ostfeld and Keesing 2000b). Masting tree species are common in boral, temperate, and tropical forests, so the cascade of effects of masting on wildlife, and the implications of this ecological cascade for both human health and forest health, may be widespread.

ECOLOGICAL CONSEQUENCES OF OAK MASTING

Effects of Masting on Rodents

Mast production provides an enormous flush of resources for seed consumers in forests dominated by masting trees. In the oak-dominated forests of the eastern United States, as well as those of eastern Europe, mast production causes high overwinter survival rates, and occasionally winter breeding, in both murid and sciurid rodents (Pucek et al. 1993, Ostfeld, Jones, and Wolff 1996, Jones et al. 1998, Wolff 1996). Abundant acorns may improve survival rates by allowing rodents to reduce foraging activities and home range sizes, thereby diminishing their vulnerability to predators. In addition, consuming an abundance of food allows storage of body fat, which may buffer rodents against harsh winter conditions. As a result of high overwinter survival and winter breeding following mast production, rodent populations begin the spring breeding season already at moderate to high density. Several studies have now demonstrated that forest rodent populations reach multiannual peaks in density in the springs or summers following mast production (Pucek et al. 1993, King 1983, Ostfeld, Jones, and Wolff 1996, Ostfeld et al. 1998, Wolff 1996, McShea 2000). Experimental simulation of masting, by providing abundant acorns on three 2.4-ha forest plots, reduced the rate of overwinter decline in white-footed mouse (*Peromyscus leucopus*) populations and resulted in spring-summer densities that were approximately

5 times higher than those on unsupplemented control plots (Jones et al. 1998). Studies in Virginia (Wolff 1996) and New York (Ostfeld et al. 1998) have demonstrated that the size of the acorn crop in the fall explains about 80% of the interannual variation in density of white-footed mice in summer.

Effects of Masting on Deer

Many studies have demonstrated the importance of acorns to the autumn and winter diets of white-tailed deer (*Odocoileus virginianus*) (Healy 1997a, Chapter 14). Although acorn availability does exert demographic effects on deer, as on rodents, population fluctuations among deer are largely independent of variations in acorn production, because of their longer mortality and natality schedules. However, recent studies have revealed a pronounced behavioral response by deer to acorn production. In the autumn of a mast year, deer are attracted to oak-dominated stands and spend a considerable portion of their daily time budgets there. In contrast, during autumns of poor acorn production, deer avoid oak-dominated stands, aggregating instead in forest of other types, such as those dominated by maples (McShea and Schwede 1993; Ostfeld et al. 1998). Because of interactions between mammals and both disease vectors and forest pests, the numerical response to acorns by rodents and the behavioral response by deer have profound implications for human health and forest health (Figure 13.1).

Interaction of Rodents and Deer with Ticks and Lyme Disease

Rodents and deer are crucial hosts for ticks of the genus *Ixodes*, which are the vectors of the Lyme-disease agent, a spirochete bacterium (*Borrelia burgdorferi*). Lyme disease is a zoonotic disease, which means that the bacterial pathogen is maintained in wildlife populations and occasionally is transmitted to humans. Unlike some other vector-borne diseases, such as malaria, humans are irrelevant to the maintenance of the Lyme-disease enzootic cycle and only become involved "accidentally" when ticks, which normally feed on wildlife, attack people. *Borrelia* infections in wildlife hosts, including rodents and deer, appear to be rather benign, resulting in no obvious symptoms and having no detectable effect on survival or reproduction. Because of the mammalian hosts' role in feeding and infecting ticks, the population dynamics and space use of the hosts are critical to the epidemiology of this expanding disease (Lane et al., 1991, Piesman and Gray 1994, Ostfeld 1997).

Lyme disease is by far the most common vector-borne disease in the United States and is increasingly common in Europe. Over the past decade in the United States, between 8,000 and 16,000 cases have been reported to the Centers for Disease Control and Prevention (CDC) each year (Ostfeld 1997). In the United States, Lyme disease is particularly common in the northeastern and northcentral states, where the vector is the black-legged tick, *I. scapularis*.

Ixodes scapularis typically undergo a two-year life cycle that includes four stages: egg, larva, nymph, and adult (Fish 1993). In autumn of each year, adult ticks feed predominantly on white-tailed deer, mating during a single 3–4-day blood meal. Females drop off after engorging and overwinter in a quiescent state on the forest floor at the site of detachment from the deer. The following late spring or early summer, engorged females produce an egg mass before dying, and the eggs hatch in mid-summer into tiny (~0.5 mm) larvae. Because the adult stage of the tick is specialized to deer, the location of deer in autumn determines the location of newly hatched larvae the following summer. Larvae remain within a few meters of the site of hatching and wait for a host to wander near enough to permit attachment, a behavior called host seeking or questing. Unlike adult ticks, larvae are not specialized in their choice of hosts and may feed from any of a wide array of mammalian, avian, or reptilian hosts.

Because transovarial transmission of *B. burgdorferi* from female to offspring is highly inefficient, the vast majority of larval ticks hatch from eggs free of the Lyme disease spirochete (Piesman et al. 1986). Therefore, larval ticks are generally harmless. Larvae may become infected with *B. burgdorferi* if they feed on an infected vertebrate host, but the probability of becoming infected varies strongly with the species of the host. In the northeastern and north central United States, blood meals taken from white-footed mice are by far the most likely to result in infection of the feeding larval tick (e.g., Levine et al. 1985, Magnarelli et al., 1988, Mather 1993). It is for this reason that *P. leucopus* is considered the principal natural reservoir for Lyme disease in North America. In Europe, several mammalian and avian hosts may be competent reservoirs, resulting in more complex ecological dynamics (Randolph and Craine 1995). Once a larval tick becomes infected, it maintains the infection through later molts and is capable of transmitting bacteria to subsequent hosts, including humans. Therefore, the population density of white-footed mice in summer, when larval ticks are active, strongly influences the number of ticks that become infected with *B. burgdorferi* (Mather and Ginsberg 1994). Because the density of infected ticks within areas that people use domestically and recreationally is the primary risk factor for Lyme disease, understanding the dynamics of white-footed mice may allow ecologists to predict and prevent human exposure to the disease (Ostfeld 1997).

After a single 2–3-day blood meal, larval ticks drop off the host and molt into the nymphal stage, which remains quiescent for 10 months or so, only becoming active the following late spring or early summer. Nymphs that acquired *B. burgdorferi* during their larval meal may transmit the disease agent to their human or nonhuman host during their nymphal meal. Because nymphs are small (~1 mm) and therefore difficult to detect, and because their season of peak activity coincides with that of humans, this life stage is probably responsible for transmitting the majority of Lyme disease cases (Barbour and Fish 1993). At forested sites in southeastern New York State, 25%–35% of nymphs are infected with the Lyme disease spirochete (Van Buskirk and Ostfeld et al., 1998, Ostfeld, unpublished data). Similar to larvae, nymphs do not specialize on any particular host species but instead feed on a wide variety of vertebrates. Feeding to repletion requires 2–3 days, after which nymphs drop off the host and molt into the adult stage, which seeks a deer host a few months later in the fall of the same year.

Acorn production influences Lyme disease risk through two different pathways, one involving deer and the other involving mice. In the autumn of a good mast year, when white-tailed deer are attracted to oak-dominated forest stands, they import their burdens of adult ticks into these habitat types, resulting in peak densities of newly hatched larval ticks the following summer (Ostfeld, Jones, and Wolff 1996, Jones et al. 1998). Because heavy acorn production also causes white-footed mouse populations to reach peaks in density the following summer, mast production results in simultaneous and syntopic peaks of ticks and the most competent natural reservoir for *B. burgdorferi*. These concurrent events result in a high probability that larval ticks will acquire the Lyme disease agent and molt into an infected nymph. The outcome is a higher than usual risk of Lyme disease during the second summer following heavy masting, given the 1-year delay before larvae that fed on abundant mice become active as infected nymphs.

Effects of Mice on Gypsy Moths

The density of white-footed mice is important not only to their parasites and pathogens but also to their prey, which include the gypsy moth (*Lymantria dispar*). The gypsy moth is a European invader of North American oak forests. In parts of the eastern United States, this species periodically undergoes population outbreaks during which it may defoliate large expanses of oak forest (Chapter 7). Gypsy moth populations tend to remain at low densities for several years before beginning a phase of rapid increase, often spanning five orders of magnitude in egg mass density over 2–3 years (Chapter 7, Campbell 1967, Ostfeld, Jones, and Wolff 1996). After one to several years of peak density, the moth populations then decline steeply, reentering a prolonged low-density phase.

Much attention has been devoted to understanding the causes of fluctuations in gypsy moth populations (Campbell and Sloan 1977, 1978, Doane and McManus 1981, Chapter 7). During the peak phase, moth populations may be regulated by their food supply, particularly when outbreaks result in massive defoliation of oak forests. Evidence indicates that the decline phase is caused by viral pathogens and parasitoids that specialize on gypsy moths and that exhibit a delayed density-dependent response to their moth hosts (Elkinton and Liebhold 1990). Other factors, such as induced chemical defenses by host trees, the use by moths of plant secondary chemicals for defense against pathogens, and delayed effects of high population density on maternal condition and fecundity, are also known to influence gypsy moth populations during both the peak and decline phases (Rossiter et al. 1988, Rossiter 1994, Hunter and Dwyer 1998).

After several larval instars, gypsy moths pupate for about two weeks in midsummer and then eclose into adults. It has long been known that white-footed mice eat gypsy moth pupae, which are a large (~ 2–3 cm), immobile, undefended food source, highly accessible to mice by virtue of their location on the forest floor or low on trunks of trees (Smith 1985, Yahner and Smith 1991). Despite their propensity to attack pupae, however, mice appear to be unimportant in regulating high-density moth populations, largely because neither the functional response nor the numerical response of mice to moths is sufficiently rapid. Nevertheless, recent research has generated strong evidence that mice, via predation on pupae, are responsible for regulating moth populations during the low-density phase (Elkinton et al. 1996, Ostfeld, Jones, and Wolff 1996, Jones et al. 1998).

Studies examining predation on freeze-dried gypsy moth pupae show that mice are the predominant predator in most years and that the proportion of pupae attacked is strongly correlated with mouse density (Elkinton et al. 1996, Ostfeld, Jones, and Wolff 1996). When mouse density exceeds $10-15$ individuals ha^{-1}, virtually 100% of the experimentally deployed moth pupae were attacked by mice within the 2-week window necessary for eclosion (Ostfeld, Jones, and Wolff 1996). In an experimental field study in which mouse density was reduced by trapping and removal, survival of both experimentally deployed and natural pupae was dramatically higher than in control sites in which mouse density was high. The result was an enormous increase in density of egg masses and caterpillars the following year on plots from which mice had been removed (Jones et al. 1998). Essentially, the reduction of mouse density during the low phase of a gypsy moth cycle released the moth population from regulation by mice and allowed it to begin a phase of rapid growth (Jones et al. 1998). This study, combined with other observational and experimental studies (e.g., Elkinton et al. 1996, Ostfeld, Jones, and Wolff 1996), indicates that moderate- to high-density mouse populations are sufficient to maintain moth populations at low densities in perpetuity, and that a crash in the mouse population when moth populations are sparse is both necessary and sufficient to cause rapid growth toward an outbreak of moths (Jones et al. 1998, Ostfeld and Keesing 2000b). Because crashes in mouse populations are predictable based on mast production (Wolff 1996, Ostfeld and Keesing 2000b), moth outbreaks and defoliation events also may be predictable well in advance.

Gypsy moths do not have a reciprocal effect on population dynamics of white-footed mice. Because moths pupate in midsummer, when food is not limiting to populations of mice (Hansen and Batzli 1978, Wolff 1996), mice do not appear to be affected by the density of gypsy moths.

The potential exists for a positive feedback loop from acorns to mice to gypsy moths to oak trees and masting (Figure 13.1). Gypsy moth defoliation of oaks may delay or prevent the production of mast crops by existing oaks and/or reduce the community dominance of oaks (reviewed by Healy et al. 1997). Temporary or long-term reductions in mast crops are expected to reduce average population densities of white-footed mice (Ostfeld, Jones, and Wolff 1996, Elkinton et al. 1996, Jones et al. 1998), which will relax the suppressive effects of mice on gypsy moths. This in turn will increase the probability of moth outbreaks and defoliation events. The existence of feedback loops adds a level of complexity to forest management, because the impacts of a particular management action may become strongly amplified.

Effects of Mice on Ground-Nesting Songbirds

In oak forests of the eastern United States, several species of songbirds, including ovenbirds, worm-eating warblers, veeries, wood thrushes, and dark-eyed juncos, nest at or near ground level. Nests of these species may be vulnerable to attack by various mammalian and avian predators, especially during incubation. Indeed, many studies using artificial ground nests suggest the potential for these predators to cause nest failure and even population declines of some passerines (e.g., Leimgruber et al. 1994, Martin 1993). Deployment of artificial ground nests, typically baited with both quail eggs and clay eggs (the latter for acquiring tooth or bill prints useful in identifying nest predators), has suggested that mammals such as raccoons and opposums, and birds such as bluejays and crows, are the principal predators. However, because quail eggs are larger and have thicker shells than typical songbird eggs, this approach may bias results against the detection of smaller predators such as mice and chipmunks (Maxson and Oring 1978), which typically are unable to handle quail eggs.

Recent studies in southeastern New York State using passerine eggs revealed that the white-footed mouse was responsible for the majority of attacks on artificial ground nests and that eastern chipmunks were the second most frequent predator. Medium-sized mammals and birds were infrequent predators on these nests (K. Schmidt, R. Naumann, J. Goheen, R. Ostfeld, E. Schauber, and A. Berkowitz, unpublished data). In oak-forest plots in which mouse populations were maintained at low densities via removal trapping, attack rates on artificial nests were significantly lower than on control plots supporting high mouse density. In contrast, experimental manipulation of chipmunk densities had no effect on nest-predation rates.

Studies of nest predation using artificial nests have a number of well-recognized potential weaknesses, including lack of parental defense, poor placement by the experimenter, and elevated attractiveness due to scent contamination. Therefore, additional studies, particularly examining attacks on natural nests, will be necessary to determine whether artificial nest experiments accurately represent rates and perpetrators of natural nest predation. However, some evidence suggests that results from artificial-nest studies may accurately reflect processes affecting success of natural nests. In a long-term study of nesting performance of dark-eyed juncos in oak forests of Virginia, Ketterson et al. (1996) showed that the proportion of nests failing to fledge young was strongly correlated with summer density of *Peromyscus* populations. At these same sites, density of mice in summer was highly correlated with acorn production the prior autumn (Wolff 1996).

The density and structure of understory vegetation may influence the survival of eggs and nestlings of ground-nesting songbirds. Because browsing by deer on forest understories may affect protective cover and the suitability of nesting sites, population size and space use of deer may also strongly influence bird populations indirectly (McShea and Rappole 1997). When acorns are abundant, impacts by deer on understory vegetation in the autumn and winter may be relaxed, due to reduced browsing, which in turn may enhance protective cover for birds the following summer. On the other hand, dense populations of deer when no acorns are available may have a strongly destructive influence on protective understory vegetation (McShea and Rappole 1997).

Potential Interactions between Mice and Their Predators

Interactions between white-footed mice and their avian and mammalian predators in oak forests have not been well studied. Despite anecdotal reports, ecologists have not yet determined whether raptor or carnivore populations experience unusually high reproductive success during years of high mouse densities. Similarly, little evidence exists to evaluate the possibility that predation by raptors and carnivores is responsible for mouse population declines from high densities. In oak-hornbeam forests of eastern Europe, acorn-caused increases in the population density of rodents, particularly *Apodemus sylvaticus* and *A. flavicollis,* appear to induce population growth by their predators, especially mustelids and owls (Jędrzejewska and Jędrzejewski 1998). These predators, in turn, attack alternative prey, such as nesting songbirds, when rodent populations collapse.

MANAGEMENT OPTIONS AND POSSIBLE OUTCOMES

The ecological studies summarized above demonstrate that a network of species of plants, vertebrates, invertebrates, and microbes interact strongly in oak forests of the eastern United States. Masting behavior by the ecosystem dominant—oak trees—sets off an ecological chain reaction mediated by behavioral and numerical responses of white-footed mice and white-tailed deer, which in turn cause changes in the dynamics of their prey (e.g., gypsy moths, songbirds), pathogens, parasites, and possibly their predators. A practical benefit of understanding the nature and strength of these ecological interactions may be the ability to predict Lyme disease risk, gypsy moth outbreaks, and nesting success of songbirds and owls, all of which are matters of practical or aesthetic importance to people. Beyond prediction, it may be possible to manage populations of deer or mice, directly or indirectly, in order to meet desired management goals. In the following section, I discuss some of the more obvious options concerning the management of vertebrate populations or the landscapes in which they occur, and speculate on the possible outcomes of these management schemes. Although the focus of management may often be a single species, the outcomes of management will almost certainly involve a number of interacting species and processes within a landscape.

Reduction or Elimination of Deer

The primary means of managing deer populations are hunting and fencing. Hunting may reduce density and population growth rates of deer, particularly if the hunting program is of appropriate scale and intensity and if it includes the taking of does (Knox 1997, Winchcombe 1993). Hunting programs rarely if ever cause dramatic reductions or local extinctions in deer populations, unless those populations are isolated, for instance, on islands. Fencing may eliminate deer completely from enclosed areas, but typically this can be accomplished only in small tracts.

Is it possible to diminish population density of black-legged ticks, and therefore the risk of Lyme disease, via moderate reductions in populations of deer? The answer appears to be no. Recent modeling efforts have addressed this question by creating mathematical simulations of tick populations that feed as adults on deer and as juveniles on rodents, and then manipulating the simulated host populations to determine potential consequences for tick density and infection prevalence (Van Buskirk and Ostfeld 1995). The model predicted that incremental reductions in deer density would not result in similar reductions in tick density. Instead, it suggested that the relationship between deer and tick abundance was strongly nonlinear, with a threshold effect on tick density only at very low deer density. The intuitive reason for this nonlinearity is that, because each individual deer may host hundreds to thousands of adult ticks in a given season, and because each female tick may lay more than 1000 eggs, even a modest population of deer may be able to support a dense population of ticks (Van Buskirk and Ostfeld 1995). Therefore, a small to moderate reduction in deer density probably will cause a trivially small reduction in tick density.

Empirical results generally are consistent with predictions of this model. Tick populations may be enormously high in areas where a hunting program results in moderate reductions in deer populations (Ostfeld, Hazler, and Cepeda 1996), as well as in places where deer populations are modest even without hunting (Wilson et al. 1988, 1990). However, on some Massachusetts islands from which deer were nearly extirpated and in mainland sites from which deer were excluded by fencing, populations of black-legged ticks experienced a dramatic reduction (results reviewed in Wilson and Childs 1997). Because of practical limitations to the installation and maintenance of fencing, this method of managing deer populations may be feasible only on a very local scale, but it could reduce Lyme disease risk around individual residential properties (Figure 13.2). In conclusion, large-scale reduction of deer density via hunting would appear to be an ineffective method of controlling Lyme disease risk, whereas fencing may be effective at small scales.

Despite the low likelihood of reducing Lyme disease risk by managing deer, there are other potential effects of deer management on the function of oak forest ecosystems, such as the nesting success of ground-dwelling songbirds.If reduction of deer density results in greater density of understory vegetation, which in turn results in higher attack rates by white-footed mice (R. Naumann, R. Ostfeld, and A. Berkowitz, unpublished data), deer management may reduce recruitment of ground-nesting songbirds. In contrast, because attack rates by raccoons and opposums often are higher in sparse than in dense understory, deer reduction could protect nests in areas where the predominant predators are medium-sized mammals (deCalesta 1994). Predicting the net effect of a deer reduction on nesting success is problematical because of incomplete knowledge of interactions between deer and other mast consumers (e.g., mice, raccoons) and of interactions between understory vegetation and susceptibility of nests to predation. For instance, if reduction of deer causes increased availability of mast for mice and raccoons, and these mast consumers increase in density, then songbird nests may be at high risk (McShea 2000). Clearly, understanding the net effects of deer management on songbirds requires additional empirical and modeling studies.

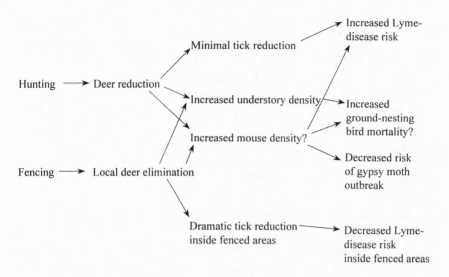

Possible deer management outcomes

FIGURE 13.2.   Methods and possible consequences of managing populations of white-tailed deer. Postulated consequences are limited to disease risk, gypsy moth dynamics, and nesting success of ground-dwelling songbirds (see Figure 13.1). Those that are highly speculative are indicated with question marks. Arrows point to the likely consequences of each phenomenon.

Reduction of White-Footed Mice

A second potential target of oak forest management is the white-footed mouse. Because white-footed mice are the principal reservoir of the Lyme disease spirochete and may be major predators on songbird nests, reducing mouse densities might have benefits for both human and forest health. On the other hand, reducing density of mice will cause relaxation of predation on gypsy moths, thereby increasing the likelihood of moth outbreaks, oak tree defoliation, and potential declines in the abundance and productivity of oaks. These conflicting probable outcomes of a single management action emphasize the crucial importance of understanding interaction webs within ecological communities.

Irrespective of the desired management outcome, managing populations of white-footed mice, particularly at scales larger than individual residential properties, probably will be much more difficult than managing deer. Reducing mouse populations directly, for instance by baiting with rodenticides, is infeasible except in small areas and may adversely affect nontarget organisms. As a consequence, the most viable means of controlling mouse populations probably will be to manage their habitats or landscapes.

Populations of *P. leucopus* often reach their highest densities in small isolated woodlots (Vessey 1987, Kaufman and Kaufman 1989, Cummings and Vessey 1994, Nupp and Swihart 1998). Several factors appear to be responsible for these high and relatively stable densities. First, in landscapes composed of woodlots within an agricultural, suburban, or urban matrix, populations of mouse predators, such as barred owls, bobcats, coyotes, foxes, and long-tailed and short-tailed weasels, may be insufficient to regulate mouse populations. Second, in some isolated forest patches, particularly those smaller than several hectares, populations of sciurid rodents (e.g., chipmunks and tree squirrels) are sparse or nonexistent, which may release mice from competitive suppression (Nupp and Swihart 1998). Third, when woodlots occur in a matrix that is inhospitable for mice, emigration is curtailed, thus negating the potential for dispersal to regulate population density. Populations of rodents in which dispersal is prevented often reach unusually high densities, which may be sustained for long periods (Ostfeld 1994).

For these reasons, it appears that management of landscapes of which oak forests are a component may be a viable option for controlling mouse populations. Larger, more continuous, or more interconnected forest patches are less likely to maintain excessively high populations of white-footed mice than are smaller, more fragmented patches (Nupp and Swihart 1996, 1998). Both extrinsic factors (predators and competitors) and intrinsic factors (dispersal) may be affected by patch size, disturbance, and the structure and compostion of the nonforest matrix (Dooley and Bowers 1996). All of these features of forested landscapes can be managed by controlling the spatial arrangement of forestry practices, agricultural fields, and residential development. It should be noted that fragmentation of forested landscapes often optimizes the population performance of deer as well as that of mice, apparently because of the juxtaposition of different food types (e.g., young vs. old forests, forest vs. oldfield vegetation; Sinclair 1997).

What are the likely outcomes for Lyme disease of reducing mouse populations? Van Buskirk and Ostfeld's (1995) model predicted that the density of juvenile black-legged ticks would be linearly (or log-linearly) dependent on density of mice. Each mouse hosts about 50–100 larval ticks and 10–50 nymphal ticks in any given season (Ostfeld, Hazler, and Cepeda 1996), and each successful blood meal for a juvenile tick promotes survival to the next stage. Questing juvenile ticks that fail to find a host within a short activity season die, and empirical evidence suggests that mouse density is an important determinant of the proportion of the juvenile tick population that is able to find a host, feed, and molt to the next stage, resulting in an approximately linear relationship between mouse density and that of juvenile ticks (Van Buskirk and Ostfeld 1995).

Because of the role of the mouse as natural reservoir for the Lyme disease microbe, their density is also a key determinant of the number of juvenile ticks that become infected with *B. burgdorferi*. Empirical studies in oak forests of both Rhode Island and New York show that the density of white-footed mice in summer is a significant predictor of the infection prevalence of nymphal ticks the following year (Mather and Ginsberg 1994, Ostfeld et al., unpublished data). Thus, the lower the density of mice, the lower the number of larval ticks that will feed on mice and molt into infected nymphs that become active the next year.

The results of the Van Buskirk and Ostfeld (1995) model suggest a means of effectively reducing the influence of mice on infection prevalence of nymphal ticks. Managing habitats to enhance populations of species that compete with or prey on mice may not only influence the density of mice but also may dilute the impact of mice on infection prevalence of ticks (Ostfeld and Keesing 2000a). Van Buskirk and Ostfeld (1995) predicted that the higher the diversity in the community of hosts for juvenile ticks, the lower the infection prevalence of ticks, and therefore the lower the risk to humans of contracting Lyme disease. In a highly diverse community of ground-dwelling vertebrates, a high proportion of larval ticks will encounter a poor Lyme disease reservoir and molt into uninfected nymphs. Thus, high vertebrate diversity dilutes the influence of *P. leucopus* on disease risk (Ostfeld and Keesing 2000a) (see Figure 13.3). Although the presence of a diverse community of vertebrates may not reduce the opportunity for ticks to successfully obtain a blood meal, the proportion of ticks becoming infected, and therefore dangerous to humans, will likely decrease with increasing host diversity (Ostfeld and Keesing 2000a).

Although directly or indirectly diminishing the density of white-footed mice reduces Lyme disease risk for humans, reducing mouse densities increases the probability of a gypsy-moth outbreak (Figure 13.2).

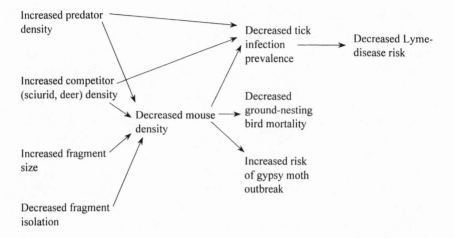

Possible mouse management outcomes

FIGURE 13.3.   Methods and possible consequences of decreasing population density of white-footed mice. Postulated consequences are limited to disease risk, gypsy moth dynamics, and nesting success of ground-dwelling songbirds (see Figure 13.1). Arrows point to the likely consequences of each phenomenon.

Experimental reductions in mouse density in 2.4-ha oak-forest plots in New York resulted in tenfold to thirtyfold increases in density of gypsy moth egg masses and caterpillars the following year (Jones et al. 1998). Densities of mice below approximately 10 ha^{-1} appear to be necessary to release gypsy moths from regulation and allow growth toward outbreak levels (Ostfeld, Jones, and Wolff 1996, Ostfeld et al. 1998). Consequently, the optimal management goal may be to avoid reducing the absolute abundance of mice but instead to reduce the abundance of mice relative to that of other species. Species diversity has two components: species richness (the number of species in the community) and species evenness (the relative abundance of each species). The higher the species diversity in the vertebrate community, by definition, the lower the relative abundance of mice will be. Thus, vertebrate diversity may be a desirable endpoint of management action for both practical and aesthetic reasons. Ecologists and managers may together be able to devise forest and landscape management options to be tested for their effectiveness in increasing diversity of vertebrates.

CONCLUSIONS

An intricate web of connections exists among species in oak forests. The nature and strength of these interactions, as well as their net effects on community and ecosystem dynamics, are still being uncovered. Our growing knowledge of these interactions facilitates scientifically sound management of oak forests and their components.

White-footed mice in the eastern United States, and their ecological counterparts in oak forests elsewhere, clearly are the hub of the highly interconnected oak-forest system. White-footed mice are among the most abundant of all vertebrates in oak forests, are quintessential generalist consumers, and may regulate populations of insects and birds, as well as of forest trees via seed predation (Ostfeld et al. 1997). They also interact strongly with parasites and zoonotic pathogens, and probably with their mammalian and avian predators. Because of this network of ecological connections, the net effects of managing mouse populations may be multifaceted and may result in conflicting outcomes for human users of oak forests. The tentative conclusion to arise from the studies described above is that high species diversity of vertebrate communities in oak forests is beneficial for both human health and forest health. High diversity of species that compete with or prey on mice will probably regulate mouse populations, directly reducing disease risk and possibly predation on songbird nests. These vertebrate predators and competitors also will dilute the influence of mice on infection prevalence of ticks (Ostfeld and Keesing 2000a, Schmidt and Ostfeld 2001).

However, the web of connections centered on white-footed mice will probably necessitate management tradeoffs. For instance, although reducing populations of mice may decrease Lyme disease risk and increase songbird nesting success, population declines of mice also increase the probability of a gypsy moth outbreak (Ostfeld, Jones, and Wolff 1996, Jones et al. 1998). Reducing the *relative* abundance of mice without dramatically decreasing their *absolute* abundance may preserve their positive effect on the regulation of gypsy moths.

Because of the benefits of reducing the abundance of white-footed mice relative to other vertebrate species, the most appropriate target of management efforts appears to be diversity rather than the population status of single species. Both species richness and species evenness appear to be enhanced in large and well-connected forested landscapes. Therefore, appropriate management efforts probably will involve features of landscapes, such as patch size, shape, and connectedness. Studies of the effects of landscape management on mouse populations and of the ecological and epidemiological consequences of any community changes that ensue will be important for wisely managing oak forests in the coming decades.

ACKNOWLEDGMENTS

I am grateful to Charlie Canham, Clive Jones, Felicia Keesing, Gary Lovett, Eric Schauber, and Jerry Wolff for many important discussions of the myriad effects of masting and for help in the creation of a research design, and to Felicia Keesing, Bill McShea, and Michael Bowers for valuable comments on this contribution. The aid of many students and research assistants has been invaluable in conducting this research. Financial support was provided by National Science Foundation (grants DEB-9306330 and DEB-9615414), the National Institutes of Health (R01-AI40076), the Nathan Cummings Foundation, the General Reinsurance Corporation, and the Mary Flagler Cary Charitable Trust. This is a contribution to the program of the Institute of Ecosystem Studies.

Chapter 14

Acorns and White-Tailed Deer

Interrelationships in Forest Ecosystems

GEORGE A. FELDHAMER

White-tailed deer (*Odocoileus virginianus*) form the most important species of cervid in the Western Hemisphere. Significant attention is paid by professional biologists and the general public to their population dynamics and control and to the sociopolitical issues surrounding recreational hunting of them, vehicular accidents involving them, and their damage to agricultural crops—the "human dimensions" noted by Warren (1997). Whitetails also are the most widely distributed New World deer species. Their geographic range extends from southern Canada south through the United States, Mexico, and Central America, to northern South America (Baker 1984). Throughout their range, whitetails occur in diverse habitats including subarctic regions, north-temperate forests, deserts, and humid tropical forests, and from wilderness areas to suburbia. Whitetails have very adaptable feeding habits and consume a wide variety of forage species. They are able to shift their use of resources depending upon availability (Anderson 1997). Oaks (*Quercus* spp.) dominate much of the deciduous forest in the eastern United States and elsewhere (Healy 1997a), and the range of oaks (see Chapter 2 and Burns and Honkala 1990) largely overlaps that of whitetails. For more than 60 years, many investigators have shown that oak mast is a preferred food of white-tailed deer and other forest wildlife species. This chapter reviews the relationship between white-tailed deer and forest communities; it focuses primarily on oak forests and oak mast. It examines the effects of acorn consumption on individual deer and on the dynamics of whitetail populations and the reciprocal impact that deer populations have on oak and other north-temperate forests in terms of regeneration rates and community ecology.

215

FORESTS AS INPUTS: EFFECTS OF ACORNS ON WHITE-TAILED DEER

It is well established that the body condition of deer has a critical effect on their over-winter survival. The nutritional value of autumn diets is a primary determinant of body condition in males and females of all age classes, but especially among fawns and yearlings. Deer that enter winter in good physical condition have a better chance of survival than those that enter this season in poor condition (Mautz 1978). In habitats where they are available, acorns have long been recognized as one of the most important autumn foods of deer (Forbes et al. 1941, Goodrum 1959, Duvendeck 1962, Korschgen 1962). A large percentage of the sound acorns available on the ground usually are taken by deer. For example, in central Pennsylvania, Steiner (1996) found that of the northern red oak (*Q. rubra*) acorns consumed by vertebrates, about 49% were removed by deer. McShea and Schwede (1993) estimated that in their study area in Virginia, in forest dominated by oak and hickory (*Carya* spp.), deer removed an average of approximately 1 kg of acorns/deer/day during autumn.

Diets rich in oak mast provide deer with more-easily digested and metabolized energy, fat, and soluble carbohydrates than do diets that are not dominated by acorns (Harlow et al. 1975, Pekins and Mautz 1987, 1988), but acorns are relatively low in crude protein (Kirkpatrick et al. 1969, Wentworth et al. 1990, and Chapter 11). Fawn growth in autumn is more dependent on available energy than on protein, however (Verme and Ozoga 1980). Likewise, dietary energy level rather than protein affects ovulation rates, fawning rates, and fat deposition in older females (Verme 1969, Warren and Kirkpatrick 1982).

Although oak mast is an excellent forage for deer, acorn availability is seasonal and production is highly variable (Chapters 9 and 10), and seed crops are produced at irregular intervals (Sork et al. 1993). Acorn yield is directly related to the age, basal area, and crown size of individual trees (Goodrum et al. 1971); thus, resource managers can help promote a steady supply of acorns by maintaining a diversity of species of red and white oak groups of different age classes within management units. Silvertown (1980) has reviewed the interesting hypothesis that irregular mast production is an evolved antipredator strategy of trees. Regardless, mast is an inconsistent source of food, and amounts available to deer differ significantly both regionally and yearly. Thus, the spatial and temporal availability of acorns in autumn has a significant impact on the nutrition, reproduction, behavior, and dynamics of white-tailed deer populations throughout the year. Acorn availability is especially critical where the quality or quantity of spring or summer forage is inadequate for deer to establish the fat and energy reserves necessary for winter survival.

Body Mass and Condition

When considering the relationship between mast availability and white-tailed deer body condition, researchers often focus on fawns and yearlings, because: (1) changes in growth and body mass are most evident at these ages, and (2) except for pregnant yearling females, reproductive activity is not likely to affect body mass. Feldhamer et al. (1989) considered mean body mass of hunter-harvested male and female fawns and yearlings during a 13-year period on the Tennessee portion of Land Between The Lakes (LBL), a 69,636-ha peninsula between Barkley Lake and Kentucky Lake in western Kentucky and Tennessee. They found that mean body mass was positively correlated with acorn yield the previous autumn. Acorn yields on the study area ranged from 0.37–55.07 kg/ha, during this period, and accounted for 42–56% of the variation in mean body mass within each age and sex group. Wentworth et al. (1992) also found that in northern Georgia, both male and female fawns, yearling males, and females ≥ 3.5 years old, weighed more in years when mast availability was fair or good. Conversely, in Virginia, Harlow et al. (1975) found no relationship between acorn availability and body mass of yearling white-tail males during the same autumn. This may have been because deer were collected too early in the winter for an effect to be manifested; also, the possible lag effect of the previous year's acorn crop was not considered.

As on LBL, body mass of males in northern Georgia was more strongly correlated with the previous year's mast index than with the current year's index. Female body mass showed a less consistent relationship with acorn production, possibly because of small sample sizes or the confounding effect of yearling reproductive activity. As noted by Wentworth et al. (1990, p. 152), "factors other than dietary quality, including reproductive performance in the previous year and stress associated with lactation, also can influence subsequent productivity of adult does (Verme 1967). Any influence of the acorn crop on adult productivity may be obscured by these confounding factors."

The importance of acorns is also reflected in other indicators of deer body condition. For example, Wentworth et al. (1990) found that, during winter, kidney fat indices for most sex and age groups were significantly higher on study sites in eastern Tennessee, western North Carolina, and northern Georgia that had high fall acorn production. Body condition of adult bucks during and immediately after the breeding season did not appear to be affected by the acorn crop. After the breeding season, however, during late winter, mast availability may have an effect on body condition.

Reproduction

As noted, body mass and condition of white-tail does are directly affected by acorn production and availability. In cervids generally, body mass and condition of maternal does as they enter the breeding season directly affect conception, neonatal development, and subsequent fawn development and body mass (Albon et al. 1983, Saether and Haagenrud 1983). Nevertheless, Wentworth et al. (1990) found that, in the southern Appalachians, adult reproduction rates, specifically the incidence of ovulation and number of fetuses in adult does, were not affected by acorn abundance. In yearling does, however, ovulation rates were significantly greater when acorns were abundant. The proportion of yearlings that bred also appeared to increase with greater acorn abundance. A similar trend occurred in the number of fetuses (from 1.04 to 1.33 per pregnant yearling) but was nonsignificant. Lack of clear relationships in these types of analyses is not too surprising, however. As Ford et al. (1997) noted, relationships between acorn abundance and white-tailed deer population characteristics in the southern Appalachians are often difficult to quantify, possibly because of small differences between study sites and low deer densities. Interestingly, Wentworth et al. (1990) found that when the acorn crop was good, mean conception date occurred approximately 9 days earlier than when acorn production was poor.

Antler Characteristics

On LBL, Feldhamer et al. (1992) considered acorn yield data and associated antler characteristics of yearling whitetails the following year. Significant linear relationships were apparent between acorn yield and mean number of antler points, mean antler beam diameter, and mean antler beam length (Figure 14.1). Similar results were reported by Wentworth et al. (1992) for all age classes of whitetails on their study area in northern Georgia.

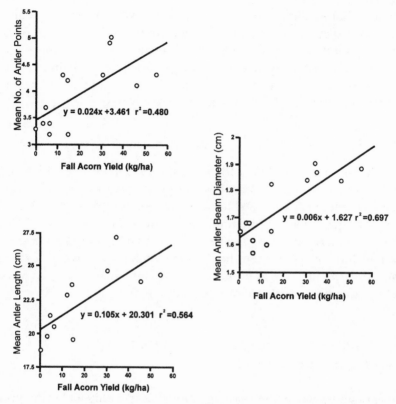

FIGURE 14.1.   Relationship between fall acorn abundance in 1976 through 1987 and (*top*) mean number of antler points, (*center*) mean antler beam diameter, and (*bottom*) mean antler beam length of yearling male white-tailed deer harvested the following year (1977 through 1988) on Land Between The Lakes, Stewart County, Tennessee. (Feldhamer et al. 1992; reprinted by permission from the *Journal of the Tennessee Academy of Science.*)

Age Structure

An increased percentage of yearlings in whitetail populations in relation to increased acorn abundance has been documented. On LBL, Feldhamer et al. (1992) reported a significant positive relationship between acorn yield and the percentage of yearling males (y = 58.93 + 0.28x, r^2 = 0.49) and yearling females (y = 23.30 + 0.21x, r^2 = 0.52) hunter-harvested two years later. A significant positive relationship also was found in Florida between acorn abundance and the percentage of yearling males harvested two years later (Harlow and Jones 1965). In the southern Appalachians, Wentworth et al. (1992) found that the percentage of all yearlings was significantly lower two years after poor acorn crops. However, they found no difference in the fawn-to-doe ratio from antlerless hunts in relation to acorn abundance, although there was a trend in mean values (0.73 fawns/doe in poor years, 0.85 in fair years, and 0.93 in years with good acorn crops). Also, the number of fawns per doe was not correlated with the previous year's mast index.

Home Range and Habitat Selection

Given the preference of deer for acorns, we would expect deer populations to respond to the patchy distribution of that resource by shifting their movement patterns. McShea and Schwede (1993) reported that in years with good mast production, the mean home-range size of white-tailed does increased from summer to fall, and there were significant autumnal shifts in habitat use, both to encompass oak-hickory habitats. Conversely, no increase in home-range size or shift in areas used was evident during a year with poor mast production.

DEER AS INPUTS: EFFECTS OF DEER ON FOREST ECOSYSTEMS

Just as quantity and quality of forage directly affect white-tailed deer populations, whitetails (like other ungulates) have a reciprocal impact on their forest communities. As noted by Nudds (1996, p. 695), "ungulates are important agents of change in ecosystems, acting to create spatial heterogeneity, modulate successional processes, and control the switching of ecosystems between alternative states." Likewise, in their review of studies on the impact of white-tailed deer browsing on forest communities in the eastern United States, Stromayer and Warren (1997, p. 227) concluded that "deer may create alternate stable states" in these communities.

White-tailed deer populations certainly create significant ecological impacts. Throughout much of the United States, population densities of whitetails have recovered from historic lows at the beginning of the twentieth century (McCabe and McCabe 1984). While this recovery reflects well on modern wildlife management practices, Waller and Alverson (1997, p. 222) suggested that "white-tailed deer have reached, and sustained, densities across much of the eastern, northern, and southern United States sufficient to cause manifold and substantial ecological impacts." White-tailed deer may therefore be considered a "keystone species" in deciduous forests, because they have a disproportionately large impact on the distribution and abundance of other species in forest communities and the resulting community structure (Paine 1995, Power et al. 1996, Waller and Alverson 1997). Long-term impacts of deer also may be evident in predominantly agricultural areas that have remnant forest communities (Nixon et al. 1991, Augustine and Jordan 1998).

REDUCED OAK REGENERATION RATES AND TREE SPECIES DIVERSITY

Oak regeneration depends on development of understory seedlings. Oak seedlings that are several years old and have well-developed root systems are more likely to successfully compete with seedlings of other rapidly growing woody species. To reach this stage of forest regeneration, oak seedlings, like seedlings of other plant species, must survive repeated deer browsing. Oaks have such a long reinitiation phase, however, that they spend a longer period in the "browsing zone" than do other species. Considering an oak forest site in central Massachusetts, Healy (1997a) suggested that regeneration was prevented at deer densities above 10/km^2. Oak regeneration and a diverse vegetative understory were maintained at deer densities of 3–6/km^2.

Impacts of deer browsing are not restricted to oak forests, of course. Hough (1965) and Whitney (1984) attributed reduced richness and abundance of woody seedlings to deer browsing in mature forest dominated by black cherry (*Prunus serotina*), sugar maple (*Acer saccharum*), and red maple (*A. rubrum*) in northwestern Pennsylvania. Working with enclosed whitetails in the same area, which had been heavily browsed for more than 50 years, Tilghman (1989) also found that tree seedling species diversity and height were reduced at higher deer densities. She felt that about 7 deer/km^2 was a reasonable management goal for this region to allow forest regeneration. Augustine and Frelich (1998) reported similar results on remnant forest patches in Minnesota dominated by sugar maple, American elm (*Ulmus americana*), slippery elm (*U. rubra*), and American basswood (*Tilia americana*). Effects of grazing by whitetails on the growth and reproduction of understory forbs were significant at high overwinter deer densities (25–35/km^2). Significantly less impact occurred at low deer densities (5–10/km^2).

Oak species are shade intolerant or intermediate in shade tolerance (see Burns and Honkala 1990). They are especially affected by deer browsing (an effect exacerbated by the repression of naturally occurring fires) as they are replaced in forest ecosystems by more shade-tolerant and browse-tolerant species, such as sugar maple. Strole and Anderson (1992, p. 143) concluded, "The fact that deer have a high preference for white oak (*Q. alba*) and a low preference for sugar maple, along with fire exclusion and competition from other low-use, browse-tolerant species, may add to the degradation of a reproducing oak forest." Whitney (1984, p. 403) felt that deer were among "the more important determinants of forest structure in the Allegheny Plateau over the past 50 years." Similarly, in a study of different forests in Ohio, including pin oak (*Q. palustris*), Boerner and Brinkman (1996, p. 309) felt that browsing by white-tailed deer "was more important than environmental gradients or climate factors in determining seedling longevity and mortality." Anderson and Katz (1993) suggested that in temperate-zone forests a period of high levels of deer browsing could be viewed as a "catastrophic event" comparable to windthrow, fire, and an insect outbreak. Regeneration failure, reduced species richness and biomass, and loss of structural diversity are long-term ecosystem impacts by whitetails; forests may take decades to recover from them, even after the reduction of deer densities (deCalesta 1997, also see Pastor and Naiman 1992). Because species composition recovers relatively faster than forest structure, Anderson and Katz (1993) suggested that a period > 70 years with minimal browsing was necessary for structural recovery. As noted previously, however, several causal factors besides deer herbivory may interact in forest damage and regeneration failure.

CONCLUSIONS

White-tailed deer are part of complex interactions in oak forest communities (see Figure 13.1, this volume). Besides altering understory vegetation, they compete for acorns with sciurids, including gray squirrels (*Sciurus carolinensis*), fox squirrels (*S. niger*), flying squirrels (*Glaucomys* spp.), and chipmunks (*Tamias striatus*), as well as woodrats, mice, ducks, woodpeckers, and other birds, such as bluejays (*Cyanocitta cristata*) and wild turkeys (*Meleagris gallopavo*) (see Chapters 9, 16, and 17).

White-tailed deer are a valued economic, cultural, and ecological resource—regardless of how *value* may be defined (Langenau et al. 1984). Despite the many positive aspects of deer, high-density deer populations in oak-dominated forests cause significant regeneration failure, reduced plant species richness and biomass, and loss of structural diversity through shifts in species composition away from preferred, browse-sensitive species. This may not necessarily be the case on rangelands or other western systems. Nonetheless, reconciling the positive and negative aspects of ungulates, especially white-tailed deer, is a necessary facet of most resource management programs.

Although in certain respects the ecological interactions of deer and acorns appear to be fairly straightforward, in reality they are complex (see Chapter 13). Multiple-use management goals also are usually complex, in that they often involve incompatible, mutually exclusive options that necessitate prioritization of competing objectives. As resource agencies such as the United States Forest Service move toward the concept of ecosystem management (U.S. Forest Service 1992), managers and biologists must set priorities. This is a difficult process, and by necessity it is always done within the context of an ever-shifting sociopolitical matrix of changing policy initiatives, legal challenges, and appeals by environmental advocacy groups for increased ecosystem protection and restoration (Grumbine 1992, 1994). Regardless, an essential part of any successful management plan must include white-tailed deer densities, in balance with both cultural and ecological carrying capacities. As we have seen, the direct and indirect effects of too many deer cascade through and change entire ecosystems, with far-reaching and potentially long-term consequences.

ACKNOWLEDGMENTS

Many thanks to D. Gibson, J. Roseberry, J. Zaczek, and the Mammal Group of Southern Illinois University (V. Barko, N. Bekiares, M. Kanekawa, A. Morzillo, C. Weickert, and J. Whittaker), T. Van Deelen of the Illinois Natural History Survey, and W. M. Knox, Virginia Department of Game and Inland Fisheries, for their critical reviews and valuable suggestions on a draft of the manuscript.

Chapter 15

Oak Trees, Acorns, and Bears

MICHAEL R. VAUGHAN

Among the three species of bears native to North America (polar bears [*Ursus maritimus*], brown bears [*U. arctos*], and black bears [*U. americanus*]), only black bears are closely tied to hardwood forests, and then only in parts of their range. In the past, black bears were distributed throughout North America, but their current distribution is significantly reduced (Figure 15.1). In the southeastern United States, in particular, black bears are limited to mountainous or coastal areas, and they are conspicuously absent from the Piedmont region. Oak distribution in the eastern United States closely matches black bear distribution (see Figures 2.1 and 2.2, this volume), suggesting a link between oak forests and black bear populations but not answering the question "Why are oak forests important to black bears?"

Almost every aspect of bear ecology is tied to diet and nutrition, including reproduction, survival, seasonal movements, growth, and harvest. This chapter discusses how bears use oak forests and why oak forests are important to bears. In particular, it focuses on oak mast (acorns) as an important food resource for bears and on oak trees as winter shelter for bears.

BLACK BEARS IN THE OAK FOREST

While black bears can and do live in a variety of habitats, they are uniquely adapted to a forest existence. Their sharp, recurved claws make them excellent climbers, and they are often seen feeding high in the tops of oak trees (Pelton 1989). Throughout much of their range, adult females den in trees and give birth to their cubs in tree cavities high off

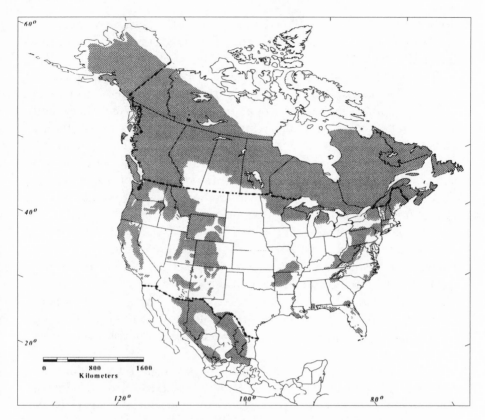

FIGURE 15.1.   Current distribution of the American black bear (*Ursus americanus*) in North America.

the ground. Bears make day beds in trees or at the base of large trees (Pelton 1989). Trees are escape cover, resting places, and playgrounds for cubs.

Although their range and distribution is not completely coincidental with the distribution of oak species in North America, black bears tend to have better success in hardwood forests dominated by oaks. For example, black bears, in general, have higher reproductive rates in the eastern deciduous forests than in the western coniferous forests (Bunnell and Tait 1981). They nearly disappeared from the Piedmont region of the southeastern United States when hardwood forests were converted to agriculture. They tend to be more abundant and exist at higher densities in hardwood forests than in any other habitat in which they are found (Bunnell and Tait 1981, Powell et al. 1997). Thus, they appear to be strongly tied to forests where their daily needs for food and cover can be met.

The food habits of black bears have been examined throughout their range. Most data on food habits come from analysis of fecal droppings collected in the wild, but occasionally stomach contents from harvested or road-killed bears are available for studies. The comparison of forage data from different studies is difficult, because often different measures are used to describe relative importance. In addition, because a food's abundance may vary annually, short-term studies may not give a true picture of food habits. Despite these difficulties, it is apparent that acorns are an important fall food for black bears (Table 15.1).

Bears are omnivorous and they are opportunistic feeders. Their diets vary seasonally and usually reflect the availability of preferred foods. However, fall is the most critical period for bears, because they have to prepare for winter dormancy and, in the case of adult females, for reproduction. Thus, fall diets must provide enough energy, in the form of fat reserves, to sustain bears for a period of 3 to 4 months during which they neither eat nor drink, but do reproduce and nourish their young (Powell et al. 1997, p. 69).

In the fall, bears go into a stage of hyperphagia (over eating), increasing carbohydrate and fat assimilation and decreasing protein intake (Pelton 1989). Since acorns provide an excellent source of fat (Chapter 11 and Wainio and Forbes 1941), and fat is bears' only source of metabolic energy during denning (Nelson et al. 1983), it stands to reason that acorns would be prevalent in the fall diet in regions of the country where oak forests and bears coincide.

Acorns are not particularly important to bears in the northwestern United States, but they are an important part of the fall diets of black bears in the southwestern, northeastern, and in particular, the southeastern United States (Table 15.1). In each of these regions the proportion of acorns in the fall diet depends in large part on acorn production; if acorns are available, they are usually the number one preference of bears. During a study in California, for instance, acorns were as little as 13% volume of the diet in a year when acorn production was low but 76% volume when acorns were abundant (Greenfell and Brody 1983). Similarly, in Pennsylvania, percent volume of acorns in the diet ranged from 24% to 66% depending on acorn productivity (Bennett et al. 1943).

In the southeastern United States, particularly in the Appalachian region, acorns always seem to be important in the fall diet of black

Table 15.1
Place of acorns in fall diet of black bears in regions of North America

	Most important fall food	Percentage acorns[a]	Source
Northeast			
Maine	Beechnuts	—	McLaughlin et al. 1994
Minnesota	Acorns	—	Rogers 1987
Pennsylvania	Acorns	23.6–66.1	Bennett et al. 1943
Vermont	Apples	2.9	Willey 1978
Southeast			
Arkansas	Acorns	34.1	Clark et al. 1987
Florida	Acorns	>50	Harlow 1961
	Saw palmetto (*Serenoa repens*)	29*	Maehr and Brady 1984
North Carolina (coastal)	Corn	17	Landers et al. 1979
Southern Appalachians (NC, GA, TN)	Acorns	48–59	Carlock et al. 1983
Tennessee	Acorns	30	Beeman and Pelton 1980
Virginia (coastal)	Inkberries (*Ilex glabra*)	9*	Hellgren and Vaughan 1988
Virginia[b]	Acorns	79*	Garner 1986
Virginia[c]	Grapes	8*	Kasbohm et al. 1995
Virginia[d]	Acorns	27*	Schrage 1994
Virginia, West Virginia	Acorns	52	Cottam et al. 1939
Northwest			
Alberta	Crowberries (*Empetrum nigrum*)	0	Raine and Kansas 1990
Montana	*Vaccinium* spp.	0	Tisch 1961
Washington	Huckleberries, apples	0	Poelker and Hartwell 1973
Wyoming	Buffalo berries (*Shepherdia canadensis*)	0	Irwin and Hammond 1985
Southwest			
California (Yosemite NP)	Apples, pears	11	Graber and White 1983
California (Tahoe NF)	Manzanitas	13–76	Greenfell and Brody 1983
California (Sequoia NP)	Acorns	55–57	Goldsmith et al. 1981
Texas (Big Bend NP)	Acorns	71	Hellgren 1993
Mexico	Acorns	34–76	Doan-Crider, unpublished data

[a]Percentage by volume unless followed by asterisk; percentage by frequency when followed by asterisk.
[b]In Shenandoah National Park (SNP) prior to a gypsy moth outbreak.
[c]In SNP during a gypsy moth outbreak that caused defoliation and a complete acorn crop failure.
[d]In SNP following a gypsy moth outbreak.

bears (e.g., Clark et al. 1987, Carlock et al. 1983). In the southern Appalachians, acorns compose a substantial proportion of the annual diet. In
coastal areas or wetland areas of the Southeast acorns are less important
but still show up in the diet (Landers et al. 1979, Hellgren and Vaughan
1988, Maehr and Brady 1984; Table 15.1).

Even in years of maximum acorn production only 70–90% of individual trees produce acorns (Chapter 10). Over a 12-year period in the
southern Appalachians, production by all oaks combined averaged 58.1
± 64.3 kg/ha (range 0.52–203.4 kg/ha); production peaks by white
oak (*Quercus alba*) averaged 4 years apart, while those for northern red
oak (*Q. rubra*) averaged 5 years apart (Beck 1977). Acorns provide most
of the calories available to bears, and during 1995, when northern red
oaks produced a bumper crop and white oaks failed, northern red oaks
produced 67.7% of all plant food calories available to bears and white
oaks contributed 5.1% (Inman and Pelton, in press). In addition, 59.3%
of all plant food calories available to bears were produced in the fall.

A good example of the importance of acorns to bears and of the flexibility of bears' diets when acorns are not available can be seen in the
food habits of bears over a 10-year period in Shenandoah National Park,
Virginia. Bear research in the park (1982–1992), coincided with the arrival of gypsy moths (*Lymantria dispar*). Gypsy moths infested the park
from 1987 through 1989, causing severe defoliation of hardwood trees
and a complete acorn failure among all species of oaks (Kasbohm 1994,
Kasbohm et al. 1995, McConnell 1988). Prior to the gypsy moth infestation, acorns were present in 79% of fall bear scats (Garner 1986). However, during the infestation period acorns were present in only 8% of
scats, and bears switched to pokeweed (*Canopholis americana*) and wild
grapes (*Vitis* spp.) as their main fall food (Kasbohm et al. 1995). Following infestation (1990–1992), when mortality of oak trees ranged from
1% to 48% of individual stands (Schrage and Vaughan 1998), acorns
were present in 27% of fall scats. Content analysis of fall diets before and
during defoliation indicated no decline in dietary nutritional quality. It
is unlikely, however, that bears were able to obtain the same quantity of
soft mast (pokeweed and grapes) in defoliation years as they were of
acorns in nondefoliation years.

It is interesting to note that the second most important food to bears
in the southern Appalachians (Powell et al. 1997, Inman and Pelton, in
press) is squaw root (*Conopholis americana*), a parasitic plant that grows
on the roots of oak trees in spring and early summer. Thus, oak trees pro-

vide, directly and indirectly, the two most important foods to bears residing in oak forest of the southern Appalachians.

Impacts of Acorns and Oaks on Bear Populations

In the southern Appalachians, hard mast, particularly red oak and white
oak acorns, is the driving force in the population dynamics of black bears
(Pelton 1989). This conclusion could be generalized to include all areas
where bear habitat is dominated by oak forests or where oak is an important component of the forest. In the fall, when bears must prepare
for the coming winter, for denning, and for producing young, they make
physiological, behavioral, and ecological adjustments to their fall food
source (Pelton 1989); and these adjustments affect every aspect of their
existence, including movements, home range, growth and condition,
vulnerability to harvest, survival, and reproduction. Thus, an acorn crop
failure or a bumper crop in a given year can alter the distribution and
abundance of bear populations on a local or regional scale.

Fall Movements

During fall, black bears depart from their normal crepuscular feeding
pattern and begin almost continuous foraging (Garshelis and Pelton
1980). In order to acquire the food necessary to accumulate the fat reserves to survive the winter denning period, bears must locate reliable
food sources, which may require extensive movements. The magnitude
of these seasonal shifts is dependent upon fall acorn production. In good
years, bears may simply move within their normal home range to elevations or aspects that produce more hard mast (Garshelis and Pelton
1981), but in poor years they may leave their home ranges entirely and
travel great distances to find areas of high food abundance (Rogers
1987).

Pelton (1989) described four types of movements or adjustments
bears in Tennessee made in response to spotty mast production: (1) general long-range movements, (2) intensive use of small areas resulting in
significant home range overlap, (3) accommodation of a prime acorn
site within the normal annual home range, and (4) departure from traditional spring and summer ranges to pockets of concentrated and abun-

dant sources of acorns in fall. In years of poor acorn production, bears moved two to four times farther than in good production years. This also was the case for bears studied in Virginia, where the movement from summer range to fall range was two to three times as great (1.7 km versus 6.7 km for solitary females and 1.2 km versus 3.4 km for females with cubs) during years of gypsy moth induced acorn crop failures as it was prior to the gypsy moth infestation (Kasbohm et al. 1998). Black bears in Minnesota traveled farther and became more attracted to human-related food sources when fall acorn crops failed (Noyce and Garshelis 1997).

While many others have reported long-range movements of bears in response to fall acorn crop failures (Jonkel and Cowan 1971, Reynolds and Beecham 1980, Garshelis and Pelton 1981, Novick and Stewart 1982, Carlock et al. 1983, Powell et al. 1997), some of the most extensive movements reported for bears in search of fall food (primarily acorns) came out of Minnesota (Rogers 1987). Rogers reported several movements of bears in excess of 90 km; 1 male moved 201 km to an area of high acorn production. It is unclear how bears find these distant food sources, but Rogers (1987) suggested they may travel to these areas as cubs, and return as adults, thus passing on this knowledge from generation to generation.

Fall movements by bears in search of food do not always result in success. Of 59 marked bears residing in gypsy-moth defoliated areas of Shenandoah National Park, 35 did not move, and only 14 of 24 that did move found acorns (Kasbohm et al. 1998). These kinds of long fall movements into marginal areas probably result in increased mortality and reduced reproduction (Carlock et al. 1983).

Home Range

In general, animal home range size varies inversely with food production (Schoener 1983), and bears are no exception. For instance, bears in Virginia increased the size of their fall ranges twofold in areas of gypsy moth–induced acorn failure (Kasbohm et al. 1998). But, home range size also may depend on the complexity of the diet as was the finding in a study in Arkansas, where fall home ranges of bears feeding on acorns were smaller than summer ranges (27 vs. 97 km^2) when bears fed on a variety of foods (Smith and Pelton 1994).

Fall home ranges of female bears in the Pisgah National Forest, North

Carolina, were found to be small when fall production of acorns was great and large when production was small (Powell et al. 1997). Powell et al. (1997, pp. 96–105) tested hypotheses concerning the relationship between three important food groups of bears (squaw root, berries, and acorns) and home range size. Acorn production explained 76% of the variation in fall home range size of females and 49% for males. Neither of the other food groups affected fall home ranges. Powell et al. also found a close relationship between fall acorn production and annual home range size, highlighting the importance of fall production of this fruit to the annual movement patterns of bears.

Habitat quality in the fall depends, in part, on the abundance of oaks at mast-producing age (Powell et al. 1997), thus a bear's fall home range tends to include a high proportion of mast-producing oak trees. Garshelis (1978), for instance, documented that fall home ranges of bears in Tennessee had significantly more oak trees than their spring and summer ranges. Garner (1986) noted the same for bears in Virginia. Bears are adapted to long periods of fasting, but their ability to locate these pockets of food abundance in fall is critical if they are to accumulate fat reserves so they can survive the winter.

Survival

Mortality rates of adults and of 1- and 2-year-old bears tend to be highest in fall, due to harvest, or in winter, due to starvation (Bunnell and Tait 1981). Cubs suffer highest mortality soon after birth (Bunnell and Tait 1981, Higgins 1997). The effect of fall mast production on cub survival is unclear. There is evidence to support a positive relationship between cub survival and fall hard mast production (Pelton 1989) and evidence to suggest that no relationship exists (Elowe and Dodge 1989). In Tennessee, cubs and yearlings suffered 80% mortality in years of poor acorn production (Pelton 1989), and females lost entire litters only in years following poor acorn production (Eiler et al. 1989). Only 2 of 16 cubs died following good years, but 7 of 8 died following poor crop years (Eiler et al. 1989), leaving Eiler to conclude that "yearly variation in cub mortality was probably related to differences in mast yields." In Minnesota, cub survival was 88% when food was abundant in the year of conception and the year of birth, but only 59% when food was poor in both years (Rogers 1987). Others studies in Tennessee (Wathen 1983) and Minnesota (Rogers 1976) found positive relationships between cub survival and fall mast production the previous year. In the latter study, cubs weighing less than 1.8 kg in late March (n = 15) had 4 times the mortality rate prior to family breakup (67% died) as heavier cubs (n = 47).

Conversely, cub survival in Massachusetts (Elowe and Dodge 1989) and Arizona (LeCount 1982) was not related to fall mast production; high and low cub survival occurred under similar forage conditions. Data collected prior to, during, and following a gypsy moth infestation in Shenandoah National Park, seemingly support these findings. Cub survival rates, and survival rates of all age classes, were unchanged following massive defoliation by gypsy moths and complete loss of the acorn crop. Prior to defoliation (Carney 1985) and following defoliation (Schrage and Vaughan 1998), acorn crops had been good. However, despite the loss of the acorn crop during years of defoliation, bears found alternative food sources and their condition remained good (Kasbohm et al. 1995).

Our work at Virginia Tech with captive bears indicates that fall nutrition for adult females does affect survival of their cubs (Vaughan, unpublished data). In food trials with some bears on ad lib diets and others on maintenance diets, all bears produced cubs, but those on maintenance diets did not have the energy to lactate, and their cubs died.

Yearling bears, on their own for the first time, likely are more vulnerable to starvation than other age classes, but not much information is available on survival of yearling bears. Yearling bears in Minnesota weighing less than 10 kg at the end of denning had higher mortality rates than those weighing more than 10 kg (Rogers 1983, 1987). Only 1 of 25 weighing less than 10 kg, but 53 of 73 weighing more than 10 kg survived. In this study, acorns were a component of the fall diet, but not the predominant food.

Most adult bears die from hunting or removal because of nuisance activity (Bunnell and Tait 1981), and there appears to be a strong relationship between fall hard mast production and both nuisance activity and harvest rates (Kasbohm et al. 1994). In Wisconsin (unpublished reports for 1954–1969 cited in Rogers 1976), failure of blueberry and red oak crops were correlated with increased damage by bears. The number of bears killed on depredation complaints was greater than 100 only when acorn and berry production fell below 20–25% of normal production. In Tennessee (Pelton 1989), game officers found 21 bears dead following the 1984 oak mast crop failure; an additional 20 had been shot after wandering onto farms and into subdivisions.

Fall food supply appears to affect the vulnerability of bears to hunting (Gilbert et al. 1978, Alt 1980, Pelton et al. 1986, Litvaitis and Kane 1994). In Pennsylvania (Alt 1980), hunting success was high when acorns were abundant, likely because the abundant food supply kept bears from entering dens until later in the fall. In New Hampshire (Kane 1989) and Massachusetts (McDonald et al. 1994), hunting success went up when fall mast (primarily acorns and beechnuts) was low. In both New Hampshire and Massachusetts, baiting was legal, thus hunters were successful in attracting bears to bait when natural food was scarce.

In Virginia, archery hunters were more successful in years when acorn production was low (Martin and Steffen 1999). During 1973–1998, the four worst years for hard mast production coincided with the four most successful years for archers. During those four years, archers accounted for 26–36% of the harvest; in all other years they never exceeded 17% of the harvest.

Noyce and Garshelis (1997) examined the relationship between natural food abundance and the bear harvest in Minnesota. They found that the percent of females in the harvest, the mean age of females killed, and hunting success were inversely related to fall production of hazelnuts (*Corylus* spp.) and acorns (Figure 15.2). Hunter success ranged from 26% to 43% and was highest when the food index (hazelnuts and acorns) was lowest ($r^2 = 0.66$; $P = 0.0013$). The percentage of females in the harvest was independent of population size, the percentage of females in the population ratio, and the number of hunters, but was in-

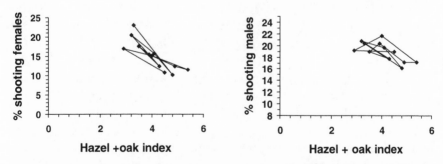

FIGURE 15.2. Percentage of Minnesota hunters shooting female and male black bears (number killed/number hunters) versus annual hazel + oak index of relative productivity of the trees. Each square represents one year of data during 1984–95. (Reprinted from Noyce and Garshelis 1997 by permission of The Wildlife Society.)

versely related to food abundance ($r^2 = 0.62$; $P = 0.002$). Both of these relationships seem to indicate that bears, particularly female bears, delayed den entry in years of poor food abundance (pregnant females are normally the first to enter dens; Kasbohm et al. 1996b). The relationship between harvest rates and food abundance was strongest for adults, less for juveniles, and not significant for yearlings, thus the average age of females in the harvest increased with decreasing food.

All of the cases cited above are clear evidence of the importance of fall production of hard mast on all age classes of bears. The relationship can be direct (starvation) or indirect (e.g., vulnerability to hunting), but the impact on the growth rate and overall health of the population can be tremendous.

Reproduction

Numerous studies indicate that every aspect of reproduction in bears, including proportion of females breeding, age of first reproduction, litter size, and interbirth interval, is affected by the nutritional condition of females in fall (Rogers 1976, Beeman and Pelton 1980, Willey 1978, Beecham 1980, Bunnell and Tait 1981, Young and Ruff 1982, Eagle and Pelton 1983, Clark et al. 1987, Eiler et al. 1989, Elowe and Dodge 1989, Clark et al. 1994). The range for each of these variables is narrow, but the impact on population growth caused by these variations can be great. Among eastern bear populations, reproductive performance differs for areas that have high versus low acorn production. In northern Minnesota, age of first reproduction was 6.3 years, average interbirth interval was 2.3 years, and average litter size was 2.4, whereas in Pennsylvania 88% of females produced young by age 4, interbirth interval was 2.0 years, and average litter size was 3.0 (Rogers 1987).

The black bear population in the Great Smoky Mountains of Tennessee and North Carolina is typical of bear populations in the mid and southern Appalachian mountains, where acorns of white and red oak are the primary fall food (Pelton 1989). There, during a 10-year period, the difference in reproductive success following years of poor acorn production versus years of good acorn production was striking. In years following poor acorn production, reproductive success was only 26% and cub mortality was 87.5% (Eiler et al. 1989). In years following good acorn production reproductive success was 89% and cub survival was high. Nine of 10 skips in the normal two-year reproductive cycle followed years of poor acorn production, and only 1 of 30 females that were trapped the summer following poor acorn production was lactating (Eiler et al. 1989). In fact, there was a linear relationship between white oak acorn abundance and percentage of females lactating the following summer (Pozzanghera 1990). These results led Eiler et al. (1989) to conclude that "the relationship between oak mast availability and black bear reproduction and cub survival emphasizes the importance of maximizing mast production from the oak component of the southern Appalachian forests."

In another region of the Appalachian range, loss of the acorn crop did not result in reproductive failure, because bears found an alternative food source of grapes and pokeweed (Kasbohm et al. 1995). During three years of a gypsy moth infestation in Shenandoah National Park, which resulted in complete loss of the acorn crop, litter size, interbirth interval, age at first reproduction, and cub survival remained unchanged between years prior to infestation (Carney 1985) and years following infestation (Schrage and Vaughan 1998). No female skipped an opportunity to breed. We concluded from this that bears have high behavioral plasticity and are extremely flexible in their diet and that acorn production alone cannot completely explain fluctuations in bear survival and reproduction; total food production and the relative nutritional value of all food items must be considered (Kasbohm et al. 1996b).

The same kind of diet flexibility was evident in Minnesota bears that fed on beaked hazel (*C. rostrata*) and red oak (Rogers 1976, 1983). When these bears had access to garbage, a higher percentage of females produced larger litters at shorter intervals. This led to the blastocyst resorption hypothesis, which hypothesized that bears in poor nutritional condition would not implant blastocysts in the uterus and would skip a year of reproduction. My students and I tested the blastocyst resorption hypothesis in captive bears and found that even bears in poor nutritional condition implanted blastocysts and in most cases brought fetuses to term. However, their inability to lactate led to starvation of the newborns, which sows devoured leaving no evidence of reproduction (Vaughan unpublished data).

Since bears breed in the summer, but implantation of the fertilized eggs and growth of fetuses is delayed until late fall or early winter (Wimsatt 1963), fall nutrition has a tremendous impact on reproductive success of female, but not male, bears. However, the idea that fall nutrition the previous year affects the breeding potential of male bears has not been examined and is a potential field for investigation.

IMPACTS OF ACORNS ON DENNING

Denning evolved as an accommodation to inclement weather and food shortage (Lindzey and Meslow 1976, Johnson and Pelton 1980, Rogers 1981). Even southern bears den, despite year-round mild weather and presence of food (Hamilton and Marchinton 1980, Novick et al. 1981, LeCount 1983). Thus, the onset of denning is likely determined by a circannual rhythm modified by weather and fall food supply (Johnson and Pelton 1980). In areas where bears are coincidental with oak forests, oak trees are related to denning behavior two ways: (1) acorn production can influence the timing of denning, and (2) bears may select oak trees as den sites.

Timing of Denning

There is ample evidence that food abundance in fall affects the timing of denning for bears (Alt 1980, Johnson and Pelton 1980, Tietje and Ruff 1980, Beecham et al. 1983, O'Pezio et al. 1983). For example, a sample of bears in Maine that fed on beechnut (*Fagus grandifolia*) in the fall denned in October when beechnuts were scarce and in November when they were abundant (Schooley et al. 1994).

There are two schools of thought regarding the denning response to fall food abundance. The prevailing idea (supported by the sources just cited) is that bears delay denning in years of good mast production, to take advantage of the ample food supply, while an alternative idea suggests that bears reach a suitable denning weight sooner in food abundant years and den earlier under those circumstances (Matson 1946). In Ontario, bears that fed on acorns denned eight days later than those that did not have acorns on which to feed (Kolenosky and Strathearn 1987b). In the southern Appalachians, where the primary fall diet is acorns, bears denned significantly later in good acorn years (Johnson and Pelton 1980, Carney 1985). An exception was in Shenandoah National Park, where bears having home ranges in areas defoliated by gypsy moths (no acorns) denned later than bears with home ranges in areas not defoliated by gypsy moths. However, the relationship between fall food abundance and den entrance is not completely clear, because other factors, such as reproductive status, weather, and hunting seasons, may influence den entrance dates.

Den Sites

The degree to which bears den in hollow tree cavities varies throughout their range. Throughout the West and most of the Northeast, most bears den on the ground (Beecham et al. 1983, LeCount 1983, Kolenosky and Strathearn 1987b, Manville 1987, Schwartz et al. 1987); but in the southeastern region of the country, a high proportion of bears den in trees (Johnson and Pelton 1981, Godfrey et al. 2000). The regional differences in den preference may be related to den site availability, security for females with cubs, or energy savings. For instance, Johnson et al. (1978) constructed a simulation model which showed that aboveground tree dens saved > 40% energy over ground dens.

Where bears prefer to den in trees, the preference is for oak trees. In Arkansas, only 2 of 48 bears (4.2%) denned in trees, but both were in white oak trees (Hayes and Pelton 1994). In Tennessee, 49 of the 95 dens examined between 1973 and 1982 (56%) were in trees. Thirteen were in chestnut oaks (*Q. prinus*) with average diameter at breast height (dbh) of 91.5 cm, 10 were in northern red oaks with average dbh of 99.4 cm, and 1 was in a scarlet oak (*Q. coccinea*) with a dbh of 81.8 cm (Wathen et al. 1986).

Two studies in Virginia, one of a protected bear population, the other of a hunted population, demonstrate the importance of large oak trees to bears as denning sites. In Shenandoah National Park, where bears are protected from hunting and where gypsy moths defoliated trees, causing loss of the acorn crop and mortality of trees that had been heavily defoliated during two to three consecutive years, 84% of pregnant females denned in trees (Kasbohm et al. 1996a). Thirty-five of 52 (67%) tree dens were in red oaks, 6 (12%) were in chestnut oaks, and 4 (8%) were in white oaks; the 7 remaining dens were in other tree species. The average dbh of all tree dens was 96.8 cm. Interestingly, during the five years of this study, 54% of the initially alive oak trees used for dens died as a result of gypsy moth defoliation, which may have resulted in long-term alteration of denning behavior (Kasbohm et al. 1996a).

In a hunted population of bears in northwestern Virginia, 109 of 171 (72%) dens were in trees (Godfrey et al. 2000). Fifty-four of the 109 (50%) tree dens were in northern red oak trees, and 43 (39%) were in chestnut oaks; the average dbh of den trees was 90.9 cm. The percentage of bears denned in trees over the four-year study period ranged from 63% to 93%, a testimony to the importance of large oak trees as den sites for black bears.

We can only speculate as to why bears in the southeastern United States, particularly pregnant females, prefer to den in trees. While heat retention and availability of large hollow trees may be important (Johnson et al. 1978), security may be an overriding influence (Godfrey et al. 2000). Hunters, dogs, and other bears (Hellgren and Vaughan 1989) are potential threats to denned bears, but the degree to which any of these threats influences denning success has not been thoroughly examined. Other potential explanations (energy savings, psychological health, comfort, thermal stability, and flexibility in denning) are discussed by Pelton (1996).

Overall Impact and Management

As noted at the beginning of this chapter, oak trees and oak mast influence every aspect of a bear's life. Throughout much of their range, bears are intimately tied to and dependent upon this resource for their food and shelter needs. Loss of this resource can and does have long-term impacts on bear populations. Reproductive rates, survival rates, and ultimately population growth rates all may depend on the year-to-year productivity of acorns and the availability of oak trees as secure den sites. One or two consecutive years of acorn crop failure can affect a bear population for years to come, as the following case illustrates.

In 1984, Great Smoky Mountains National Park, Tennessee, experienced an acorn crop failure. As a result, cub recruitment in 1986 was almost nil (Pelton 1989). This led to synchronized breeding in 1986 (i.e., by females that had failed to reproduce in 1985 plus females that had yearlings in 1985), and a high cub recruitment rate in 1986. By 1989 this bumper crop of young bears dispersed throughout the park and the number of panhandler bears and nuisance complaints rose sharply, as did the harvest rate around the park. This scenario likely played out for several more years until alternate-year breeding returned, with about 50% of adult females producing every year, or until another crop failure caused another spike in the population. In either case, this example illustrates the need for long-term studies of how yearly fluctuations in acorn production drive the dynamics of bear populations.

Because oak trees and oak mast are so important to black bears, it is critical that forest be properly managed to insure a continual supply of mature, mast-producing oak trees, along with buffer species, such as wild grapes, to reduce the severity of periodic acorn crop failures (Eiler et al. 1989). Management practices should be specific to the area or particular habitat conditions. For the southern Appalachians, Pelton (1989) recommended long cutting rotations, to take advantage of the most productive years for oak trees, and careful selection of the configuration and oak regeneration potential for proposed clearcuts. He also noted the importance of maintaining mixed oak-pine forest types rather than converting to pure pine stands, and research to determine the usefulness of shelterwood, seed tree, and group selection methods in maintaining the stability of the oak mast potential of an area. In forested wetlands of the Mississippi Alluvial Valley, where availability of dry, secure den sites is a concern, maintenance of large-cavity trees of overcup oak (*Q. lyrata*) and bald cypress (*Taxodium distichum*) is important (White et al. 1996).

In Louisiana, the Black Bear Conservation Committee (BBCC), an organization formed to manage and recover the endangered Louisiana black bear (*U. a. luteolus*), published a black bear management handbook and a restoration plan specific to that area (Black Bear Conservation Committee 1996, 1997). Their management recommendations for bottomland hardwoods include:

—Maintenance of large tracts of bottomland hardwood forest with a mixture of tree species
—Uneven-aged management through group selection and small patch harvest cuts, which research has shown is best for oak regeneration
—Maximization of tree vigor for best hard mast production
—70–100 year rotation (minimum 50 years) for best hard mast production
—Intermediate cuts to improve tree species composition, remove poor quality trees, and promote oak regeneration
—Identification and maintenance of large-cavity trees with minimum 91 cm dbh
—Bear-friendly harvest strategies, such as small clear cuts, maintenance of travel corridors, maintenance of clumps of hard mast producing trees
—Maintenance of a diverse mixture of mast species to reduce annual fluctuations in mast abundance
—Stand improvement for high-graded stands (from which all big trees have been removed) and other poor-quality stands on sites capable of substantially greater mast production

CONCLUSION

Bears are dependent on oak trees and their energy-rich fruits throughout much of their range. Adult bears feed in and around oak trees, den in oak trees and give birth to their young there, rest in day beds at the base of oak trees, and seek shelter from hunters and other threats in oak forests. Cubs and yearlings play in oak trees and feed on their fruits. Bears in the southern Appalachians have adapted well to the change in forest composition from domination by American chestnut to oak domination, and maturation of oak forests over the past 50 years likely has resulted in improved conditions for bears and increased numbers of bears (M. R. Pelton, University of Tennessee, personal communication). Continued success of bear populations, particularly in the southeastern United States, is dependent upon wise management and conservation of oak forests.

Chapter 16

Turkeys, Acorns, and Oaks

DAVID E. STEFFEN, NELSON W. LAFON,

AND GARY W. NORMAN

The wild turkey (*Meleagris gallopavo*) is uniquely adapted to a wide variety of habitats and landscapes throughout North America. The present distribution of five different subspecies includes locations from southern Canada, northern Mexico, and all states except Alaska. Primarily associated with forested habitats in the eastern half of the United States, the eastern subspecies (*M. g. silvestris*) is the most widely distributed wild turkey. The Florida wild turkey (*M. g. osceola*) is confined to peninsular Florida. The Rio Grande turkey (*M. g. intermedia*) is found in the more open habitats of Kansas, Oklahoma, and Texas. These open habitats generally include limited tree cover or cultivated land. The Merriam's turkey (*M. g. merriami*) is found primarily in ponderosa pine–oak forest habitats at higher elevations of the western mountains. Gould's turkeys (*M. g. mexicana*) occupy habitats associated with oak-pine forests in northwestern Mexico, southern Arizona, and southern New Mexico. Restoration efforts involving the eastern, Merriam's, and Rio Grande subspecies (and hybrids of these subspecies) have established many populations outside the historic ranges, primarily in the far western states.

A flexible diet has helped the wild turkey adapt to many habitats. Wild turkeys are opportunistic and omnivorous feeders, consuming a widely diverse diet that reflects the availability of many plant and animal foods (Hurst 1992). Despite an ability to consume many food types, acorns and other hard mast generally are considered important food items for turkeys in all seasons throughout their range. When available, acorns may be the primary food consumed during fall, winter, and spring (Mosby and Handley 1943, Wheeler 1948, Korschgen 1967, Schroeder 1985, Hurst 1992).

ACORNS IN THE DIET

The literature is rich with documentation of turkey food habits. Food habit studies have been thoroughly summarized by Schemnitz (1956), Schorger (1966), Korschgen (1967), and Hurst (1992). The studies in these reviews are based on field observations and foods identified from crops, stomachs, gizzards, and droppings (Hurst 1992). Data are usually reported as frequency of occurrence (e.g., percent occurrence in droppings), percent composition (by volume or weight), or qualitative criteria.

Eastern Turkey

Based on percent of the diet, acorn use for the eastern subspecies is generally higher than that for other subspecies (Table 16.1). On average, percent acorn composition of the diet of eastern birds is highest in the winter (33.2%), followed by spring (20.5%) and fall (20.4%). Average summer use of acorns represent only 1.2% of the eastern diet. Acorns represent the highest volume of food in the eastern turkey diet during the winter and spring (Hurst 1992).

A large variation in acorn use is found among years, seasons, and studies. Depending on the particular study, acorn composition of the eastern wild turkey's diet may be nearly nonexistent at anytime during the year. However, average seasonal acorn composition has been shown to be as high as 62.8%, 11.4%, and 56.9% during the spring, summer, and fall, respectively, in Pennsylvania (Kozicky 1942), and 68.6% during the winter in Michigan (Korschgen 1967). Acorns accounted for 73.3% of the January foods for Missouri turkeys (Korschgen 1967).

Florida Turkey

Studies indicate that acorn composition of the diet for the Florida subspecies is also high during the fall months (29.3%) (Table 16.1). However, the few quantifiable studies for the Florida subspecies were conducted almost exclusively during the fall period. Other work suggests that the Florida subspecies relies on acorns as heavily during the winter months as it does in the fall (Korschgen 1967, Williams 1992). Acorns are also eaten during spring and summer but form a less important share of the diet then (Barwick et al. 1973, Williams 1992).

Table 16.1
Mean seasonal acorn composition of wild turkey diets by subspecies

Subspecies	Acorn composition of diet (%)				Sources
	Spring	*Summer*	*Fall*	*Winter*	
Eastern					
Mean	20.5	1.2	20.4	33.2	Bailey et al. 1951, Baughman and Guynn 1993, Bennett and English 1941, Christisen and Korshgen 1955, Culbertson 1948, Exum et al. 1987, Good and Webb 1940, Hayden 1979, Kennamer and Arner 1967, Korschgen 1973, Kozicky 1942, Lewis 1962, Martin et al. 1939, Moore 1964, Mosby and Handley 1943, Rivers 1940, Tabatabai and Kennedy 1984, Webb 1941
Number of studies	12	8	14	13	
Data years	30+[a]	19	43	38	
Merriam's					
Mean	6.4	1.1	12.8	6.4	Burget 1957, Hoffman 1962, Laudenslager and Flake 1987, Mackey and Jonas 1982, Murie 1946, Peterson and Richardson 1973, Reeves and Swank 1955, Schemnitz et al. 1985, Scott and Boeker 1973, Smith and Browning 1967, Wakeling and Rogers 1995
Number of studies	5	4	13	6	
Data years[b]	14+	9+	30+	13+	
Rio Grande					
Mean	0.0	0.2	3.5	0.0	Korschgen 1967, Litton 1977, Quinton and Montei 1977
Number of studies	3	2	3	2	
Data years	5	3	5	3	
Gould's					
Mean	0.1	2.5	1.0	3.0	Lafon and Schemnitz 1995, Potter et al. 1985
Number of studies	2	2	1	1	
Data years	2	2	1	1	
Florida					
Mean	—	—	29.3[c]	—	Barwick et al. 1973, Powell 1967, Schemnitz 1956
Number of studies	0	0	3	0	
Data years	0	0	11	0	

Notes: Means from individual studies were weighted by the data years in each study. Seasonal boundaries are: spring, April–June; summer, July–September; fall, October–December; winter, January–March.

[a]Because of a small sample of birds (n = 16), 12 springs from Korschgen (1973) were weighted as 1 spring.
[b]Because of a small sample of birds (n = 29), 6 years of data from Scott and Boeker (1973) were weighted as 1 year.
[c]Generally an aggregate of fall and winter samples.

Merriam's Turkey

Of the western wild turkey subspecies, only the food habits of the Merriam's turkey have been well documented (Table 16.1). Acorn composition of the diet is highest in the fall (12.8%) and lowest in the summer (1.1%). Spring and fall acorn composition of the diet is 6%. Hoffman (1962) considered acorns least important during the spring in Colorado. Although less prevalent when compared with the diet of the eastern subspecies, acorns are still considered a key component of the turkey diet across much of the Merriam's range (Ligon 1946, Hoffman 1962, Korschgen 1967, Smith and Browning 1967, Mackey and Jonas 1982, Schemnitz et al. 1985, Wakeling and Rogers 1995). As with the eastern subspecies, there is a wide range in acorn use by Merriam's turkeys among seasons and areas.

Rio Grande and Gould's Turkeys

Because studies of the subject are few, the full spectrum of acorn use throughout the year may not be adequately represented for the Rio Grande and Gould's subspecies. Relative to the other subspecies, diet composition of acorns is much less for both Rio Grande and Gould's turkeys (Table 16.1). Although little acorn utilization was observed in an area with only infrequent live oaks (*Quercus virginiana*), Litton (1977) noted that acorns played an important role in the Rio Grande diet in other areas of Texas. Other researchers also regard acorns as an important component of the Rio Grande diet (Walker 1941, Beck and Beck 1955, Korschgen 1967, Beasom and Wilson 1992). Although a poor acorn production year in New Mexico resulted in low utilization by Gould's turkeys (Potter et al. 1985), Leopold (1959) suggested that acorns might be the single most important food item in the fall and winter.

FACTORS AFFECTING ACORN USE

The dietary proportion of acorns is influenced by a combination of factors, including acorn availability, food preference, and habitat structure. Acorn availability is probably the most influential factor in determining use. Availability is primarily related to oak composition of the habitat, acorn production, snow depth, edibility, and acorn persistence.

Turkey Age and Sex

Regardless of age, there is little difference between the diets of male and female turkeys (Korschgen 1973, Beasom and Wilson 1992). Juveniles (July–September) have a diet similar to that of adults (Hurst 1992). Healy (1992a) observed that young birds ate acorns as soon as they were available in the fall, and by mid-September poults could even swallow large northern red oak (*Q. rubra*) and chestnut oak (*Q. prinus*) acorns. Acorns were part of the summer diet for poults (2–24 weeks old) in Florida (Barwick et al. 1973).

Acorn Availability

Acorn availability inevitably depends in part on the species of oak and their abundance in the habitat. As would be expected, low acorn use was found on South Carolina pine plantations (Baughman and Gyunn 1993) and in a Texas habitat with few oaks (Litton 1977). Even where oaks are common, acorn composition of the diet may be low, because acorn production is often poor, making acorns an unpredictable food source (Dickson 1990, Hurst 1992, Porter 1992). As opportunistic feeders, turkeys readily adapt to the changing and uncertain food conditions that result from periodic mast failures.

The persistence of acorns following seedfall is influenced by the size of the seed crop and the rate of consumption by wildlife. In northern ranges, deep snows will restrict the ability of turkeys to locate and exploit food (Porter et al. 1980). Although turkeys can scratch through 30 centimeters of snow, depths greater than 10 centimeters greatly inhibit foraging and encourage turkeys to seek alternative foods or to feed in agricultural areas (Vander Haegen et al. 1989, Healy 1992a). Heavy infestations of weevil larvae also limit the edibility and availability of acorns (Schorger 1966).

Many other species of wildlife also consume acorns and potentially reduce acorn availability to turkeys (Good and Webb 1940, Bailey et al. 1951, Latham 1956, Schemnitz 1956, Bailey and Rinell 1967, Hurst 1992). In the past, acorn shortages in some winters were attributed to consumption by passenger pigeons (*Ectopistes migratorius*) (Schorger 1966). In years of average or poor acorn production, most of the acorn crop is consumed by the end of seedfall; acorns persist into the following spring and summer only after bumper crops (Goodrum et al. 1971). Korschgen (1973) observed a general association between spring consumption of acorns and fall production in Missouri for both the white oak group and black and red oak groups. We conducted an additional analysis of Korschgen's (1973) data (n = 4 years) and found a perfect Spearman's correlation between percentage of use in the spring and production from the previous fall (r_s = 1.00, $P < 0.001$). In contrast, Good and Webb (1940) found very limited acorn supplies during the spring in Alabama, even after heavy mast crops.

Preference for Acorns

When available, acorns of all oak species are readily utilized by turkeys throughout their range. Oak mast appears to be preferred by wild turkeys over most other naturally occurring foods. In contrast to deer and other mammals (see Chapters 11, 12, and 14) turkeys do not prefer acorns of the white oak group over those of the red oak group. The only rigorous evaluation of preferences among oak species was conducted by Minser et al. (1995). In trials among 12 oak species, Minser et al. (1995) showed that swamp white oak (*Q. bicolor*) (a relatively large acorn) was least preferred among the oaks and had a preference ranking comparable to pecans (*Carya illinoensis*). Most species of oak mast were preferred over pecans, hickory (*C. tomentosa*), walnuts (*Juglans nigra*), and Chinese chestnuts (*Castanea mollissima*) (Minser et al. 1995).

Color and shape might play a more important role in food selection for turkeys than taste or smell (Pelham and Dickson 1992). Birds can probably distinguish between the 4 primary tastes of salt, sweet, acid, and bitter (Portmann 1961), but compared to mammals they have a relatively poor sense of taste (Pelham and Dickson 1992). Minser et al. (1995) concluded that nut size was an important factor in food preference by turkeys. Species with larger nuts (including swamp white oak) were not utilized as readily as oak species that produced smaller acorns. Dalke et al. (1942) speculated that the small acorns of post oak and blackjack oak were especially appealing to turkeys. On the other hand, Wakeling and Rogers (1995) hypothesized that preference by Merriam's turkeys for acorns, over the much smaller seed of ponderosa and pinyon pine, was consistent with predictions from optimal foraging theory.

While nut size influences mast preferences, turkeys readily substitute less preferred foods in their diet to conform to the most available food sources. Row crops only become important in the turkey diet when mast production is limited (Lewis and Kurzejeski 1984, Vander Haegen et al. 1989, Kurzejeski and Lewis 1990, Lewis 1992). When winter mast is abundant, corn and other crops are seldom used (Schemnitz 1956, Lewis and Kurzejeski 1984). However, feeding trials in Tennessee showed that corn, northern red oak acorns, and white oak acorns were all highly and equally preferred by turkeys (Minser et al. 1995).

Habitat Structure

The features of habitat structure associated with food resources may be compelling factors in turkeys' final selection of habitat and food. Predator avoidance may be as important as finding food in influencing where turkeys forage (Williams 1992). If row crops (e.g., corn) are equally preferred to acorns (Minser et al. 1995), the ultimate food selection criterion for turkeys may be linked to the security of forested cover. This may help explain why turkeys prefer to remain in wooded areas when acorns are abundant and why row crops become important only during poor mast years (Kurzejeski and Lewis 1990, Lewis 1992). Wunz (1979) also observed that clearings were used more in years of poor oak mast.

Turkeys also are reluctant to forage in habitats with dense, woody understories (Bailey and Rinell 1967). As a result, dense stands of bear oak, which exclude herbaceous plants, are rarely used by turkeys except at the edges (Schorger 1966).

IMPACTS OF ACORNS AND OAKS

Because acorns are a preferred and highly utilized food source, their availability might be an important factor in turkey population processes. Mast availability and use may ultimately affect the condition, behavior, survival, and productivity of wild turkey populations.

Movements and Behavior

The home range size, movements, and habitat use of wild turkeys are apparently influenced by food availability, and winter foods exert the greatest effect (Lewis and Kurzejeski 1984, Kurzejeski and Lewis 1990, Lewis 1992). Wunz and Pack (1992) observed smaller annual home ranges in superior habitats, and winter home ranges appear to be smaller during years with abundant acorn crops (Lewis and Kurzejeski 1984, Kelley et al. 1988).

Because turkeys favor mast crops (especially acorns) as winter foods, the locations of wintering areas are determined by acorn availability (Ellis and Lewis 1967, Hurst 1992). In movements that coincide with mast availability and acorn drop, turkeys typically shift their ranges from fields and pastures to forests in fall and winter (Eaton et al. 1970, Barwick and Speake 1973, Speake et al. 1975, Eaton et al. 1976, Everett et al. 1985, Porter 1992). In a study of wild turkeys in managed pine forests, the turkeys responded to mast availability by using bottomland hardwoods more in fall and winter (Exum et al. 1987). Gould's turkeys have moved as much as 12.7 km between summer and fall ranges, probably to take advantage of mast crop availability (Schemnitz et al. 1990). When mast crops are good, these shifts in habitat can be very distinct and abrupt (Eaton et al. 1970, Barwick and Speake 1973, Healy 1992a).

Seasonal shifts in home range are reduced when food resources are abundant and distributed throughout the range (Korschgen 1967, Kurzejeski and Lewis 1990). When mast crops are scarce, flocks move to areas where other food sources (especially row crops) are abundant (Dalke et al. 1942, Ellis and Lewis 1967, Kurzejeski and Lewis 1990, Healy 1992a). As turkeys search for food resources, shifts from summer range to fall and winter range tend to be greater during years with poor mast production (Kurzejeski and Lewis 1990, Healy 1992a). In northern Missouri, turkeys traveled up to 4.8 km to use row crops in winter when acorn production was poor (Kurzejeski and Lewis 1990). When acorns are still available in the spring, after good mast crops, spring movements remain small (Godwin et al. 1994).

Winter flocking behavior seems to be influenced by the availability of acorns and other foods. When natural foods are scarce or unavailable, winter flocks usually are composed of fewer than 10 birds (Wunz and Pack 1992). As winter progresses and acorn supplies become depleted, flocks reduce their size, to facilitate food acquisition (Wheeler 1948). Larger flocks occur with abundant mast conditions (Wheeler 1948) or when winter foods are abundant and localized (e.g., crops) (Wunz and Pack 1992).

Condition of Turkeys

Acorns, with their high crude fat content and high metabolizable energy (Billingsley and Arner 1970, Decker et al. 1991), become available in the fall when energy demands are increasing. Despite their energy richness, their low protein and phosphorous contents suggest that acorns are largely inadequate to supply the nutritional needs of breeding or growing turkeys (Beck and Beck 1955, Short and Epps 1976, Pattee and Beasom 1979).

Dietary energy requirements increase during the colder fall and winter months throughout much of the wild turkey range. Standardized for a 4.23-kg turkey hen, two separate studies estimated that 247 kcal/day (New Hampshire) and 347 kcal/day (Minnesota) were required to maintain body weight at an ambient temperature of 0°C (Decker et al. 1991, Haroldson 1996). If only acorns were consumed to meet these winter energy demands at temperatures of 0°C, 4.23-kg hens from these respective studies would require 77.9 g/day (about 26 acorns) and 109.5 g/day (about 37 acorns). An extra 19 grams of acorns (about 6 acorns) would be required per hen for every 10° decrease in ambient temperature (Haroldson 1996). An 8-kg turkey gobbler would need 512 kcal/day (239.9 g/day or about 80 acorns/day) for a maintenance diet (Decker et al. 1991). A full turkey crop, which can easily contain > 100 acorns (Mosby and Handley 1943, Schemnitz 1956), would satisfy the daily energy requirement. Because winter diets of turkeys typically are not 100% acorns (Table 16.1), fewer acorns would actually be required to maintain thermoregulation in most areas. Although individual acorns vary widely in size among and within species, we have assumed that an average acorn weighs approximately 3 grams (Christisen and Korschgen 1955, Decker 1988).

Availability of oak mast seems to influence turkey condition, but direct measurements of food availability and turkey condition have been difficult to obtain. Low body weights and weight loss were assumed to be the result of an absent acorn crop in a study in Mississippi (Seiss 1989) and of deep snows in a study in Minnesota (Porter et al. 1980). Merriam's turkeys appeared to enter the winter in better condition after good acorn crops (Korschgen 1967).

Mortality

Because turkeys have a flexible diet and the ability to fast for long periods without starving, it is assumed that mast failures are of little significance to turkey survival. Other useful foods are usually available and good populations exist where few oaks occur (Bailey et al. 1951, Korschgen 1967, Markley 1967, Ignatoski 1973). Although body weights may be reduced during years with poor mast, adaptability to food shortages generally prevents starvation (Uhlig and Bailey 1952, Schorger 1966, Powell 1967). Substitutes for hard mast include ferns, bulbs, tubers, spore heads of club moss (*Lycopodium* sp.), seeds of forbs and grasses, and crops (Wunz and Pack 1992). In northern areas, where all foods may become unavailable because of deep and continuous snow cover, turkeys are capable of surviving without food for two weeks while losing > 40% of their body weight (Hayden and Nelson 1963, Markley 1967).

During unfavorable winter weather conditions, however, fall and/or winter survival of turkeys can be related to the availability of food resources, especially acorns (Porter et al. 1980, Porter et al. 1983, Vander Haegen et al. 1989, Vangilder 1995). In areas with limited fall hunting mortality, higher fall survival rates of hens generally were associated with better mast production (Vangilder 1995). No association was apparent between mast abundance and winter mortality rates. Even so, Vangilder (1995) suggested that acorn availability might be important for winter hen survival during adverse weather conditions.

Unfavorable ambient conditions (e.g., severe winters at northern latitudes, persistent deep snows), coupled with poor acorn production, can result in negative energy balances and starvation. Access to agricultural landscapes and food plots has been shown to mitigate the combined effects of mast failures and severe winters on mortality rates (Vander Haegen et al. 1989, Roberts and Porter 1995, Vangilder 1995).

Turkey populations also may experience greater fall hunting mortality during years when mast is scarce than when it is abundant (Menzel 1975, Wunz 1979). Hunter success may improve in the fall during mast shortages because turkeys concentrate activity at localized food sources (e.g., row crops, fields), making them more visible and vulnerable (Healy 1992a, Wunz and Pack 1992). Additional analyses of a five-year turkey population dynamics study in Virginia and West Virginia (Pollock et al. 1997, Pack et al. 1999) suggested that hunting mortality rates of adult hens during the fall tended to be higher under poor mast conditions ($P = 0.01$) (Table 16.2). Unlike for adult hens, fall hunting mortality rates for juvenile hens did not vary ($P = 0.80$) as a function of mast production.

Fall hunting mortality may be disproportionately magnified when heavy hunting pressure is coupled with mast failure. When acorn crops were abundant, regions with heavy (Virginia) and light (West Virginia) fall hunting pressure had similarly low adult hen hunting mortality rates (0.05 and 0.04, respectively); but when acorn availability was scarce, hunting mortality rates of adult hens increased much more ($P = 0.05$)

Table 16.2
Fall hunting mortality rates of wild turkey hens based on mast conditions and hunting pressure 1990–1993 in Virginia and West Virginia

Age	Hunting pressure	Mast conditions Good–Excellent[a] Mean	SE	Poor–Fair[b] Mean	SE	Overall mortality rate Mean	SE
Adult[c]	Low[e]	0.04	0.02	0.07	0.01	0.06	0.01
	High[f]	0.05	0.03	0.17	0.01	0.11	0.04
	Overall	0.04	0.01	0.12	0.03	0.08	0.02
Juvenile[d]	Low	0.10	0.03	0.11	0.06	0.10	0.03
	High	0.16	0.08	0.18	0.06	0.17	0.04
	Overall	0.13	0.04	0.14	0.04	0.13	0.02

Sources: Pollock et al. 1997, Pack et al. 1999. Although study began in 1989, due to small sample sizes in the fall of 1989, hunting mortality rates from that year were not considered.

[a]Based on mast surveys, the two best mast years (1991 and 1993).
[b]Based on mast surveys, the two worst mast years (1990 and 1992).
[c]Hens >1-year-old.
[d]Hens <1-year-old.
[e]In a 4-week fall season in West Virginia with no concurrent firearms deer hunting.
[f]In an 8- or 9-week fall season in Virginia with 1 or 2 weeks of concurrent firearms deer hunting.

in heavily hunted areas than in the lightly hunted areas (0.17 vs. 0.07, respectively). Regarding the combined effects of hunting pressure and mast conditions, no differential change in fall hunting mortality was observed for juveniles ($P = 0.93$). In Pennsylvania, fall turkey harvests were noticeably greater on a heavily hunted area during years with poor acorn production than on an area with low hunting pressure (Wunz 1979).

In addition to mast abundance, fall harvest is also influenced by a variety of other factors, including the availability of alternate foods, turkey population size, juvenile recruitment during the year, and hunting pressure. In West Virginia, there was actually a positive correlation between regional indices of mast abundance and fall harvest during the years 1943 through 1950 (Uhlig and Bailey 1952, Uhlig and Wilson 1952). These studies concluded that cold weather in May resulted in both poor mast production and lower turkey productivity and that warmer May temperatures resulted in good mast years and higher turkey productivity. High productivity yielded population increases for larger fall harvests (Uhlig and Bailey 1952).

Reproduction

Many studies have indicated a strong relationship between wild turkey reproductive success and nutrition. Nutritional deficiencies and poor habitat quality may lead to fewer hens' laying (Pattee and Beasom 1979), later initiation of laying (Billingsley and Arner 1970, Pattee and Beasom 1979, Porter et al. 1983), earlier end to nesting attempts (Pattee and Beasom 1979), fewer eggs being laid (Hayden and Nelson 1963, Gardner and Arner 1968, Billingsley and Arner 1970), fewer subadult nesters (Rumble and Hodorff 1993), lower renesting rates (Vander Haegen et al. 1988, Hoffman et al. 1995), fewer poults hatching (Pattee and Beasom 1979), and lower brood survival (Porter et al.1983).

Few studies have specifically addressed the role of acorn availability prior to breeding efforts. Palmer et al. (1993) noted later nest initiation following acorn crop failures in Mississippi. However, Wunz (1979) found no clear evidence that acorn production had any effect on the following summer's reproduction. Despite wide differences in mast production over a five-year period (Pollock et al. 1997), recruitment rates did not vary among years in Virginia and West Virginia (Norman et al. 2001). Considering the importance of acorns to winter diets and that reproductive success is dependent on the energy budget and endogenous reserves of females during the weeks prior to breeding (Porter et al. 1983), winter acorn availability is likely important to turkey reproduction.

In addition to hen productivity, nutrition also affected gobbler breeding capabilities and behavior. The nutritional quality of the habitat affected the growth and testes development of subadult gobblers (Lewis and Breitenbach 1966). Small gobblers, those weighing less than 6.35 kg, rarely exhibited advanced stages of spermatogenesis (Lewis and Breitenbach 1966, Blankenship 1992). Winter food restrictions that caused weight losses of 20–30% before breeding season delayed strutting and gobbling (Hayden and Nelson 1963). Among environmental factors, available mast might influence energy reserves and subsequent gobbler investment in gobbling and breeding activities (Lint et al. 1995, Miller et al. 1997).

Population Impact

While acorn production and availability appear to exert a great influence on wild turkey diet, movement, behavior, survival, and productivity, the ultimate role of oak mast as a population limiting factor is unclear. Because good wild turkey populations exist where oaks are rare (Markley 1967, Powell 1967, Kothmann and Litton 1975, Dickson et al. 1978), and turkeys generally do not starve during mast failures (Korschgen 1967, Markley 1967), acorns appear not to be a necessity for survival. Despite the apparent influences on population parameters, winter foods seldom act as an ultimate limiting factor for the omnivorous and opportunistic wild turkey. Less obvious population influences of winter mast probably are manifested by changes in carrying capacity and population growth rates.

Winter foods, oak mast, and other habitat factors are probably most important at the northern and southern edges of turkey range and in more heavily forested habitats (Healy 1981, Lewis 1992, Porter 1992). Populations have been observed to fluctuate with changes in food resources and mast abundance (Ligon 1946, Porter 1992, Vangilder 1995), especially when deep snows influence food availability (Healy 1981, 1992b).

Other Values of Oak Trees

In addition to their value as acorn producers, oak trees and forests have other uses for turkeys. In parts of the wild turkey range, especially in the Central Plains and the West, oak trees are more valuable as roost sites than as food sources (Crockett 1973, Healy 1992b). Because their roosting options sometimes are limited, Texas Rio Grande turkeys, New Mexico Gould's turkeys, and Oklahoma Merriam's turkeys commonly roost in live oaks (Markley 1967, Beasom and Wilson 1992), Emory oaks (*Q. emoryi*) (Potter et al. 1985), and bur oaks (*Q. macrocarpa*) (Crockett 1973), respectively. Turkeys prefer coniferous trees, which usually provide better thermal cover and roosting structure than oak trees (Hoffman 1968, Kilpatrick et al. 1988, Porter 1992, Flake et al. 1995). Depending on the associated understory characteristics, oak habitats are often selected by wild turkeys for nest sites (Hoffman 1962, Shaw and Mollohan 1992) and brood rearing (Mackey and Jonas 1982, Ross and Wunz 1990, Schemnitz et al. 1990). Rio Grande gobblers use shaded groves of live oaks to escape the summer heat in Texas (Beasom and Wilson 1992). Scrub oak (*Q. ilicifolia*) thickets provide escape cover for turkeys in West Virginia (Glover 1948). Bur oak habitats provide important forest corridors between roosting sites in South Dakota (Flake et al. 1995).

MANAGEMENT OF OAKS FOR TURKEYS

While turkeys are usually associated with extensive, mature hardwood forests (Schorger 1966), mast-producing forests are not absolutely essential to wild turkey populations. As a striking example, Rio Grande turkeys have expanded their range into treeless areas of west Texas (Kothmann and Litton 1975, Litton 1977). Even for the eastern subspecies, extensive stands of timber are not necessary (Little 1980). Eastern turkeys in Minnesota occupy habitats with as little as 12% forested cover (Hecklau et al. 1982).

Although vast forests may not be critical to turkeys, habitat management should generally encourage mature, mast-producing forests that incorporate adequate nesting, brood, and winter-survival cover (Williams 1992, Wunz and Pack 1992). Management for forest diversity helps provide more consistent mast production (Minser et al. 1995) and suitable vegetative structure (Williams 1992) to accommodate the diverse diets and other ecological needs of wild turkeys. Dutrow and Devine (1981) concluded that timber production was not compromised by managing habitat for turkeys in Virginia. Specific oak management strategies should primarily focus on maximizing acorn production, promoting other mast-producing species, providing open areas, and encouraging a desirable herbaceous understory.

CONCLUSIONS

Many wild turkey studies have been influenced by the environmental perturbations caused by changes in acorn availability. The unpredictability of acorn production has provided turkey managers and researchers with numerous, albeit unplanned, opportunities to observe the likely impacts of changes in food availability on wild turkey populations. Due to variation in alternative food availability, weather conditions, hunting pressure, and presence of competitors, it is difficult to make generalizations about the impact of a particular change in oak mast availability. Even so, all these studies have contributed to our empirical knowledge about the relationship of acorns to wild turkey food consumption, condition, reproduction, survival, behavior, and populations. Recognizing that oaks are an important component of turkey habitat, this collective assessment of acorn-related impacts has enhanced our overall understanding of wild turkey population ecology.

ACKNOWLEDGMENTS

This work is a contribution of Pittman-Robertson Federal Aid in Wildlife Restoration Project WE-99. The administrative support of R. W. Duncan and R. W. Ellis is gratefully acknowledged.

Chapter 17

Squirrels and Oaks

CHRISTOPHER C. SMITH AND MARTIN A. STAPANIAN

Squirrels of the genus *Sciurus* are diurnal arboreal rodents that feed to a large extent on the seeds of trees and the fruiting bodies of fungi, most of which have a relatively high lipid content (Smith 1995). Among the trees upon which they feed, some oaks (*Quercus* spp.) have the largest, least well protected edible seed kernels. Acorns also have some of the lowest lipid and protein contents of any of the seeds upon which squirrels feed (Wainio and Forbes 1941, Baumgras 1944, King and McClure 1944, Ofcarcik and Burns 1971, Smith and Follmer 1972, Short 1976, Short and Epps 1976). Management plans for forests in the United States, particularly in the Southeast and Northwest, involve conifers, hardwoods, and, recently, vertebrate species, including squirrels. In the Ohio River drainage area, pine plantations are less important than in the Southeast and Northwest, and forest management is concerned with creating a balance of hardwoods that will feed wildlife and supply firewood and lumber (Nixon and Hansen 1987). In order to evaluate existing management plans and to create sound new ones, it is necessary to understand the relationships that have evolved among squirrels, hardwoods, and conifers. These relationships include the choices squirrels make among species of tree seeds to eat, cache, and remove from caches at various seasons of the year.

FRUITING BODIES OF TREES IN TEMPERATE AND BOREAL FORESTS

In boreal forests, both angiosperm and gymnosperm canopy trees are wind pollinated. Further, both hold seeds behind many woody scales within one catkin. In temperate deciduous forests, most wind-pollinated trees form fruiting bodies that contain one seed. These seeds may be within hard, relatively large nuts (e.g., Fagaceae and Juglandaceae) that are dispersed by scatter-hoarding birds and rodents, attached to wings for wind dispersal (e.g., *Ulmus* spp., *Fraxinus* spp., and *Acer* spp.), or in pulp swallowed by and dispersed through the guts of birds and mammals (e.g., *Celtis* spp., and *Juniperus* spp.). Smith et al. (1990) and Smith (1998) provide evidence for the hypothesis that the difference in packaging of seeds in boreal and temperate forest trees is the result of a difference in the probability of successful cross-pollination in relation to regional tree species diversity. Tree species diversity is relatively low in boreal forests. Annual flower production at the landscape scale varies greatly from year to year as a physiological response to regional weather conditions (mast flowering). This mast flowering results in a high annual variance in the density of conspecific pollen at the landscape scale. In years when the trees are committing a large amount of their resources to both male and female catkins, they have a high percentage of viable seed. In years of low reproductive effort, few male and female catkins are produced and seed set is low. Trees can commit a high percentage of their female reproductive effort to protective woody scales, because they are assured a high frequency of cross-pollination and seed set during the years that they expend the vast majority of their reproductive resources. In temperate forests, the higher species diversity makes the density of conspecific pollen too low to assure a high frequency of cross-pollination, and trees make many small flowers with no commitment to protective or dispersal tissues until after pollination.

This delayed commitment to protection is possible because each seed is made by a separate fruit. Most wind-pollinated species in the temperate forests, including oaks, initiate their flowers with the bud that makes new leaves and stems. Therefore, there is little annual variation in flower production (Smith 1998). The annual variation in oak seed crops in the temperate forest is a result of high variation in the success of pollination and the availability of resources for fruit growth after pollination (Sork et al. 1993). Unlike Smith (1998), who counted flowers on the tree at the time of pollination and nearly mature acorns before they were being harvested by vertebrates, Sork et al. (1993) measured flower production by early stages of fruit abortion and mature acorns in seed traps below the tree. Thus, early flower failure may have been missed and acorn production underestimated by predispersal predation.

TREE REPRODUCTIVE PRODUCTS
AND THE CACHING AND SOCIAL
BEHAVIOR OF SQUIRRELS

Differences in tree reproduction in boreal and temperate forests in North America have had a profound effect on the nature of food caching and social behavior in the diurnal tree squirrels found in those two forest biomes (Smith 1995). In boreal forests, red squirrels (*Tamiasciurus hudsonicus*) and Douglas squirrels (*T. douglasii*) are parapatric and store large numbers of closed conifer cones in a damp cache near the center of a territory defended by one adult squirrel (Smith 1968). Individual territories with larder hoards of cached cones allow more effective harvesting and storing of cones than would a territorial system defended by mated pairs (Smith 1968). The two species in the genus are interspecifically territorial (Smith 1968). They cannot effectively divide up their food resources to be sympatric, because the food-caching behavior would be much less efficient if each species used only half of the potential resources (C. C. Smith 1981).

In temperate deciduous forests, two sympatric species in the genus *Sciurus* store individual nuts in a scattered pattern (scatter-hoards) in the ground, usually around their individual nest trees (Stapanian and Smith 1978). Squirrels in the genus *Sciurus* are not territorial, but they do demonstrate a social hierarchy around a concentrated food source (Pack et al. 1967). The system of scatter-hoarding individual nuts reduces the probability that larger animals, such as deer (*Odocoileus virginianus*) and turkeys (*Meleagris gallopavo*), will be able to effectively exploit the nuts falling under the tree of origin (Smith and Reichman 1984). The ranges of gray squirrels (*S. carolinensis*) and fox squirrels (*S. niger*) are broadly sympatric in the temperate deciduous forests of North America. The two species use the same foods with the same relative efficiencies (Smith and Follmer 1972), but they avoid competitive exclusion by being different in their efficiency of utilizing different habitats. Fox squirrels, which have about double the mass of gray squirrels, are found in open habitats where there is space between individual tree canopies and scarce underbrush, while gray squirrels frequent densely forested habitats with abundant underbrush (Nixon and Hansen 1987). The individual trees in open habitats can generally have larger fruit crops, because they benefit from reduced competition for light; and fox squirrels can escape predation in the open habitats better than gray squirrels, because of their

larger size. It is these two squirrel species that are most closely tied ecologically and evolutionarily with oaks and acorns.

Species in the genus *Sciurus* have the potential to exploit pines, and
tassel-eared squirrels (*S. aberti*) even limit their use of trees mainly to
ponderosa pine (*Pinus ponderosa*). It would appear that gray and fox
squirrels have not evolved the ability to digest cambium while resisting
the effects of resins to the extent that has been reached by other sciurids,
such as tassel-eared squirrels and red squirrels. In the natural vegetation
of the eastern deciduous forests there would have been little selective
gain in such an ability when seeds of angiosperms were commonly available.

The Nutritional Value of Tree Seeds

There is a marked difference in nutritional content of seed kernels in
the families Fagaceae and Juglandaceae. The kernels of seeds in hickory
nuts (*Carya* spp.) and walnuts (*Juglans* spp.), among the Juglandaceae,
have a relatively high content of protein (> 20%) (Wainio and Forbes
1941, Baumgras 1944) and lipids (> 20%) (Wainio and Forbes 1941,
Baumgras 1944, Smith and Follmer 1972), giving the kernels a high energy content per gram of dry weight (> 25,200 joules) (Smith and
Follmer 1972). Among the Fagaceae are acorns and chestnuts (*Castanea*),
which are relatively low in protein (< 10%) (Wainio and Forbes 1941,
Baumgras 1944, King and McClure 1944, Ofcarcik and Burns 1971,
Short 1976, Short and Epps 1976) and are variable by oak subgenus in
lipids (Baumgras 1944, King and McClure 1944, Ofcarcik and Burns
1971, Smith and Follmer 1972, Short 1976, Short and Epps 1976, and
Chapter 11), giving them a relatively low, but variable, energy content
per gram of dry weight (Smith and Follmer 1972, Havera and Smith
1979). Acorns in the subgenus of white oaks (*Leucobalanus*) and chestnuts (Wainio and Forbes 1941) have a low lipid content (< 10%) and
energy content per gram of dry weight (< 21,000 joules). In contrast,
acorns in the subgenus of red oaks (*Erythrobalanus*) have a relatively high
lipid content (varying around 20%) and energy content per gram of dry
weight (> 21,000 joules). Although smaller, the seed kernels of species
in the Pinaceae are generally similar to those in the Juglandaceae, having proportionately high concentrations of protein (> 20%) and lipids
(> 20%) (King and McClure 1944) and energy content per gram of dry

weight (> 25,200 joules) (Smith 1968). Kernels of seeds in the Fagaceae are thus less nutritionally desirable for tree squirrels than are the seed kernels from the Juglandaceae and Pinaceae.

The food value that squirrels obtain from different species of nuts is often difficult to measure, because when squirrels feed in the field they sometimes leave behind, uneaten, part of the kernel of hickory nuts when they discard the remains of the nut and they often leave part of the kernel of walnuts. The shells of walnuts and hickory nuts are convoluted and partition the kernel into lobes separated by hard shell. When opening a walnut, squirrels will gnaw through a thin part of the shell and scrape out the part of the kernel that they can reach with their incisors through the hole. Over half of the kernel of some nuts may be discarded. Gray squirrels eating black walnuts in captivity spend an average of 15 minutes on a walnut before discarding it with kernel left uneaten (Smith and Follmer 1972), while a fox squirrel observed on a golf course in late spring took 45 minutes to consume all the kernel from a black walnut (C. C. Smith, personal observation). Nixon and Hansen (1987) calculate that fox squirrels require 16 walnuts or 3 bur oak acorns as a daily ration in winter on the basis of observation of captive squirrels. Smith and Follmer (1972) calculated the caloric content of black walnut kernels to be approximately 60% of the caloric content of bur oak acorn kernels. It is likely that squirrels, as their cached stores run low, will put more effort into finishing the kernel of a walnut.

The efficiency with which red, gray, and fox squirrels metabolize the energy of seed kernels correlates closely with the lipid content of the kernels (Smith and Follmer 1972, Havera and Smith 1979). When presented with shelled seed kernels, gray and fox squirrels show a preference for the kernels of walnut and hickory nuts over those of both red and white subgenera of acorns. They also show a preference for the kernels of one species of red oak over those of two species of white oak acorns (Smith and Follmer 1972). Another factor that influences preference is the extra mass of gut content that results from an acorn diet. This can be an adaptive problem for tree squirrels, which must lift their body weight up trees to acquire food and escape some enemies. Stalheim-Smith (1984), comparing the forces exerted by the muscles of the forelimbs of fox squirrels and of Gunnison's prairie dogs (*Cynomys gunnisoni*) of similar mass, determined that those of fox squirrels are slightly greater. However, the muscles of fox squirrels fatigue slightly faster than those of Gunnison's prairie dogs. Extra weight is therefore a handicap for tree squirrels but an advantage for prairie dogs, which are moving soil with their forelimbs. It is significant that tree squirrels store very little body fat, cache food, and do not hibernate, while semifossorial sciurids store considerable body fat, usually do not cache food, and undergo periods of torpor.

PATTERNS OF SEED USE BY TREE SQUIRRELS

Tree squirrels in the genera *Sciurus* and *Tamiasciurus* cache seeds as the seeds mature in the late summer and autumn. For *Tamiasciurus*, which make a central cache within a territory, caching is most efficient if those conifer cones containing the most food content are the first to be carried to the cache. This priority is true both among species of conifers (Smith 1968, 1970) and within a single species (Elliott 1988). However, this type of caching strategy is not used by *Sciurus* in making their scattered caches. When presented with two adjacent piles, one each of bur oak acorns (*Q. macrocarpa*) and black walnuts (*J. nigra*), fox squirrels removed and cached almost equal numbers of the two species (Stapanian and Smith 1984).

Removing the walnut shell requires much more time than removing acorn caps and shells (Smith and Follmer 1972). In the autumn and spring, when weather conditions are on average more favorable than those in winter, squirrels can afford to take the time to eat hard-shelled walnuts and hickory nuts. In winter, however, when minimizing the time out of the nest is critical, acorns should be the food of preference. Moreover, the extra weight in the gut from eating acorns would not be as significant a problem for squirrels in the nest in winter as it would be when harvesting nuts in the autumn or mating in the spring. The stomach contents of gray and fox squirrels in Ohio support this general pattern, although the volume of kernels of species in the Juglandaceae exceeded that in the Fagaceae by about 50% for the year (Nixon et al. 1968).

Because gray and fox squirrels cache species in both the Juglandaceae and the Fagaceae simultaneously, they can act as dispersal agents for species in both families when the trees are in mixed stands. How can a seed predator function as a dispersal agent for species the seed of which it consumes? By spreading and caching nuts away from their tree of origin, squirrels reduce the chance that the seeds will be consumed by larger animals, but scattering caches does not prevent other squirrels from locating the buried seeds (Cahalane 1942, Stapanian and Smith 1978, 1984). Lewis (1980) has even observed gray squirrels finding nuts below 70 cm of snow. Because the individual seeds are dispersed in many caches they cannot all be watched and defended.

The density of nuts cached in the fall is insufficient to last all winter. C. C. Smith and J. M. Briggs (unpublished data) found that fox squirrels near Manhattan, Kansas, removed very few nuts buried by investigators over a 6-week period of intense cold and snow cover, but during the following spring and summer when nuts were buried at the same locations most were removed and at a rate that was more than a population of squirrels could eat. It seems, therefore, that these squirrels were depending on their own caches during the poor winter weather and were replenishing their caches when the weather allowed them to spend more time foraging. Such behavior would allow seed in scattered caches to begin to germinate and thus would explain how squirrel behavior can result in effective seed dispersal.

SOUTHERN PINES AS A FOOD SUPPLY FOR SQUIRRELS

Pine seeds are a seasonal staple in the diet of tassel-eared squirrels in the southern Rocky Mountains (Farentinos 1972) and also of fox squirrels in southeastern United States (Loeb and Lennartz 1989, Weigl et al. 1989, Wigley et al. 1989). Even in Kansas, where pines are not native, fox squirrels feed on seeds from young cones when they can find them, and do so as early as June (C. C. Smith personal observation). Pine seed kernels are high in protein and lipid content and energy concentration. Although pine seeds are an effective source of highly nutritious food, they are available to squirrels only until the seeds are shed from the cones. The seeds are small enough that an animal as large as a squirrel is unlikely to be able to compete with smaller mammals and birds to exploit individual seeds on the ground once they have been shed from the cones, and squirrels have not evolved cheek pouches that would allow them to carry and cache seeds after extracting them from cones. Squirrels therefore extract the seeds from closed pine cones. They eat pine seeds before the kernel is completely mature (C. C. Smith personal observation), although they do not eat immature acorns, which, unlike immature pine seed kernels, are a poor quality food (Smith and Follmer 1972). Because pine seeds provide this early nutrition, pines may be a particularly valuable food source for squirrels in a mixed pine and hardwood forest, but in a pure stand of pines there would be no food supply once the seeds were shed. Mixed pine and hardwood forests may be more suitable to fox squirrels than to gray squirrels, because pines open up the understory in forests (Wigley et al. 1989). Fox squirrels evolved with acorns and hickory nuts as an effective winter food supply and apparently have not evolved the tassel-eared squirrel's ability to use pine cambium as a winter food supply.

FLUCTUATIONS IN FOOD SUPPLY AND SQUIRREL POPULATIONS

Large fluctuations in gray and fox squirrel populations have been reported from early in the spread of European settlers in North America (Schorger 1949) until as recently as autumn 1965 in the Smoky Mountains. There have been large-scale emigrations of gray and fox squirrels from temperate forests in response to failure of forest seed crops as there have been of red squirrels from boreal forests under the same circumstances (Smith 1968, 1970). The crop failures in boreal forests result from the concurrent variation in mast flowering and fruiting of the few species of trees in boreal forests. These crop variations benefit the trees, both in more efficient pollination (Smith et al. 1990) and in reduced seed predation (Smith 1970, Smith and Balda 1979). Temperate forests do not experience mast flowering (Smith 1998), so the failure of mast crops is not the result of the masting behavior found in boreal forests. The decimation of squirrel populations by mast failures in temperate forests may not be an example of positive selection, since it eliminates one of the main types of seed disperser. Only the white oaks, whose acorns germinate early, would profit by the demographic change to a younger-aged squirrel population that would leave a higher percentage of viable embryos in cached acorns the year after a general mast crop failure in temperate forests (Fox 1982).

In temperate forests, the probability of a forestwide mast failure caused by poor pollinating conditions should be reduced by the fact that species of red and white oak acorns that mature in the same year are pollinated a year apart. There would have to be successive years of poor pollinating conditions for pollination to be the cause of a general failure in the acorn crop. In Kansas, oaks pollinate one to two weeks earlier than black walnuts. If that difference is general between the families of Fagaceae and Juglandaceae, there would have to be three different periods of poor pollinating conditions to cause local mast crop failure. Conditions such as drought, which might affect the maturation of all mast crops, may be a more frequent cause of general mast crop failure in temperate forests.

MANAGEMENT CONSIDERATIONS
FOR SQUIRRELS AND OAKS

The natural life span of individual trees is much longer than that of an individual squirrel. Trees are much easier to find and kill than squirrels and are of much greater economic importance than squirrels, which is why human activity can directly change forest composition and, in turn, indirectly change squirrel populations. The ideal forest for gray and fox squirrels would have a mix of species of red and white oaks and some species of the Juglandaceae. For fox squirrels, conditions would be improved by having a mix of pines and hardwoods. The economic yield from the rapid growth and turnover in trees has favored planting monocultures of pines as a silvicultural practice.

Millions of acres of hardwood forests in the southeastern United States have been converted to agriculture and pine plantations since the 1930s, particularly in the oak-pine uplands and the mixed hardwood bottomlands in major river valleys (Smith and Linnartz 1980 and references therein). Clearcutting and subsequent planting of pine plantations is expected to increase in the future (U.S. Forest Service 1988b, Alig et al. 1990, Boyce and Martin 1993). Competing woody vegetation is typically suppressed in the established plantation by controlled burning (Richter et al. 1982, Binkley et al. 1992), application of synthetic hormones and herbicides (Walker 1980), and other techniques (Swindel et al. 1984). As the planted pines become more dominant, diversity trends in the maturing plantations "are expected to be less ecologically favorable" (Swindel et al. 1984, p. 19). Other studies suggest that when stands of mixed-species forests are cleared and replaced by pine plantations, the result is a locally impoverished flora and fauna relative to the "original" stand (Atkeson and Johnson 1979, Repenning and Labisky 1985, Childers et al. 1986, Felix et al. 1986, Skeen et al. 1993).

Stapanian et al. (1997, 1999) and Stapanian and Cassell (1999) conducted a large, probability-based synoptic survey of forests in Georgia and Alabama. They found that silviculturally disturbed forested lands in the region had significantly lower tree species richness than in undisturbed forests. Wyant et al. (1991) obtained similar results for North

Carolina. In particular, species richness of trees was significantly lower in disturbed areas of the loblolly pine (*Pinus taeda*), shortleaf pine (*P. echinata*), oak-pine, and oak-hickory forest cover types (Stapanian et al. 1997). These three forest cover types account for approximately 71% of the total forested area of Georgia and Alabama (Stapanian et al. 1997). Thus, changes detected in their species composition and structure have important ramifications for conservation and forest management for the entire region.

Frequencies of individual species in managed and unmanaged areas of these forest types have been quantified (Stapanian and Cassell 1999) for the region. In the oak-hickory forest type, the regional frequencies of chestnut oak (*Q. prinus*) and flowering dogwood (*Cornus florida*) are significantly less in disturbed than in undisturbed areas. Loblolly pine, which is not a common associate of undisturbed oak-hickory forests (Eyre 1980), occurs more frequently in disturbed areas of oak-hickory in the region. Loblolly pine is planted for economic reasons in some managed oak-hickory stands in the region. Chestnut oak and loblolly pine typically occupy the overstory at maturity, and flowering dogwood is a mid-canopy species.

Wigley et al. (1989) found that cover in mid- and upper-canopy layers is greater in mixed pine-hardwood stands than in pure pine stands in the Southeast. Wildlife species such as fox squirrels which depend on both layers occur less frequently in pine plantations (Loeb and Lennartz 1989). At least three of the five subspecies of fox squirrels known to occur in the Southeast have experienced significant decreases in the past century, and at least one more appears to be declining in the Carolinas (MacClintock 1970, Loeb and Lennartz 1989, Weigl et al. 1989). This decline has been attributed to the decline of the mature mixed pine and hardwood forest, replaced by the pure pine forest (Loeb and Lennartz 1989).

The western gray squirrel (*Sciurus griseus*) has been designated a threatened species in Washington State. *S. griseus* is closely associated with communities of Oregon white oak (*Q. garryanna*) and conifers. Populations of this squirrel are small, fragmented, and declining, primarily because of loss and fragmentation of suitable habitat (Ryan and Carey 1995). The oak woodlands in the forest-prairie-wetland interfaces of the Pacific Northwest are being reduced and fragmented by human development (Ryan and Carey 1995) and are often replaced by monocultures of conifers, especially Douglas-fir (*Pseudotsuga menziesii*). Ryan and Carey's (1995) management objectives for the region include maintaining large, continuous open-form oak stands with adjacent intergrading conifers, prairies, and wetlands.

From a purely economic and management standpoint, a mixture of conifers and hardwoods in stands could mean a higher net rate of return, because of reduced start-up and maintenance costs (Lentz et al. 1989) and other resource and economic benefits (Leopold et al. 1989), including wildlife diversity. This is especially important for private landowners, because mixed conifer-hardwood stands also meet multiple-use objectives (Leopold et al. 1989). However, the volume of growth of conifers per unit area would be lower in mixed stands. The short-term rate of economic return may be less in mixed stands, because it takes most hardwoods longer to reach marketable volume. However, the lower volume may be offset by the higher values per board foot associated with selected hardwood species (Leopold et al. 1989). The long-term economics, including those from multiple use of mixed stands, warrant renewed investigation.

ACKNOWLEDGMENTS

The U.S. Geological Service/Biological Resources Division provided support to work on this manuscript.

Part III

Management of Oaks for Wildlife

Chapter 18

Fire and Oak Management

DAVID H. VAN LEAR AND PATRICK H. BROSE

The oak-hickory-pine, Appalachian oak, and the oak-hickory forest types described by Kuchler (1964) occupy most of the oak-dominated area of the eastern United States. In these types it is often difficult for oaks to regenerate, especially on better-quality sites (Loftis and McGee 1993). Oak regeneration failures are generally attributed primarily to poor initial establishment of oak seedlings and to their slow juvenile growth when they are present. Of these two causes, the slow juvenile growth of oak is considered more serious (Lorimer 1993).

In the dense shade of mature mixed-hardwood stands, oak seedlings and seedling sprouts do not develop into competitive stems. Overstory removal by either partial or complete cuttings fails to stimulate height growth of oak reproduction but does release regeneration of well-established shade-tolerant species and facilitates establishment of fast-growing shade-intolerant seedlings. Subsequent stand development is to a mixed mesophytic forest with oak as a minor component or altogether absent (McGee 1979, Abrams 1992, Lorimer et al. 1994). This successional trend is a relatively recent phenomenon, having developed during the past 75 years, and is tied to the exclusion of fire from eastern hardwood forests (Little 1974, Van Lear and Johnson 1983, Lorimer 1993).

Foresters have been admonished for most of the twentieth century to avoid burning in hardwood stands, primarily because it is well documented that wildfires cause a marked increase in heartrots in hardwoods (Nelson et al. 1933, Wendel and Smith 1986). As a result, fire research in hardwoods has lagged far behind that in pines. Occasional studies, however, have suggested that fire historically has played an important ecological role in oak-dominated forests. In the northeastern United States, Swan (1970) and Niering et al. (1970) noted that oak seedlings resisted root kill from fire better than their competitors did. Waldrop et al. (1987) and Augspurger et al. (1987) found similar results in the southeastern United States. Thor and Nichols (1974) described results of a long-term study in central Tennessee in which oak regeneration doubled following 6 years of annual burning or periodic burning at 5-year intervals. These studies, coupled with reviews of the fire history of the eastern United States (Little 1974, Van Lear and Johnson 1983, Crow 1988, Van Lear and Waldrop 1989, M. Williams 1989, Abrams 1992), strongly suggested that fire had played an important role in the development and maintenance of oak forests.

In this chapter, we discuss the ecological role of fire in oak-dominated forests and present guidelines for using prescribed fire to regenerate oaks. Particular emphasis is placed on a new regeneration method that utilizes partial harvesting followed by prescribed fire.

FIRE ECOLOGY OF OAKS

Oaks have a number of adaptations that enable them to survive in regimes of frequent fire. Mature oaks have thicker bark than most other hardwood species and, even when wounded, resist rot and are capable of living for decades after bole injury (Lorimer 1989). Oak regeneration increases its root-to-shoot ratios following disturbance (Barnes and Van Lear 1998, Hane 1999), which at least partially explains the fact that oaks are more tenacious sprouters than other hardwoods following repeated topkills (Waldrop and Lloyd 1991). Acorns have hypogeal germination, which results in the root collar and accompanying dormant buds being located well below the soil surface (Burns and Honkala 1990), protected from the heat of a surface fire. These adaptations allow upland oaks to survive in regimes of frequent fire, not so much because oaks thrive on fire but because they can withstand the potentially lethal effects of fire better than most species.

Frequent surface fires create environments that perpetuate oaks on better-quality sites. Surface fires remove much of the mid- and understory strata in mature mixed hardwood stands, reducing shading (Barnes and Van Lear 1998). Fire prepares a favorable seedbed for caching of acorns by squirrels and jays by removing excessive litter from the forest floor (Darley-Hill and Johnson 1981, Galford et al. 1988). Fire reduces surface soil moisture, discouraging seedling establishment of mesophytic species (Barnes and Van Lear 1998). It may also control insect predators of acorns and new seedlings (Galford et al. 1988).

As is true of many fire-adapted plants, oaks promote the occurrence of fire. An oak-dominated overstory adds about 4.5 mg/ha/yr of leaf litter to the forest floor (Loomis 1975). This litter remains curly, creating a porous fuelbed for autumn fires. Unlike leaf litter of mesophytic species, which forms a flat mat upon compaction and decays rapidly, oak leaf litter undergoes little decay during the winter. In regions where snowpacks are heavy, oak litter recurls after snowmelt, once again creating a porous fuelbed capable of carrying a surface fire during the spring fire season (Lorimer 1989).

Based on these characteristics and the historic methods of land use by Indians and settlers, many researchers have suggested that fire played a major historical role in the establishment and maintenance of oak forests. However, the nature of that role is not well established, and until recently no silvicultural prescriptions for using fire to regenerate oak had been rigorously tested.

EFFECTS OF FIRE INTENSITY AND SEASON OF BURNING

Literature concerning effects of fire in hardwood stands is often conflicting. For example, some researchers have reported that fires have occasionally created oak-dominated stands (Roth and Hepting 1943, Carvell and Maxey 1969, Ward and Stephens 1989), probably because intense fires controlled competition and stimulated rapid growth of oak reproduction (Johnson et al. 1989, Lorimer 1989). Conversely, others have reported that fires have had little effect on species composition in young stands (Johnson 1974, McGee 1979, Waldrop et al. 1985, Augspurger et al. 1987).

Because of these conflicting results, it has been difficult to predict effects of fires on young hardwood reproduction. The variation in results is due in large part to differences in season of burning and intensity of fire. Season of burning is important because it affects the physiological condition of the plant and the ability of species to resprout. Fire intensity is critical because certain species, such as the oaks, can survive higher intensity fires than can their competitors (Brose and Van Lear 1998). Fire intensity can be thought of as the dosage rate, that is, how much fire is applied (Pyne et al. 1996). To correctly study fire effects, one must document the preburn conditions of the plant community, thoroughly monitor behavior of the fire (especially fire intensity), and then reinventory the burn site for several years.

PREHARVEST BURNING IN MIXED HARDWOOD STANDS

More than 70 years of fire exclusion have altered stand structure and composition of eastern hardwood stands. These changes are manifested in increased stand density, greater canopy stratification, and a higher proportion of shade-tolerant and fire-intolerant species, especially in the understory. Although one understory burn is unlikely to markedly modify these conditions, repeated understory burning over long periods of time creates a stand structure on drier sites that favors oaks (Thor and Nichols 1974).

Van Lear and Watt (1993) developed a theoretical prescription to encourage oak regeneration by burning the understory of mature mixed hardwood stands near the end of the rotation. They hypothesized that fire would reduce understory and midstory competition and enhance the relative position of the more fire-resistant oaks in the advance regeneration pool. Barnes and Van Lear (1998) tested this hypothesis in the Piedmont of South Carolina and found that one spring fire, that is, one burn early in the growing season when leaves were expanding, was as effective as three winter fires in reducing density of understory and midstory stems in mature mixed hardwood stands. Understory and midstory density was reduced by almost 50% (Figure 18.1). They also found that winter and spring understory burns increased the number of oak rootstocks in the regeneration layer and improved their root-to-shoot ratios. Damage to overstory oaks was much less than that to yellow-poplar and was confined mostly to the smaller diameter stems with thinner bark (Table 18.1).

The main drawback with understory burning is that the overstory canopy remains unbroken, which slows development of oak regeneration into competitive stems. Several burns spread over at least 10 years will likely be needed to establish competitive oak regeneration in this manner. Such an approach is handicapped by the expense and risks of multiple prescribed fires, making it a rather unattractive option for landowners and managers.

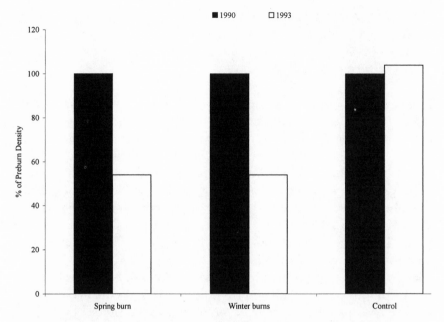

FIGURE 18.1.   Reduction of understory and midstory densities (trees from 2.5–10 cm dbh) in mature, mixed-hardwood stands following three winter burns and one spring burn in the Clemson Experimental Forest in the Piedmont of South Carolina.

Table 18.1
Extent of damage to boles of trees >11 cm dbh one year after the last of the three winter burns and two years after the spring burn, by percentage of species

	Spring burns			Winter burns		
Species	No damage	Bark scorched	Bark split	No damage	Bark scorched	Bark split
Oak species	83	10	7	70	26	4
Yellow-poplar	33	33	33	8	42	50
Dogwood	56	—	44	16	8	76
Combined competition[a]	67	—	33	57	21	21

Source: Barnes and Van Lear 1998.

[a]Numerous species which occurred in relatively minor numbers and included red maple, American beech, black gum, sourwood, sassafras, and others.

Shelterwood Cutting with Prescribed Fire to Control Competition

The Shelterwood-Burn Method

Frequently oak-dominated stands on better-quality sites have abundant but noncompetitive oak reproduction in the regeneration layer. These oak stems are generally less than 30 cm tall and have root collar diameters less than 13 mm. When such stands are harvested by either clearcutting or the shelterwood method, oak reproduction is out-competed by rapidly growing shade-intolerant species and/or well-established shade-tolerant species. The result is a species conversion of the site (Loftis 1983, Schuler and Miller 1995). In the southeastern United States, yellow-poplar is an especially vigorous competitor of oaks on better-quality sites and usually dominates following an overstory harvest.

A promising regeneration technique has recently been developed in the Piedmont of the southeastern United States to enhance the competitive position of oak regeneration in such stands (Keyser et al. 1996, Brose and Van Lear 1998, Brose, Van Lear, and Cooper 1999). This shelterwood-burn method is a three-step process (Figure 18.2) based on the silvics and fire ecology of oak and yellow-poplar regeneration. It involves an initial shelterwood harvest that removes roughly half of the overstory basal area, leaving the best dominant and codominant oaks in the shelterwood. In this first cut of the two-cut shelterwood method, all yellow-poplars are removed and logging slash is kept away from bases of residual oaks by directional felling. This partial harvest is followed by a 3-to-5-year waiting period, after which a relatively hot prescribed fire is run through the advance regeneration during the growing season.

Oak reproduction, if it is to become a dominant member of the next stand, must be relatively free to grow. Such oaks were defined as straight vigorous stems at least 1.3 m tall with no major competitors within 3 m (Nix 1989). Oaks are much more resistant than yellow-poplar to rootkill by seasonal prescribed fires, especially as fire intensity increases. In areas that were burned in the spring with high-intensity flames, density of free-to-grow oaks exceeded 800 stems/ha (Figure 18.3), because yellow-poplar competition was controlled. In contrast, winter burns provided little control of yellow-poplar and, even with a high-intensity fire, oak density did not reach 300 stems/ha. Summer fire resulted in substantial numbers of free-to-grow oaks at all fire intensity levels, but especially in the two medium intensity levels. High-intensity summer fires killed many

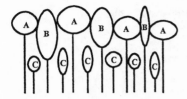

Typical upland mixed-hardwood stand.

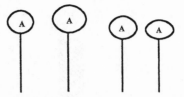

2. Initial cut to a shelterwood (40%-60% basal area reduction).

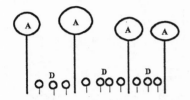

3. After 3-5 years, yellow-poplar dominates the advance regeneration pool.

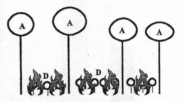

4. Prescribed fire topkills the advance regeneration, forcing rootstocks to sprout. Overstory damage and mortality limited to trees with slash at their bases.

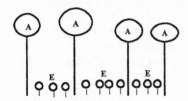

5. Oak now dominates the advance regeneration pool. Three management options available.

6a. Harvesting overstory without additional burning creates a new oak forest.

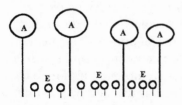

6b. Retaining overstory without additional burning creates a two-age stand.

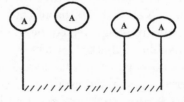

6c. Retaining with repeated burning either stockpiles oak sprouts or creates an oak savanna.

FIGURE 18.2.   Schematic diagram of the shelterwood-burn technique. A = high-quality oaks; B = hickories, poor-quality oaks, and yellow-poplars; C = American beech, flowering dogwood, and red maple; D = mixed-hardwood regeneration dominated by yellow-poplar; and E = mixed-hardwood regeneration dominated by oaks.

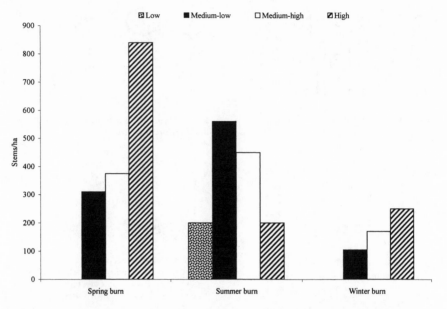

FIGURE 18.3.   Density of free-to-grow oaks in the advanced regeneration pool under Piedmont shelterwoods after fires of various intensities ignited during different seasons. (Adapted from Brose, Van Lear, and Cooper 1999.)

of the smaller oak seedlings while low-intensity summer fires failed to control competition.

Not all of a tract will burn with the same intensity, nor will it have uniform oak reproduction over the entire area. We suggest that if free-to-grow oaks exceed 370/ha and 60% or more of the stocking plots have at least one free-to-grow oak, the stand can be considered regenerated with the likelihood that oaks will be a dominant component of the next stand. If oak dominance is not a management objective, the stand may be considered regenerated even if these criteria are not met. If more oaks are desired, additional fires may be prescribed as dictated by leaf litter accumulation.

Why the Shelterwood-Burn Method Is Successful

Oaks differ from yellow-poplar, their major competitor on most good Piedmont sites, in their sensitivity to fire. Because of epigeal germination at the surface of the mineral soil, root collar buds of yellow-poplar seedlings are easily killed by surface fires (Kolb et al. 1990). Oak seedling survival is highest when acorns are buried about 3 cm deep in mineral soil (Sander et al. 1983). Because many acorns are buried by wildlife and germination is hypogeal, root collar buds of oaks are protected from the heat of surface fires by several cm of mineral soil.

The length of the waiting period between the initial shelterwood cut and the prescribed burn is critical. The shelterwood continues to produce litter, which creates the continuous bed of fine fuel necessary to carry the fire. In the 3-to-5-year period following the initial cut, logging slash settles and loses its foliage, minimizing the risk of bole damage to residual trees caused by flareups in heavy fuels (Brose and Van Lear 1998, Brose, Van Lear, and Cooper 1999). Although the yellow-poplar was removed in the initial shelterwood cut, yellow-poplar seed remains viable in the duff for up to 10 years (McCarthy 1933). Most of this stored seed germinates during the waiting period, and seedlings are killed in the subsequent fire. The waiting period also allows residual overstory trees to recover from the shock of the initial cut before they will be stressed again by the prescribed fire.

Shading from the shelterwood prevents yellow-poplar regeneration from growing so large during the waiting period that it could not be killed by fire. Hane (1999) found that shoot biomass of yellow-poplar regeneration in 7-year shelterwoods was only 30% of that in clearcuts of similar age. In shelterwoods, root-to-shoot ratios of oak reproduction increased for at least 7 years as the advanced regeneration developed. In contrast, yellow-poplar's shoot-to-root ratios declined during this period. This contrasting pattern of development in shelterwood stands makes oaks more vigorous sprouters than yellow-poplar after stems are killed by prescribed fire (Brose and Van Lear 1998).

After the waiting period, a growing season burn alters composition of the advance regeneration to favor oaks. Spring fires that move at about 1 m/min and have flame lengths > 1m give best results. In the Piedmont, a prescribed fire of this type will likely result in an advance regeneration pool with sufficient free-to-grow oaks to dominate the next stand. Fires that are of lower intensity and/or conducted during other times of the year will result in regeneration pools with fewer oaks (see Figure 18.3). If care is taken to prevent accumulation of slash at the bases of residual overstory trees or to remove it before burning, then fire-induced damage and mortality will be minimal (Brose and Van Lear 1999).

Applications of the Shelterwood-Burn Technique

This technique has many potential uses. Where timber management is the goal, it will be useful for perpetuating oak stands on better-quality sites for high-quality sawlogs. The shelterwood-burn method is attractive to land-owners because the initial cut of the shelterwood method produces income immediately. A small portion of the profit is then used to pay for the pre-scribed burn. Careful planning for the burn before the initial timber harvest can reduce the costs of the prescribed fire. The second shelterwood removal cut, if taken, is at least as profitable as the initial cut, because the best oaks are retained and some probably increase in value. In addition, landowners have the option of maintaining the classic structure of the shelterwood while stockpiling oak regeneration with periodic burns (Brose, Van Lear, and Keyser 1999). Many landowners may appreciate the aesthetic appearance of the shelterwood in contrast to that of young clearcut stands.

If wildlife management is the goal, the shelterwood-burn method can be used to sustain hard mast production while providing a source of palatable browse during the regeneration period. Many upland game species, for example, deer, grouse, rabbit, squirrel, and turkey, utilize the mast, browse, and cover in a regenerating shelterwood. Nongame species, including neotropical migrant songbirds, may benefit from various options available with the technique (Brose, Van Lear, and Keyser 1999, Lanham et al., in press).

Closely related to wildlife management is the concept of ecosystem restoration. If frequent burning is prescribed, especially growing season burns, after the initial shelterwood cut, a hardwood savanna can be created. Repeated burning at short intervals favors grasses and herbaceous plants over woody species. Use of this variation of the technique could restore some of the hardwood savannas frequently described by early explorers and settlers in the Southeast (Buckner 1983, Pyne 1982, Van Lear and Waldrop 1989, Abrams 1992). If the cutting regime were varied and additional burning was conducted, this technique could restore the vanishing oak savannas and open hardwood forests of the eastern United States (Brose, Van Lear, and Keyser 1999).

Research suggests that the shelterwood-burn technique may be successful in various physiographic regions and sites. It has been tested in the Piedmont of Virginia. Arthur et al. (1998) noted that repeated burning in oak-pine communities on xeric sites in the Cumberland Plateau reduced regeneration of red maple and other non-oak species and promoted chestnut oak regeneration. On mesic sites in Wisconsin, Kruger

(1992) found that two spring burns reduced densities of sugar maple and hop hornbeam by 80%, while density of northern red oak increased. Ward and Gluck (1999), in Connecticut, observed that burning several years after a shelterwood harvest favored oaks and reduced competition from birch and shrubs. In the northeastern United States, hot fires in mountain laurel thickets created a disturbance similar to a shelterwood harvest and opened canopies, allowing oaks to escape the dense shrub layer (Moser et al. 1996).

In areas where deer density is extremely high and where areas of early successional habitat are relatively rare, the shelterwood-burn method probably will not work. Shelterwood stands in such areas would be magnets for deer. Where high deer populations are not a problem and prescribed burning is feasible, oaks are likely to benefit from the technique.

Conclusions

Fire has played a major role in the regeneration and development of oak-dominated forests in the eastern United States. However, silvicultural research to develop effective ways to use fire for oak regeneration has been rather piecemeal and has generally lacked the long-term commitment necessary to document the potential of prescribed fire as an oak regeneration tool. Recent studies indicate that understory burning in mature mixed hardwood stands creates environmental conditions (e.g., reduced midstory/understory density, larger root/shoot ratios for oak, reduced forest-floor thickness, and more xeric conditions in the surface soil) that should favor oak regeneration. However, the drawback of understory burning in mature mixed stands with unbroken overstory canopies is that developing oak reproduction does not reach sufficient size to be capable of dominating the new stand when overstory trees are harvested.

A more efficient and successful way to enhance oak regeneration with fire is to combine burning with an earlier shelterwood cut. This technique has the potential to regenerate competitive oak stems on good-quality sites by reducing the vigor of oaks' competitors, especially if growing season burns of relatively high intensity are used. It works because it mimics a disturbance pattern (i.e., partial overstory disturbance followed by fire) that has shaped the composition of eastern forests for millennia. It is also economically attractive, because the initial shelterwood cut yields an immediate income and the subsequent prescribed burn is low in cost compared to alternative treatments.

Chapter 19

California's Oak Woodlands

RICHARD B. STANDIFORD

California's oak woodlands, also known as hardwood rangelands, cover 10 million acres, or 10% of the state (Bolsinger 1988, Greenwood et al. 1993, Pacific Meridian Resources 1994). These areas have an overstory tree canopy predominantly in the oak genus (*Quercus* spp.) and an understory of exotic annual grasses and forbs and occasional native perennial grasses (Griffin 1973, Bartolome 1987, Holmes 1990, and Allen et al. 1991).

Since European settlement of California, oak woodlands have been managed primarily for livestock production. These areas have taken on a new importance since the recognition that they have the richest species abundance of any habitat in the state, including more than 300 vertebrate species, 5,000 invertebrate species, and 2,000 plant species (Verner 1980, Barrett 1980, Garrison 1996). Oak woodlands also enhance the water quantity and quality, outdoor recreation, and aesthetic effect of the region. Over 80% of California's oak woodlands are in private ownership (Greenwood et al. 1993).

OAK SPECIES AND CLASSIFICATION

The five major oak species occurring in oak woodlands include three deciduous white oak species—blue oak (*Q. douglasii*), valley oak (*Q. lobata*), and Engelmann oak (*Q. engelmannii*)—and two evergreen oaks—coast live oak (*Q. agrifolia*) and interior live oak (*Q. wislizeni*). Three additional species are found in these oak woodlands, namely California black oak (*Q. kelloggii*), Oregon white oak (*Q. garryana*), and canyon live oak (*Q. chrysolepis*). However, these three are more typically found on moister, more productive conifer-dominated sites. Table 19.1 briefly describes the general characteristics the most important of these oak species.

Various systems have been used to classify oak woodlands in California. Table 19.2 provides a general cross-reference of these classification systems. The distribution, density, and abundance of the various oak species, together with other tree, brush, and herbaceous species, form the basis for evaluating the potential of a hardwood rangeland site for providing economic and ecological utility. Although there are numerous ways of classifying California's oak-dominated woodlands, the five vegetation types used in the California Wildlife Habitat Relationships System will be used in this discussion (Mayer and Laudenslayer 1988). These names for California oak woodlands are based on the dominant tree species and include valley oak woodland, blue oak woodland, blue oak–foothill pine woodland, coastal oak woodlands, and montane hardwood forest (Table 19.3).

SPATIAL AND TEMPORAL ASPECTS OF OAK WOODLAND SUSTAINABILITY

Landscape-Level Sustainability

Landscape factors affecting oak woodland distribution include long-term climatic factors and, more recently, human-caused events. Pollen analysis shows shifts in distribution of oak stands along altitudinal gradients (Byrne et al. 1991). Over the past 40 years, California's oak woodlands have decreased by over one million acres on a statewide scale (Bolsinger 1988) due to human-induced factors. Major losses from 1945 through 1973 were from rangeland clearing for enhancement of forage production. Major losses since 1973 have been from conversions to residential and industrial development. Regionally, some oak woodlands have decreased due to urban expansion (Doak 1989), firewood harvesting (Standiford et al. 1996), range improvement (Bolsinger 1988), and conversion to intensive agriculture (Mayer et al. 1985). The results have been habitat fragmentation, increased conflicts between people with different value systems, predator problems, and soil and water erosion.

Stand-Level Sustainability Considerations

From 1932 to 1992, blue oak woodland canopy density and basal area increased under typical livestock grazing and influenced by fire exclusion policies (Holzman 1993). This indicates that many oak stands were sta-

Table 19.1
General characteristics of California's important hardwood rangeland oak species

Characteristics	Blue oak	Interior live oak	Coast live oak	Valley oak
Scientific name	*Quercus douglasii* Hook. & Arn.	*Quercus wislizeni* A. DC.	*Quercus agrifolia* Nee	*Quercus lobata* Nee
Common names	blue, white, mountain, rock, iron, post, jack, Douglas	Interior live oak, highland live oak, Sierra live oak	Coast live oak, California live oak, encina	Valley, white, California white, mush, water, swamp, roble
Height	Usually 20–60 ft.; tallest > 90 ft.	Usually 30–75 ft.; shrub form 8–10 ft.	Usually 20–40 ft.; may reach 80 ft.	40–120 ft.
Mature tree dbh	Less than 1 ft, up to 2 ft.; largest > 6 ft.	1–3 ft.	1–4 ft.	1–4 ft.; largest > 8 ft.
Longevity	Long-lived, 175–450 yrs.	150–200 yrs.	Long-lived, 125–250 yrs.	Long-lived, 200–250 yrs.
Sprouting	Variable sprouter; not vigorous on dry sites	Very vigorous sprouter	Very vigorous sprouter	Not a vigorous sprouter
Acorn	Matures first year; variable in shape; warty scales; cup very shallow	Matures second year; very slender, pointed, 1 in. long; cup over half the nut	Matures first year; 3/4 to 2 3/4 in.; cup over 1/3 of nut and not warty	Matures first year; size variable but large and tapered; cup over 1/3 of nut, warty
Foliage	Deciduous; blue-gray color; smooth or slightly to deeply lobed edges; 1–3 in. long and 1/2– 2 in. wide	Evergreen; smooth to very spiny-toothed; dark green above and lighter below with waxy/shiny surface; 1–4 in.; flat	Evergreen 1–3 in.; roundish; dark and shiny above with gray or rusty fuzz underneath; cupped or spoon-shaped	Deciduous; leaves leathery with shiny, dark green-yellow above and grayish below; deep irregular lobes; 2–4 in.

Shade tolerance	Seedlings not tolerant	Somewhat shade tolerant	Shade tolerant throughout life	Seedlings somewhat tolerant; mature trees intolerant
Fire tolerance	Tolerates grass fires, not hot brush fires	Not very tolerant, but sprouts well after fire	Very tolerant of hot fires, due to thick bark	Not tolerant of fires
Elevation	500–2000 ft. in north; up to 5000 ft. in south	Below 2000 ft. in north; above 6200 ft. in south	Below 3000 ft. in north; up to 5000 ft. in south	500–800 ft. in north; up to 5600 ft. in south
Associates	Grades into open valley oak stands at low elevations, into denser live oak stands at higher elevation; foothill pine common	In pure stands or mixed with blue and/or coast live oak; valley oaks in south	Forms pure stands; also grows with interior live oak and coast live oak	Blue and Oregon white oak; sometimes interior live oak
Sites	Hot, dry sites with rocky soils, 12–40 in. deep; can't compete with live oak on better sites	Wide range, from valleys to foothills; moister areas than blue oak	Common on valley floors and not-too-dry fertile slopes	Prefers fertile, well-drained bottomland soils, streambeds, and lower foothills
Notes	Confused with valley oaks when leaves are dusty	Confused with coast live oaks; distinguished by flat leaves	Confused with interior live oak but rounded and cupped leaves	Confused with Oregon white oak but acorns pointed with warty cups

Continued

Table 19.1 (Continued)

Characteristics	Engelmann oak	California black oak	Oregon white oak	Canyon live oak
Scientific name	Quercus engelmannii Greene	Quercus kelloggii Newb.	Quercus garryana Dougl.	Quercus chrysolepis Liebm.
Common names	Engelmann, mesa	Black, California black oak	Garry oak, white oak, Oregon oak	Canyon live oak, canyon oak, gold cup oak, live maul, maul oak, white live oak
Height	20–50 ft.	60–90 ft.	50–80 ft.	60–80 ft.
Mature tree dbh	1–2 ft.	1–4 ft.	2–3 ft.; largest > 5 ft.	1–4 ft.; largest > 5 ft.
Longevity	100–200 yrs.	100–200 yrs.; occasionally up to 500 yrs.	100–200 yrs.	Up to 300 yrs.
Sprouting	Variable sprouter	Excellent sprouter	Excellent sprouter	Variable sprouter
Acorn	Matures first year	Matures second year; 1 1/2 in. long; thin cup over half the nut	Matures first year; 1 in. long with shallow cup	1 1/2 in. long; thick, shallow cup
Foliage	Considered deciduous but foliage may persist during winter; similar to blue-gray color of blue oak	Deciduous; 5 in. long; 5–7 lobed; spiny leaf tips; dark yellow-green above and pale yellow-green below	Deciduous; 4–6 in. long; evenly and deeply lobed with rounded leaf tips; lustrous dark green and shiny above and pale green below	Evergreen; 3 in. long; persists 3 or 4 seasons on tree; usually not lobed; leathery

Shade tolerance	Seedlings tolerant, mature trees intolerant	Seedlings intermediately tolerant; mature trees intolerant	Seedlings intermediately tolerant; mature trees intolerant	Tolerant
Fire tolerance	Very tolerant of hot fires	Very sensitive to cambium being killed in hot fires	Maintained in open stands by regular, low-intensity fires	Sensitive to hot fires
Elevation	Under 4000 ft.	200–6000 ft.	500–3000 ft.	300–5000 ft.
Associates	In pure stands and with coast live oak	Most common with tanoak, madrone, mixed conifer forest species; also with coast live oak, interior live oak, and blue oak	Douglas-fir and mixed evergreen forests; Pacific madrone and tanoak	Mixed conifer, chaparral, and woodland species; tanoak, Douglas-fir, Pacific madrone, coast live oak
Sites	Warm, dry fans and foothills	More common on forest sites; found on moister hardwood rangelands; well-drained soils	Cool, humid sites near coast to hot, dry sites inland	Most widely distributed oak in state; sheltered north slopes and steep canyons
Notes	Very limited range in south makes protection a high priority	Protected by Forest Practice Act on timberlands; commercial properties for finished lumber	Protected by Forest Practice Act on timberlands	Both a shrubby and tree form; very dense wood

Table 19.2
Cross-reference for hardwood rangeland classifications in the Sierra Nevada

Allen et al. 1991	CALVEG[a]	Griffin 1977	Munz and Keck 1973	Kuchler 1988	Eyre 1980	Mayer and Laudenslayer 1988
Blue oak series	Blue oak	Foothill woodland–blue oak phase	Foothill woodland	Blue oak–digger pine	Blue oak–digger pine	Blue oak woodland
Blue oak series	Blue oak–digger pine	Foothill woodland–blue oak phase	Foothill woodland	Blue oak–digger pine	Blue oak–digger pine	Blue oak–digger pine
Interior live oak series	—	Interior live oak North slope phase	Foothill woodland	—	—	Montane hardwood
—	Canyon live oak	Blue oak phase, interior live oak phase	Foothill woodland	Sierran montane forest	Canyon live oak	Montane hardwood conifer, montane hardwood
Black oak series	Black oak	Black oak phase	Foothill woodland	Sierran montane forest	California black oak	Montane hardwood conifer, montane hardwood
Valley oak series	Valley oak	Valley oak phase	Foothill woodland	Valley oak savanna	—	Valley oak woodland
—	—	Riparian forest	—	Riparian forest	—	Valley foothill riparian
—	—	Northern oak woodland	Northern oak woodland	Oregon oak	Oregon white oak	Montane hardwood

[a]U.S. Forest Service and California Department of Forestry and Fire Protection Land Cover Mapping and Monitoring Program.

Table 19.3
Acreage of California hardwood rangeland habitat types

Habitat type (CWHR)	Central Coast[a]	San Joaquin Valley/ Eastside[b]	Sacramento Valley/ North Interior[c]	Central Sierra[d]	North Coast[e]	Southern California[f]	Total
Blue oak woodland	1,096,990	1,078,080	945,170	365,920	75,900	34,000	3,596,060
Blue oak–foothill pine woodland	283,180	332,090	458,620	230,530	0	0	1,304,420
Valley oak woodland	54,600	16,870	1,760	0	2,230	1,000	76,450
Coastal oak woodlands	1,277,630	24,710	20,790	0	217,650	399,000	1,939,770
Montane hardwood	632,880	775,450	1,087,910	1,019,910	539,020	86,000	4,141,170
Total	3,345,270	2,227,200	2,514,240	1,616,360	834,800	520,000	11,057,870

Source: California Department of Forestry and Fire Protection database.

[a]Alameda, Contra Costa, Lake, Marin, Monterey, San Benito, San Luis Obispo, San Mateo, Santa Barbara, Santa Clara, Santa Cruz, Solano, Sonoma, and Ventura counties.
[b]Fresno, Kern, Kings, Madera, Merced, San Joaquin, Stanislaus, and Tulare counties.
[c]Butte, Colusa, Glenn, Lassen, Modoc, Plumas, Sacramento, Shasta, Sierra, Siskiyou, Solano, Sutter, Tehama, Trinity, Yolo, and Yuba counties.
[d]Amador, Calaveras, Eldorado, Mariposa, Nevada, Placer, and Tuolumne counties.
[e]Del Norte, Humboldt, and Mendocino counties.
[f]Imperial, Los Angeles, Orange, Riverside, San Diego, and San Bernardino counties.

ble or increased over a moderately long period, despite perceived natural regeneration problems (Muick and Bartolome 1987, Bolsinger 1988, Swiecki et al. 1997). However, more than 20% of the study sites were converted to other land uses, primarily residential subdivisions, during this period (Holzman 1993). A similar study of changes in total woody cover of blue oak–foothill pine woodland from 1940 to 1988 found these areas were relatively stable (Davis 1995).

Pollen analysis studies document the dynamics of oak woodland composition over a very long term and highlight the changing influence of human populations (Byrne et al. 1991). Oak woodlands were relatively stable during the long period of use by Native Americans. Beginning approximately 150 years ago with European settlement, introduction of livestock, and clearing for intensive agriculture, oak densities declined.

Exotic annuals first show up in the pollen record at that same time. After that initial exploitation of the oak resource in the early settlement period, since about 1850, oak cover has increased dramatically. Current oak densities, as determined from the pollen record, are at their highest level, due to fire exclusion policies of the last 50 years and the management practices associated with ranching that aim for low intensity but high extensivity.

OAK WOODLAND ECOSYSTEM PROCESSES

Since the introduction of domestic livestock and exotic annuals by European settlers, oak woodland ecosystems have changed dramatically. Herbaceous composition has changed from perennials to annuals (Holmes 1990). Fire intervals and intensity have increased (McClaren and Bartolome 1989b). Overstory cover, where not converted to another land use, has generally increased (Holzman and Allen-Diaz 1991). Soil moisture late in the growing season has decreased, and soil density has increased due to compaction from higher herbivore densities (Gordon et al. 1989). Riparian zones are now less dense and diverse (Tietje et al. 1991). (See Table 19.4.) These ecosystem process changes are discussed below.

Herbaceous Composition

The pre-European herbaceous community in oak woodland understory included native perennial bunchgrasses and forbs (Holmes 1990). Native species were displaced by alien annuals from Europe, Asia, Africa, and South America with the arrival of European settlers (Burcham 1970). Urbanization is accelerating the invasion of exotic plant species, although mitigation projects are under way that often require restoration of native grasses.

Grazing Processes and Forage Production

Livestock grazing has had a major impact on California's oak woodlands. By 1880, Spanish coastal missions had four million sheep and one million cattle (Holmes 1990), fostering a large demand for forage and oak browse. Currently, two-thirds of all woodlands are grazed (Huntsinger 1997). In addition to domestic livestock grazing, feral hogs consume

Table 19.4
Comparison of oak woodland conditions before European settlement, during extensive-ranching period, and in urban interface areas

Pre-European settlement	Extensive-ranching period	Current urban
Perennial herbaceous layer	Invasion of exotic annuals	Increasing annuals invasion, especially noxious weeds
Regular fire interval	Continuation of regular fire interval	Fire suppression policies, long fire interval, increased fire intensity
More-open overstory layer	Range clearing and tree thinning	Increased overstory layer of unconverted stands
Soil moisture higher and later into growing season	Soil moisture late in growing season decreased due to exotic annuals	Decreased soil moisture late in growing season due to exotic annuals
Lower soil bulk density	Increased soil bulk density	Increased soil bulk density
Snags, large woody debris	Snags and woody debris cleaned up in typical management activities	Less attention to clean-up; increased snags and woody debris
Dense, diverse riparian zone	Riparian zones less dense and diverse	Higher human use of riparian zones, increased storm runoff from urban areas
Lower herbivore densities	Higher herbivore density, primarily domestic livestock	Decrease in domestic livestock

acorns, while rodents such as ground squirrels and pocket gophers utilize large quantities of acorns and seedlings.

Grazing has both positive and negative effects on oak woodland sustainability. Positive grazing effects include: reduced moisture competition between oaks and herbaceous material (Hall et al. 1992); reduced leaf area in seedlings, which may help conserve moisture late in the growing season (Welker and Menke 1990); habitat for rodents who consume acorns and young seedlings may be reduced; and fuel ladders are eliminated, reducing the probability of crown fires in grazed woodlands. Negative effects of livestock grazing include: livestock and other grazing animals consume oak seedlings and acorns (Swiecki et al. 1997, Adams et al. 1992, Hall et al. 1992); grazing may increase soil compaction, making root growth for developing oak seedlings more difficult (Gordon et al. 1989); and soil organic matter may be reduced.

The oak canopy has an effect on forage production, composition, and quality that varies around the state depending on precipitation, oak species, and amount of oak canopy cover. Oaks compete with the forage understory for both sunlight and moisture, and they alter the nutrient status of the site because of the deep rooting of oaks and nutrient cycling from litter fall.

Oak removal was historically recommended as a means of increasing forage production on hardwood rangelands (George 1987). For the deciduous blue oak, most studies have demonstrated increased forage production following tree removal on areas previously containing over 25% canopy cover and receiving over 20 inches of rain (Kay 1987, Jansen 1987). Conversely, where there is less than 20 inches of rain, areas with low blue oak canopy (less than 25% cover) consistently had higher forage yields than adjacent open areas (Holland and Morton 1980, Frost and McDougald 1989). In areas with moderate blue oak canopy cover (25% to 60%), there was a variable canopy effect on forage production (McClaren and Bartolome 1989a). In areas with less than 20 inches of rainfall, zones under the canopy had consistently higher forage production throughout the growing season. In areas with over 20 inches of annual rainfall, the open areas had higher forage production than the areas under the moderate oak canopy.

Blue oak, in the southern and central portion of its range, provides green forage earlier (in the presence of adequate rainfall) and in higher quantities (15% to 100% greater) compared to the forage in open areas (Holland 1980, Frost and McDougald 1989, Ratliff et al. 1991); the difference in forage quality and quantity (though not necessarily timing of initial growth) may be even more pronounced during drought, due to the shading provided by tree canopies and the consequent reduction in moisture loss through evapotranspiration (Frost and McDougald 1989).

In evergreen live oak stands, with leaves that shade forage growth during the winter and early spring months, the few studies that have been carried out show a larger competitive effect of oaks on forage production (Ratliff et al. 1991). In general, live oaks stands with over 25% canopy cover will have lower forage growth than cleared areas. However, one study in the southern foothills of the Sierra Nevada showed that in drought years, live oak shading helped conserve soil moisture, resulting in higher forage production than on open sites (Frost and McDougald 1989). Table 19.5 summarizes the results of research studies of the relationship between oak canopy and forage production (Frost et al. 1997).

The increase in forage production beneath blue oak canopies, or in

Table 19.5
The effect of density of oak canopy on hardwood rangeland
forage production

Species group	Canopy cover	Winter forage production	Spring forage production
Live oaks	Scattered (< 10% cover)	Varies	Varies
	Sparse (10–25% cover)	Varies	Varies
	Moderate (25–60% cover)	Inhibited	Inhibited
	Dense (> 60% cover)	Inhibited	Inhibited
Deciduous oaks	Scattered (< 10% cover)	Enhanced	Enhanced
	Sparse (10–25% cover)	Enhanced	Enhanced
	Moderate (25–60% cover)	Varies	Varies
	Dense (> 60% cover)	Inhibited	Inhibited

Source: Adapted from Frost et al. 1997.

areas previously beneath blue oak canopies, is attributed, in part, to increased soil fertility caused by leaf fall and decomposition (Jackson et al.
1990, Frost and Edinger 1991, Firestone 1995). Enhanced soil fertility
also improved forage quality beneath blue oaks or where blue oaks had
been removed. However, since the nutrient input from leaf litter ceases
after tree removal, forage production increases will be temporary, gradually declining until soil fertility reaches levels similar to those in adjacent open areas. Long-term studies have found that it may take 15 years
for this nutrient effect from oak cover to dissipate after tree removal (Kay
1987).

Oak canopies also have an effect on forage species composition. Studies have found that understories of both blue and live oak stands favor
later successional herbaceous species, such as wild oats (*Avena fatua*), soft
chess (*Bromus mollis*), and ripgut brome (*Bromus diandrus*). Clovers (*Tri-
folium* spp.), annual fescues (*Vulpia* spp.), filaree (*Erodium* spp.), and soft
chess account for more of the total herbage biomass in open areas than
under oak canopy (Ratliff et al. 1991, Holland 1980).

In general, managers of livestock enterprises on hardwood range-lands should consider the following general guidelines when managing their oaks (Standiford and Tinnin 1996):

There is little or no value in removing blue oaks in areas with less than 20 inches of annual precipitation.

On areas with over 20 inches of annual rainfall, thinning oaks where the canopy exceeds 50% will have the greatest effect on forage production.

In areas thinned for forage enhancement, residual tree canopies of 25% to 35% are able to maintain soil fertility and wildlife habitat and to minimize erosion processes.

Tree removal activities should always be planned, considering all values of the trees, including wildlife habitat, soil stability, etc. in addition to the possible forage production benefits.

Soil Processes and Nutrient Cycling

In an investigation of soil-associated characteristics under canopies of different tree species and in open grassland sites, Frost and Edinger (1991) found higher organic carbon levels, greater cation exchange capacity, lower soil bulk density, and greater concentrations of some nutrients (at a soil depth of 0–5 cm) under blue oak canopies than in open grassland. Organic matter input from blue oak leaf litter primarily accounts for this finding; leaching of nutrients from rainwater drip may also make a significant contribution. The soil conditions beneath interior live oak and blue oak are similar; shading from the evergreen canopy, therefore, is thought to primarily account for the reduced total annual herbage production below interior live oaks under moderate environmental conditions (Frost and Edinger 1991). This may at least partially account for the higher production under blue oak canopies as compared to open grassland sites in the central Sierra Nevada foothills. Upon removal of overstory blue oak species, there is a gradual decline to levels comparable to the open grassland (Holland 1980, Kay 1987). Frost and Edinger (1991) attribute this to the reservoir of nutrients in blue oak litter, which is gradually depleted over time.

Oak woodlands with perennial grasses keep soil moisture later into the growing season than do woodlands with annual grasses (Gordon et al. 1989). This difference in soil moisture may partially explain the observed lack of sapling recruitment in oak woodlands. Evaluation of hardwood rangeland soil bulk density shows that areas with livestock grazing have a higher density than ungrazed areas.

Working in the foothills of the Sierra Nevada, Jackson et al. (1990) found that soils under blue oak canopies have higher nitrogen turnover rates and inorganic nitrogen contents than surrounding open grassland soils, due primarily to the higher nitrogen content from mineralization of oak leaf litter. There was no difference in soil water potential between the understory and the open grassland. The increased fertility under the blue oak canopy did not result in enhanced forage productivity; however, blue oaks do maintain a reservoir of soil organic nitrogen that could be rapidly depleted if the oaks were removed.

Oak Regeneration and Recruitment Processes

One of the key concerns that landowners, policy makers, and the public have about the state's hardwood rangelands is whether there is adequate oak regeneration to sustain current woodlands and savannas. Several surveys of oak regeneration (Bolsinger 1988, Muick and Bartolome 1987, Standiford et al. 1991, Swiecki et al. 1997) have shown a shortage of trees in the sapling size class for certain species (especially blue oak, Engelmann oak, and valley oak) in certain regions of the state (sites at low elevation, on south- and west-facing slopes, on shallow soils, or with high populations of natural or domesticated herbivores). If this shortage of small trees continues over time, then the oak stands may gradually be lost as natural mortality factors or tree removal take their toll on the large, dominant trees in the stand and woodlands convert to other vegetation types such as brushfields or grasslands.

Local deciduous oak regeneration was abundant prior to 1900. Present stand structure suggests that oak regeneration was more frequent in the past. However, deciduous oaks have reproduced poorly in the past 50 years (Griffin 1977). While seedlings do become established, few develop into saplings. Live oaks, whose seedlings may be more resistant to grazing and browsing than are those of the deciduous oaks, have produced saplings with more success over the past few decades. Pocket gophers, a significant seedling predator, may prefer deciduous oak roots to those of live oaks. Much of the failure of deciduous oak seedling establishment may be attributed to acorn and seedling damage by cattle, deer, rodents and insects (Griffin 1977).

According to demographic studies and experimental research, older individuals dominate most extant blue oak populations; current recruitment levels are insufficient to maintain the present distribution of blue oak (Davis 1995, Swiecki et al. 1997). Valley and blue oak are not regenerating in sufficient numbers to maintain existing stands (Muick and Bartolome 1987). However, the causes and mechanisms, which seem to vary according to species, region, and site, are still under investigation (Bartolome 1987).

Current research indicates that present recruitment of blue oaks may arise from a gap mechanism by which an understory seedling bank persists until a moderate stand disturbance (such as clearing, fire, or natural tree mortality) occurs, after which sapling recruitment proceeds (Swiecki et al. 1997). Natural blue oak regeneration is thus a multiple-step process that requires many years for completion. When grazing pressures and plant competition are minimal (such as along roadsides beyond pastures) or where microhabitat is favorable, pioneer establishment of blue oaks in open sites can occur. However, the seedling bank–gap–sapling recruitment mechanism, or seedling advance regeneration, seems to be the most pervasive mode of natural blue oak recruitment and regeneration. Undisturbed sites with moderate to dense oak canopies are unlikely sites for sapling recruitment. Since there are various steps and life history stages involved in the seedling advance regeneration mechanism, several variables affect regeneration at different stages of the recruitment process.

Swiecki et al. (1997) evaluated the status of stand-level blue oak regeneration and examined how management tools, environmental factors and site history affect blue oak sapling recruitment at 15 different blue oak–dominated locations. Saplings (basal diameter of 1 cm or greater and a diameter at breast height [dbh] of 3 cm or less) were found on 15.3% of the plots; the majority of these had grown from seedlings as opposed to sprouting from cut stumps. Based on observed mortality and sapling recruitment, 13 out of 15 sites had a net loss in blue oak density and canopy cover. These results differ from the earlier discussions on stand-level stability and demonstrate the need for long-term regeneration monitoring. Statistical analysis showed variables such as topographic position, browsing intensity, recent canopy gaps and clearings, and total overstory and shrub cover were associated with sapling recruitment at most of the study locations. Insolation, soil water holding capacity, and repeated fire were important at some locations.

Swiecki et al. (1997) propose the following recommendations to enhance blue oak recruitment:

Minimize understory shrub clearing in blue oak communities.

Reduce the intensity and duration of browsing pressure on woody vegetation.

Employ fire to manipulate understory vegetation so as to favor recruitment.

Leave intact at least 20% overstory canopy cover within 0.1 ha unit of regeneration sites.

Following gap creation, minimize livestock use until blue oak saplings are taller than browse level.

Minimize "potentially adverse impacts" in sites near the limits of blue oak range.

Valley oak has experienced inadequate regeneration during the past century (Griffin 1973, Bernhardt and Swiecki 1991, and Danielsen and Halverson 1991). Seedlings do become established, but few develop into saplings. Invasion of alien annual grasses, which make less water available to oaks than do native perennial grasses, may be one cause of this effect.

Stump sprouting has been widely observed in most oak woodland species. Studies have shown a high probability of achieving stump sprouting for blue and live oak species. This observation reduces the concern expressed over a lack of sapling trees in these woodlands (McCreary et al. 1991, Standiford et al. 1996).

Figure 19.1 is a chart that may help hardwood range landowners and managers assess oak regeneration (Standiford and Tinnin 1996). It shows that there are several questions to be raised in considering the process of oak regeneration (Bartolome et al. 1987). First, there needs to be an assessment of the current stand structure and whether it is consistent with the objective for the stand. Second, the health and vigor of the existing trees need to be assessed to determine if recruitment of small trees is needed to offset future tree mortality. Third, the number of seedling and sapling trees should be evaluated. In areas where trees have died, seedlings and saplings will be needed to replace the lost trees. When overstory density is below the desired level for management objectives, seedlings and sapling trees will be needed to increase tree density.

Oak Restoration and Planting

On areas that have been determined to have a regeneration problem, it may be necessary to establish more oaks. Planting acorns or seedlings

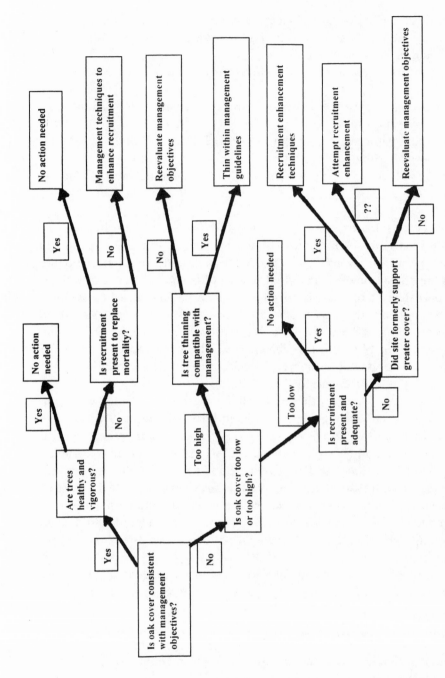

FIGURE 19.1.   Decision key to aid evaluation of oak regeneration on hardwood rangeland.

may be necessary where recruitment is inadequate to maintain oak cover on a long-term basis. However, the same factors that limit natural oak regeneration can inhibit artificial regeneration of native oaks. In general, substantial care must be taken to plant, protect, and maintain young oaks in the field to ensure success.

Riparian Management Processes

Although a small percentage of the state's water supply originates in hardwood rangelands, virtually all of it flows through oak woodland riparian zones (California Department of Forestry and Fire Protection 1988). Also, most of the state's major reservoirs are located on oak woodlands. Riparian zones provide important habitat for wildlife and aquatic organisms. Management activities influence water quality and wildlife and fisheries habitat. Yet, removal of up to one-third of the oak canopy had little effect on water quality and yield in one regional study (Epifanio et al. 1991). As part of the state's water quality management plan, new rangeland management practices are being developed (Humiston 1995). In urban interface areas, riparian zones are often subject to very high levels of human use for recreational purposes. Scott and Pratini (1997) documented how urban development increases human use of riparian areas, lowering the habitat value for various wildlife species and decreasing overall biological diversity.

Fire Ecology Processes

Fire is an important ecosystem process and management tool in oak woodlands. It affects oak woodland stand structure, oak regeneration processes, wildlife habitat, nutrient cycling, and economic uses for domestic livestock. The ecological effects of fire depend on its frequency, timing, and intensity, as well as the patch size from fire-induced mortality. The burning intensity and timing of adjacent vegetation types, such as chaparral and montane forests, influence fire effects in oak woodlands. Recent increases in the acreage damaged by stand-destroying fires in oak woodlands—the result of decades of attempting to exclude fire from wildland areas—point to the need to develop strategies that include fire in management activities if we are to sustain the economic and ecological values of our oak woodlands.

Fire Frequency. Because of the long period of human habitation of oak woodlands, it is extremely difficult to separate the "natural" role of fire from the human use of fire as a management tool. Lightning fires have helped shape oak woodlands. Lightning-caused fires originate from major storms coming northward from Mexico. It is speculated that decades may pass between major lightning-caused fire events in oak woodlands (Griffin 1977). Oak woodlands are extremely well adapted to hot summer fires (Mooney 1977). Mature oaks can survive regular low-intensity ground fires, and the young seedlings and saplings of most woodland oak species have the capacity to resprout after being top-killed by fire.

Native Americans made frequent use of fire in their stewardship of oak woodlands (Holmes 1990). There are numerous accounts of Native Americans using burning in woodlands to enhance habitat for game species, to improve access for hunting and gathering of acorns, and to maintain plant materials in an appropriate growth form for crafts (Jepson 1910, Cooper 1922). However, it is almost impossible to document the frequency, intensity, and extent of burning by Native Americans from existing fire ecology studies.

The first European settlers in oak woodlands continued to use fire as a management practice, to keep stands open for livestock production and to encourage forage production. Records indicated oak woodland burning intervals of 8 to 15 years by ranchers (Sampson 1944). Local prescribed burning associations were set up in various locations around the state, where neighbors came together annually to help conduct burns in the highest priority areas.

The use of burning as a management tool to mimic the effects of nature ceased on the state's conifer forest lands in the early part of the century. However, ranchers continued the extensive use of planned burning until the 1950s. At that time, the use of fire in oak woodlands declined, curtailed by negative urban attitudes towards fire, increasing housing density in rural areas of the state, concerns about liability from escaped prescribed fires, and air quality concerns. Fire suppression became the standard management strategy on oak woodlands, as it had become decades earlier on conifer lands.

McClaren and Bartolome (1989b) evaluated fire frequency in Central Sierra oak woodlands. Fire frequency had been around 25 years prior to settlement by Europeans in the mid-1800s. After settlement by Europeans, fire frequency became about every 7 years. No fires were observed from 1950 until the mid-1980s, when fire suppression was the dominant practice. Stephens (1997) observed similar fire frequencies in the central Sierra Nevada.

Effects of Fire on Oak Woodland Sustainability. Higher fire frequencies in the past may have created conditions more conducive to oak regeneration. McClaren and Bartolome (1989b) compared the age structure of oak stands with fire history and showed that oak recruitment was associated with fire events. Most oak recruitment in their Central Sierra study area occurred during periods of high fire frequency during the 1880s to 1940s. Oak recruitment has been rare since fire suppression.

The reasons that higher fire frequency enhances oak regeneration are not entirely clear. Allen-Diaz and Bartolome (1992) looked at blue oak seedling establishment and mortality and the practices of grazing and prescribed burning in coastal areas of hardwood rangelands. Neither grazing nor burning significantly affected oak seedling density nor the probability of mortality when compared to ungrazed and unburned areas, suggesting that seedling establishment is compatible with grazing and fire. Lawson (1993) studied the effect of prescribed fire on coast live oak and Engelmann oak in southern California and found higher seedling mortality in areas of prescribed fire.

Perhaps the importance of fire to oak regeneration is explained by its enhancement of oak sprout growth, which has been documented by Bartolome and McClaren (1989b). They concluded that in areas of moderate grazing with fire intervals of around 7 years, seedlings taking up to 18 to 20 years to exceed the livestock browse line (around 5 feet) would survive to become saplings and persist in the stand. In heavily grazed areas, only those trees that exceeded the browse line in 10 to 13 years would be recruited. Other factors affecting oak regeneration which would be influenced by the timing of fire events include the seedbed for acorns, the competition for moisture from herbaceous species, and the habitat for wildlife species that feed on acorns and seedlings.

Fire also has a major effect on the structure and composition of oak woodland stands. Lawson (1993) showed the differential effects of fire on coast live oak and Engelmann oak stands. Coast live oak had a higher mortality than Engelmann oak following fire. Coast live oak had greater height growth in unburned areas, while Engelmann oak had greater height growth following fire. The thicker bark of Engelmann oak provided more protection for the Engelmann oak sprouts. The study concluded that concerns about the decline of Engelmann oak habitats in southern California might be mitigated by reintroduction of fire to encourage Engelmann oak in these mixed stands.

Fire also kills diseases and pests, such as the filbert weevil (*Cucurlio occidentalis*) and the filbert worm (*Melissopus latiferreanus*), both of which can infest the acorn crop (Lewis 1991); and it reduces fuel ladders under oak canopies, preventing high-intensity crown fires.

Wildlife Habitat and Biodiversity Processes

California's hardwood rangelands provide habitat for over 300 vertebrate wildlife species, more than 2,000 plant species, and an estimated 5,000 species of insects. Figure 19.2 illustrates the diversity of vertebrate wildlife species predicted by the California Wildlife Habitat Relationships model for each of the five major hardwood rangeland habitat types (Mayer and Laudenslayer 1988). The management and long-term sustainability of California's hardwood rangeland habitats will best be served if ecological components and their interrelationships are recognized and addressed by owners and managers.

Wildlife are abundant inhabitants of the oak woodlands; the persistence of this varied wildlife can only be assured by maintaining the diverse habitats contained within the Sierra Nevada oak woodlands. These wildlife species depend on the oak trees, shrubs, grasses, forbs, seeds,

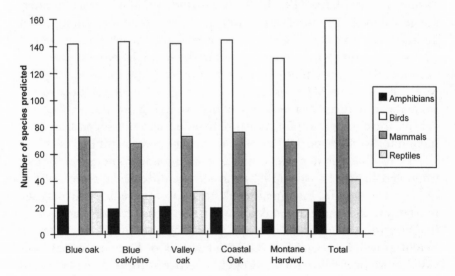

FIGURE 19.2. Numbers of amphibians, birds, mammals, and reptiles estimated to occur in the five California hardwood rangeland habitats described in the California Wildlife Habitat Relationships System. This list includes only those species in the system that are estimated to use one or more tree size and canopy cover classes for breeding, feeding, and/or cover.

fruits, insects, and other components of the oak woodland system. Much of the wildlife diversity is directly related to the diversity of trees, shrubs, logs, leaf litter, grasses, forbs, and other habitat components (Block 1990). A change in one of these components (such as alteration in tree density through urbanization or fuelwood removal) changes other factors (light regime, shrub layer, leaf litter, etc.).

In a three-year study of nongame wildlife populations at the Sierra Foothill Research and Extension Center, Block and Morrison (1990) found 113 bird species (at least 60 of which bred at the site), including 43 year-round residents, 11 winter residents that bred elsewhere, 17 breeding species that wintered elsewhere, 21 migrant birds, and 21 incidental species (Block 1990). Much of the bird species diversity is directly related to the plant diversity in oak woodlands. For example, Hutton's vireo (*Vireo huttoni*), orange-crowned warbler (*Vermivora celata*), and Wilson's warbler (*Wilsonia pusilla*) are closely associated with interior live oak. Birds such as the white-breasted nuthatch (*Sitta carolinensis*) and western bluebird (*Sialia mexicana*) are closely associated with blue oak (Block 1990). Wintering and migrant birds rely on the woodland resource for survival. Moreover, the specific habitats utilized by the birds change seasonally. For example, many resident birds obtained insects from foliage of blue and interior live oaks during the breeding season but were restricted to live oaks when the blue oaks had shed their leaves.

Block and Morrison (1990) recorded five small mammal species, including the brush mouse (*Peromyscus boylii*), pinyon mouse (*P. truei*), deer mouse (*P. maniculatus*), dusky-footed woodrats (*Neotoma fuscipes*), and ornate shrews (*Sorex ornatus*). They found low numbers of small mammals in their study sites and hypothesized that the indictment of small mammals as a major inhibitor of white oak regeneration is premature.

Block and Morrison (1990) found one amphibian species (the California slender salamander [*Batrachoseps attenuatus*]) and three reptile species (the western fence lizard [*Sceloporus occidentalis*], western skink [*Eumeces skiltonianus*], and the southern alligator lizard [*Gerrhonotus multicarinatus*]) in their Sierra Nevada sites. Distribution of the California slender salamander was limited to interior live oak stands, while western fence lizard and the southern alligator lizard and western skink were found in both live and blue oak stands.

Favorable hardwood rangeland habitats supply food, water, and cover to sustain wildlife species. Each habitat element provides unique niches, favoring particular wildlife species. Conversely, the absence of a particular element in a habitat may limit species diversity. Examples of elements of a hardwood rangeland habitat that are important to consider include riparian zones, vernal pools, wetlands, dead and downed logs and other woody debris, brush piles, snags, rock outcroppings, and cliffs.

Riparian habitat elements are used by almost 90% of all hardwood rangeland wildlife species, illustrating the importance of conserving this habitat element where present. Over one-third of all bird species on hardwood rangelands make use of snags, suggesting that management strategies maintaining an appropriate number of snags will result in greater wildlife species diversity. Downed woody debris from fallen limbs or dead trees provide an extremely valuable habitat for most reptiles and amphibians, as well as for many bird species. Oak woodland management for wildlife must retain dead trees along with trees in various stages of vigor in order to maintain critical wildlife habitat (Block et al. 1990). Mid-elevation hardwood rangeland habitats with several oak species, vertical diversity in vegetation structure, and diverse riparian zones have the richest diversity of wildlife (Motroni et al. 1991).

The threats to continued high biodiversity on hardwood rangelands include (1) fragmentation of large blocks of extensively managed hardwood rangelands; (2) reduction in important habitat elements such as snags, woody debris, and diverse riparian zones; and (3) increasing interface with urban areas, bringing household pets, humans, and fire suppression policies into contact with hardwood rangeland habitats. These threats to biodiversity can be reduced by encouraging cluster development and conservation of connecting corridors between large hardwood rangeland habitat blocks (Giusti and Tinnin 1993).

CONCLUSION

Oak woodlands are an important ecological component in California. Sustainability of ecological values is threatened by rapid population growth and the resulting conversion and fragmentation of woodland habitats. Many counties have started adopting local conservation strategies to conserve oak woodlands. Education and research have played a major role in conservation. Significant accomplishments have been made in rural areas of the state, where livestock and natural resource management are the predominant land uses. Where individual landowners have the ability to implement management activities that affect large acreage, education and research have contributed to decisions that favor conservation of oak woodlands.

However, for much of California, conversion of oak woodland habitats to urban or suburban land use is having a dramatic impact on sustainability of resource values. Incorporation of ecological information into land use plans adopted by county governments is only beginning. Since conversion to residential and industrial uses is ultimately a land use decision, it is a political process involving action by elected officials and input from different constituencies. The political and economic forces vary greatly in different parts of the state. Since success in this area will require that multiple individuals agree on a political course of action, this issue will present a large challenge.

Chapter 20

Ecology and Management of Evergreen Oak Woodlands in Arizona and New Mexico

PETER F. FFOLLIOTT

Evergreen oak woodlands—also referred to as encinal woodlands, from the Spanish word *encinal* meaning wholly or mostly oak, the western live oak type in the Society of American Foresters' classification (Ffolliott 1980), and the Madrean evergreen woodland formation (Brown and Lowe 1980, Brown 1982—are valuable environmental and economic resources in Arizona and New Mexico. These woodlands provide preferred habitats for a variety of wildlife species (including a large number of species listed as threatened, endangered, or sensitive by the federal and state governments), forage for livestock, wood products, watershed protection, and recreation and tourism (Ffolliott and Guertin 1987, Ffolliott 1992). A key to sustaining these multiple-use values lies in finding a balance between conservation and sustainable development.

CHARACTERISTICS OF EVERGREEN OAK WOODLANDS

Emory oak (*Quercus emoryi*) is the most common tree species throughout evergreen oak woodlands. It is associated in varying intermixtures of Mexican blue oak (*Q. oblongifolia*), gray oak (*Q. grisea*), silverleaf oak (*Q. hypoleucoides*), and Arizona white oak (*Q. arizonica*) (Ffolliott 1980, Brown 1982, McPherson 1992, 1997, McClaran and McPherson 1999). Overstory associates are alligator juniper (*Juniperus deppeana*) and border pinyon (*Pinus discolor*). Scattered ponderosa pine (*P. ponderosa*) occurs with Emory oak at higher elevations.

Distribution

Evergreen oak woodlands are concentrated in the Sierra Madre Occidental of Mexico, from where they reach northward into southeastern Arizona and southwestern New Mexico; scattered stands are also found in Texas (Ffolliott 1980, 1989, McPherson 1997, McClaran and McPherson 1999). These woodlands cover approximately 80,300 km^2 in aggregate, although their delineation is difficult, as inconsistent criteria have been used in their classification. The woodlands are found in foothills, bajadas, barrancas, and sierras of the lower elevations of mountain ranges, between 1,200 and 2,200 m. The soils that dominate in these woodlands have been formed mostly from mixed sedimentary and igneous rock (Hendricks 1985, McClaran and McPherson 1999). Structural development of the woodlands is determined by soil type and depth; most stands are found on rocky soils less than 6 m in depth (Hendericks 1985, Medina 1987, Pollisco et al. 1995, McPherson 1997). Annual precipitation ranges from 300 to 560 mm, of which about one-half falls during the growing season of May through early September. Winter precipitation, mostly rain but occasionally snow, falls from late November through the middle of March.

Habitat Types

Evergreen oak woodlands in Arizona and New Mexico are classified into 13 habitat types (plant associations), named after the dominant overstory tree and the dominant herbaceous understory plant (U.S. Forest Service 1987). A predominant habitat type of the evergreen oak woodlands in southeastern Arizona is *Q. emoryi/ Bouteloua curtipendula*. There are two habitat types dominated by Mexican blue oak, two by gray oak, four by Emory oak, three by Arizona white oak, and one by silverleaf oak. There are also six habitat types which are dominated by border pinyon, most of which include oak in the overstory or understory.

Structure, Density, and Growth

The woodlands contain small, multiple-stemmed, irregularly formed trees (Brown 1982, Ffolliott 1989, Ffolliott and Gottfried 1992, Gottfried

Table 20.1
General morphological characteristics of evergreen oak trees in Arizona and New Mexico

Characteristic	Emory oak	Mexican blue oak	Gray oak	Silverleaf oak	Arizona white oak
Height	12–15 m	8–10 m	15–20 m	18–20 m	15–18 m
Mature tree dbh	15–25 cm	10–15 cm	18–20 cm	15–20 cm	18–20 cm
Maximum tree dbh	60–75 cm	35–45 cm	60–75 cm	50–75 cm	60–75 cm
Mature tree crown diameter	1–1.5 m	0.75–1.25 m	1.25–2 m	1.25–1.75 m	1.25–2 m
Leaves	Broadly lance-shaped, 2.5–6 cm long, short spiny point with a few short spiny teeth, leathery, flat, shiny yellowish green on both sides	Oblong 2.5–5 cm long, rounded at both ends, edges rarely toothed, thin and firm, without hairs at maturity, blue-green	Elliptic to ovate, 2–5 cm long, blunt or sharp-pointed at apex, rounded at base, edges with few teeth toward apex, thin, gray-green or blue-green	Lance-shaped, 5–10 cm long, 1.2–2.5 cm wide, sharp-pointed, narrow base, edges rolled under with a few small spiny teeth, thick and leathery	Obovate or oblong, 2.5–7.5 cm long, short-pointed or rounded at apex, heart-shaped or rounded at base, edges wavy-lobed and toothed toward apex, thick and stiff
Acorns	1.2–2 cm long, rounded, one-third or more enclosed by cup, sweetish and edible	1.2–2 cm long, rounded, one-third enclosed by cup	1.2 cm long, rounded, one-third enclosed by cup	1–1.5 cm long, pointed, one-third enclosed by cup, hairy inside	2–2.5 cm long, shallow cup
Bark	Thick, divided into plates, black	Fissured into small squarish plates, gray	Fissured into shaggy plates, light gray	Deeply furrowed into ridges and plates, blackish	Fissured into thick plates, light gray to whitish

et al. 1995, McPherson 1997). General morphological characteristics of the evergreen oak trees in Arizona and New Mexico are described in Table 20.1. Species composition and stand structure and densities depend largely on inherent site characteristics and past tree harvesting activities. One-, two-, or three-aged stand structures are common. Intermingled with the trees are a variety of grasses and grasslike plants, forbs, and succulents, often in parks and savannalike mosaics.

Common graminoids include gramas (*Bouteloua curtipendula, B. gracilis, B. hirsuta,* and *B. radicosa*), lovegrasses (*Eragrostis intermedia* and *E. mexicana*), and muhlys (*Muhlenbergia emersleyi* and *M. longiligula*) (Brown 1982, McClaran et al. 1992, McPherson 1992, 1994, 1997, McClaran and McPherson 1999). Several species of perennial and annual herbaceous dicots emerge in early spring following the cessation of winter rains; a second group emerges coincident with the summer monsoon (McPhearson 1994, 1997). Openness of the landscapes is related to soil properties, site characteristics, and history of land use.

Numbers of trees range from a few scattered individuals to several hundred stems per hectare. Volumes of stemwood vary from less than 2 m^3/ha to over 100 m^3/ha (Ffolliott 1989, Conner et al. 1990, Van Hooser et al. 1990, Gottfried et al. 1995). Annual growth rates for stemwood are relatively low, ranging from 0.25 to about 1 m^3/ha. Production of branchwood, leaves, and roots has not been measured. Natural mortality is low, likely because the long history of tree harvesting activities (over 100 years) has reduced the number of older trees (80 years or older) at high risk. Nevertheless, all species of evergreen oak in Arizona and New Mexico are susceptible to infection by *Inonotus andersonii,* a major cause of wood decay (Fairweather and Gilbertson 1992).

ECOLOGICAL RELATIONSHIPS

Much of what is known about ecological relationships in the evergreen oak woodlands in Arizona and New Mexico has been derived from studies on two species: Emory oak and Arizona white oak. Emory oak is a red oak while Arizona white oak belongs to the white oak subgenus (Kearny and Peebles 1960). An overview of autecology of these species precedes a consideration of synecological relationships.

Autecology

Little is known about the reproductive cycles of evergreen oak species in the region (McPherson 1997). Analysis of growth-rings indicates that Emory oak produces acorns at 40 years of age, and acorn production in Arizona white oak begins at about 80 years (Sanchini 1981). Personal observations by the author suggest acorn production at earlier ages on more productive sites in southeastern Arizona. Acorn production for both species generally varies among years and individual trees, however. Periodicity of acorn production in unknown. Predation by vertebrates and invertebrates can result in large losses of acorns from the soil surface (Nyandiga and McPherson 1992, McClaran and McPherson 1999).

An absence of seed dormancy in Emory and Arizona white oak, combined with the tendency of the species' acorns to mature coincident with the onset of summer rains, suggests that the acorns either utilize soil moisture to germinate immediately upon dropping, or the acorns do not germinate at all (McPherson 1992, 1997). Adequate burial in mineral soil enhances the likelihood of acorn germination. While the environmental conditions necessary for acorn germination and seedling establishment occur infrequently in evergreen oak woodlands of Arizona and New Mexico (Pase 1969, Borelli et al. 1994), such germination and establishment does not appear to be the life-history stage that limits replacement of the oak (Sanchini 1981, Nyandiga and McPherson 1992, McPherson 1992). Nevertheless, a high proportion of oak plantlets originate vegetatively by stump sprouting or root suckering in dry years (Borelli et al. 1994).

Seedling mortality can approach 50% within the first year, with nearly all of the mortality being attributable to the frequent drought conditions (Pase 1969, McClaran and McPherson 1999). Seedling survival rates generally flatten out at about 25% by the third year after acorn germination. Annual height growth of the surviving seedlings often reaches 2.5 to 3 cm by the third year.

Growth and mortality of mature Emory oak and Arizona white oak (60 years old) are both low. The low productivity of these oaks and of other evergreen oaks in the region is not surprising, considering the dry climate and the shallow, rocky, and poorly drained soils where these trees are found (Ffolliott and Gottfried 1992, McPherson 1994, 1997, Gottfried et al. 1995). Analysis of life-tables for Emory oak and Arizona white oak indicates that the risk of mortality remains relatively constant throughout their expected lifespans (Sanchini 1981), implying that the causes of mortality for both species are stochastic events such as drought, wildfire, and the occasional ice storms.

Synecology

While the community-level processes in the evergreen oak woodlands of Arizona and New Mexico have been poorly studied (McPherson 1992), the type has been extensively described (Axelrod 1958, Brown 1982, Goldberg 1982, McPherson 1997). Plant communities have been described generally along elevational gradients (Wallmo 1955, Marshall 1957, Whittaker and Niering 1964, 1965, Niering and Lowe 1984). The woodlands at elevations between 1,220 and 1,520 m are "open oak woodlands" and those between 1,520 and 2,130 m "pine-oak woodlands." The woodlands merge with oak savanna and eventually semi-desert grassland ecosystems at lower, and therefore drier, elevations. Emory oak or Mexican blue oak dominate in these savannas; at higher elevations, the woodlands grade into pine-oak woodlands and, eventually, ponderosa pine forests.

The species composition and the structure of evergreen oak stands are related to a number of site factors in addition to elevational gradients, primarily slope-aspect combinations and incidences of disturbance (McPherson 1992, 1997). Stand-level disturbances by fire, disease, and plant-control and land-clearing activities are relatively minor in occurrence (Kruse et al. 1996), but the absence of observed disturbances does not necessarily mean that disturbance does not influence stand structure. Disturbances are likely to affect stand structure and productivity (Ffolliott and Gottfried 1992, Gottfried et al. 1995, McClaran and McPherson 1999).

On some sites, there is little evidence of high-intensity, stand-replacing wildfire having occurred for 100 years ago or more (Swetnam et al. 1989). Nevertheless, past fires have affected species composition, size-class distributions, and stand density (Niering and Lowe 1984, Barton 1991, Kruse et al. 1996). Grazing by livestock and wildlife ungulates can also alter stand structure while again leaving little evidence of disturbance (Sanchini 1981). Grazing favors oak establishment if competing grasses are consumed. Harvesting trees, mostly selection-cutting of fuelwood, is a more evident disturbance. Stand densities are reduced and individual tree size-class distributions changed in accordance with the intensity of harvesting, although growth of residual trees is often increased. Mature trees with a few scattered stump sprouts and root suckers normally compose stands that have not been harvested in the past 50 years (Bennett 1992, 1995, Ffolliott 1992, Ffolliott and Gottfried 1999). More-recently harvested sites characteristically have fewer mature trees and a few pre-harvest but numerous postharvest stump sprouts.

Structural diversity of evergreen oak woodlands, that is, the abundance of foliage in vertical layers, is also changed by harvesting timber (Sharman and Ffolliott 1992). With increasing intensities of harvesting, increasingly fewer trees in the tallest height classes remain in the stand. Taller trees provide more habitat niches for nongame birds than do shorter trees (Balda 1969), and, therefore, the diversity of bird habitats can be adversely affected by timber harvesting.

WILDLIFE MANAGEMENT PRACTICES

Wildlife management in the evergreen oak woodlands, and elsewhere in Arizona and New Mexico, includes practices centering on population management and practices focusing on habitat management (Krausman and Smith 1990). Population management practices involve, but are not limited to, formulating hunting regulations to keep game populations in balance with often fragile habitats, establishing land preserves (refuges) to reduce the impacts of decimating factors, and replenishing wildlife populations by stocking when appropriate. Habitat management practices are largely based on knowledge of species' habitat requirements, plant successional patterns, and management tools that modify composition and density of plants on a site.

Preferred habitats of many wildlife species occur in areas that are intermixtures of tree, shrub, and herbaceous cover. However, removal or modification of vegetation as part of any management practice changes landscape diversity, which can affect the habitats for wildlife. Vegetation management, therefore, is intrinsically linked to the well-being of deer, Gould's turkey, small game, and nongame bird species in these ecosystems.

Deer

White-tailed, Coues, and mule deer inhabit a variety of vegetative types in Arizona and New Mexico, but their greatest densities are found in types that include mountain browse, chaparral shrublands, and evergreen oak woodlands (Anthony and Smith 1977, Krausman and Smith

1984, Smith and Anthony 1992). Deer most often use evergreen oak habitats that contain varying combinations of mountain mahogany (*Cercocarpus breviflorus*), alligator juniper, border pinyon, kidneywood (*Eysenhardtia polystacha*), snowberry (*Symphoricarpos oreophilus*), grama grasses (*Bouteloua* spp.), and ocotillo (*Fouquieria splendens*). Habitats within core areas of home ranges are mostly oak-dominated regardless of the associated vegetation.

Fire, tree harvesting, and livestock grazing can alter habitats preferred by deer. Low-intensity, prescribed fire has relatively little impact, but high-intensity, uncontrolled wildfire can detrimentally change structural complexity, species composition, and richness of the habitats (Ffolliott et al. 1996). Postharvesting stump sprouts increase the availability of browse (Gottfried et al. 1995); however, excessive tree harvesting can turn an open woodland into a dense brush field. Effects of livestock grazing on deer habitats depend largely on the intensity, timing, and location of the grazing, the habitat condition and seral stage of the vegetation, and the prevailing climatic conditions (Anthony and Smith 1977, Krausman and Smith 1984, Smith and Anthony 1992). Overgrazing reduces the ground cover and alters plant species composition, particularly of forbs, detrimentally altering habitat conditions on many sites. It is necessary, therefore, that management of fire, tree harvesting, and livestock grazing be integrated into the management of deer habitats.

Gould's Turkey

Management guidelines developed for Merriam's turkey (*Meleagris gallopavo*), which inhabit the higher-elevation ponderosa pine forests (Hoffman et al. 1993), have been modified to suit Gould's turkey, although the latter guidelines are based on a more limited research base. Mast-producing evergreen oak trees ranging from 20 cm to 40 cm in diameter, along with border pinyon and canyon grape (*Vitis arizonica*), provide food for Gould's turkey in Arizona and New Mexico (Schemnitz and Zornes 1995). Therefore, evergreen oak stands are maintained with a minimum of 100 mast-producing trees where possible. Available water, frequently a limiting factor in turkey habitats, should be within 1.6 km. Priority is given to developing water sources (by the installation of metal tanks, cement troughs, etc.) when they are lacking, to enhance sites that are otherwise good brood-rearing and nesting habitat.

Mature and overmature oak trees with open crowns and large, horizontal branches, are preferred turkey roost trees (Schemnitz and Zornes

1995). Younger trees have a potential to become roost trees as they increase in size if their crowns remain open, for entrance to the trees. Chihuahua pine (*P. leiophylla* var. *chihuahuana*), sycamore (*Platanus wrightii*), and cottonwood (*Populus fremontii*) of mature status are also roost trees.

Openings created by managers to improve livestock forage production in turkey brood habitats should not exceed 150 to 200 m in width and should be irregular in shape (Schemnitz and Zornes 1995). These openings should be dispersed to make up no more than 15% of turkey habitats. Such openings usually have abundant insects, an essential part of poult diets (Schemnitz and Zeedyk 1992). Livestock stocking levels should be adjusted to allow a maximum of 35% forage utilization. Livestock grazing within regularly used brood habitats should be deferred until September 1.

Small Game

The presence of large, mature evergreen oaks is universal throughout the range of Arizona gray and Chiricahua fox squirrels and, quite likely, is essential in preferred habitats. Oaks furnish acorns, a key food composing up to 30% of squirrels' diet in the autumn (Brown 1984). Therefore, a sufficient number of mast-producing trees should be retained in a stand as a food source for these two species of squirrels. These squirrels are largely foragers; they do not cache food or regularly bury mast like their eastern cousins. There is little reason for them to do so, because their habitats are relatively snow-free.

Squirrels build several leaf nests or none (Brown 1984). When they are built, two types of nest are constructed: a flat, platformlike structure for resting in the summer months and a more substantial, covered, bolus nest that is used as a nursery and in lieu of a den site. These nests are found in oak trees more than 12 m tall; squirrel nests are also observed in sycamore, walnut (*Juglans major*), and cottonwood trees, and Apache pine (*P. engelmannii*). They are situated at a fork of two or more large branches. Trees affording these characteristics must be a component of a stand if it is to sustain squirrel populations.

Preferred habitat conditions for lagomorpha inhabiting evergreen oak woodlands in Arizona and New Mexico are unknown. A need for diverse food sources, preferentially small stems of shrubs with high moisture content, and sufficient protective cover to hide from predators, are apparently major determinants of habitat selection. Lagomorpha seemingly prefer shrub-grassland mosaic habitats within the woodlands.

Nongame Birds

Diverse nongame birds inhabit the oak evergreen woodlands. Specific needs provided for by the oak and associated vegetation include food and cover, and nest, roost, and perch sites (Block et al. 1992). The species of oak and the proportion of the tree foraged upon are species-specific (Marshall 1957, Balda 1969, Block et al. 1992, Morrison 1999). Therefore, a diversity of tree species and sizes is needed in a stand to accommodate the bird species permanently or temporarily inhabiting woodlands. Many species of neotropical birds that typically breed in temperate environments of North America and winter in tropical environments of Central and South America also inhabit the evergreen oak woodlands, extensively for foraging (Hutto 1985, Block et al. 1992). With conservation of neotropical birds becoming increasing important in the region, a better understanding of the birds and their habitat requirements is critical.

Mexican spotted owl, a year-round resident of the evergreen oak woodlands, is listed as threatened by the federal government, because of concerns over habitat loss and population declines. Mexican spotted owls are found in stands dominated by evergreen oak species, sometimes co-occurring with alligator juniper and border pinyon (Marshall 1957, Ganey and Balda 1989). However, habitat use by this bird has not been intensively studied, and, therefore, the importance of evergreen oak habitats to this owl species is largely unknown (Ganey et al. 1992). Information on habitat requirements is critical to implementing multiple-use management strategies for maintaining viable populations.

OTHER MANAGEMENT PRACTICES

Values of evergreen oak woodlands in Arizona and New Mexico for livestock and wood production, watershed protection, and recreation and tourism are incorporated into ecosystem-based management plans to accommodate people's increasing desires for these multiple benefits.

Livestock Production

Evergreen oak woodlands provide forage for livestock production. Situated between the higher montane forests and lower desert ecosystems, these woodlands frequently hold a key to balancing livestock numbers with available forage resources in Arizona and New Mexico (Ffolliott

1989, McClaran et al. 1992, McPherson 1997, McClaran and McPherson 1999). Composition of livestock herds in the southwestern United States generally differs from that of herds in northwestern Mexico. A predominance of cattle is found in the former, while higher proportions of sheep and goats graze in the latter (Downing and Ffolliott 1983). These differences in herd composition are attributed to economic and sociocultural orientations in the two countries.

Production of forage appears to be unrelated to the density of overstory oak trees, although the reverse might not be true. Perennial grasses can restrict the ability of Emory oak to become established on ungrazed grasslands (McPherson 1992, McClaran and McPherson 1999). However, once overstory trees have been cleared, annual forage production can increase 100 to 200 kg/ha. The question of whether a sacrifice in wood resources warrants obtaining increased forage remains unanswered (Allen 1992, McClaran et al. 1992).

Wood Production

Fuelwood and fenceposts are the wood products most commonly obtained from evergreen oak woodlands. Other wood products are processed from smaller, irregular stems and capitalize on physical characteristics of the wood that offer opportunities for furniture and novelty items (Ffolliott 1989, Ffolliott and Gottfried 1992, Maingi and Ffolliott 1992). Supplies of wood from the woodlands are projected to increase at a constant rate into the future; demands will also increase, but (in contrast to supplies) at continuously increasing rates (Conner et al. 1990, Van Hooser et al. 1990, Ffolliott and Gottfried 1992, Gottfried et al. 1995).

Extrapolation of anticipated supply and demand trends suggests that demands for wood could exceed supplies by the first part of the twenty-first century. Therefore, management agencies have begun to limit the amount of wood that can be harvested, through cutting restrictions and, in some instances, total cutting exclusions (Bennett 1992, Gottfried et al. 1995). Regional harvesting restrictions have been reenforced by environmental pressures to manage the woodlands on an ecosystem basis.

Watershed Protection

Watershed management practices that protect fragile watershed resources are of paramount necessity in the evergreen woodlands of Arizona and New Mexico (Lopes and Ffolliott 1992, Baker et al. 1995).

These practices include establishing tree, shrub, or grass covers on severely eroded or otherwise degraded sites; controlling gully formation and soil mass wasting by constructing check dams and other control structures; and limiting tree harvesting, livestock grazing, and road construction on sites undergoing or potentially subjected to excessive soil loss.

Investigations of hydrologic relationships in the evergreen oak woodlands have only recently begun. A better understanding of interception, transpiration, and infiltration processes has been obtained from these investigations (Lopes and Ffolliott 1992, Haworth and McPherson 1994, Baker et al. 1995, Ffolliott and Gottfried 1999). This information is crucial to prescribing effective management practices to improve conservation, protection, and sustainable development of watershed resources in the evergreen oak woodlands (Ffolliott et al. 1993).

Recreation and Tourism

Recreation and tourism activities in the evergreen oak woodlands are regional income generators. Deer and quail hunting has long been a dominant recreational use (McClaran and McPherson 1999). Nonconsumptive recreation includes hiking, camping, sightseeing, bird watching, and picnicking, although comprehensive statistics on these recreational experiences are incomplete (Conner et al. 1990). Nevertheless, increasing numbers of people from within and outside of the region are spending considerable time enjoying the unique settings found in the evergreen oak woodlands (McClaran et al. 1992).

Silvicultural Prescriptions

Natural regeneration from seeds (acorns) of evergreen oaks is episodical. As a consequence, evergreen oak woodlands in Arizona and New Mexico might not reproduce in numbers sufficient to sustain themselves for wood production if sexual reproduction is to be depended upon (Borelli et al. 1994). Poor regeneration from acorns has been attributed to livestock grazing (Phillips 1912) and the common early-summer drought (Pase 1969, Nyandiga and McPherson 1992). Therefore, vegetative reproduction by sprouting from stumps and roots could be a more important regenerative mechanism in the long run.

Because evergreen oak trees sprout vigorously after they are cut, silvicultural prescriptions based on coppicing treatments can form a basis for obtaining regeneration following the harvesting of trees for fuel-wood and other wood products (Gottfried and Ffolliott 1993, Gottfried et al. 1995). Furthermore, subsequent harvesting intervals might be shortened through the proper timing of selective thinning of post-harvesting stump sprouts, to increase growth of the residual stems. A preliminary guide for managers planning future fuelwood harvesting is to thin clumps of sprouts five years after initial harvesting, leaving one to three residuals that represent the largest and most vigorous of the postharvesting stump sprouts (Touchan and Ffolliott 1999). The sustainability of regenerative patterns by vegetative reproduction is unknown, however.

Silvicultural prescriptions based on clearcutting and shelterwood, seed tree, and selection cuttings have been listed as possibilities for the evergreen oak woodlands by the U.S. Forest Service (1987) in their classification of habitat types. However, widespread applications of these silvicultural prescriptions have not been tested.

SUMMARY

The need for more holistically conceived, ecosystem-based management of the evergreen oak woodlands in Arizona and New Mexico is intensifying with increasing demands for the multiple-use values of these woodlands. Most of these woodlands are located in the public domain; consequently, public land management policies dictate much of what can be done on these lands. A limited harvesting of trees will likely continue. However, inherently low levels of growth, irregular stem forms of the trees, a lack of markets for local wood products, and increasing environmental concerns restrict the harvesting opportunities. Wildlife values, livestock production, watershed-protection considerations, and increasing recreation and tourism are heightening pressures on decision makers and managers to ensure that each of these other uses is given due consideration in the future.

Chapter 21

Managing Eastern Oak Forests for Wildlife

WILLIAM M. HEALY

The rich literature on the ecology and silviculture of oaks has been thoroughly reviewed by Dey (Chapter 5). Both Dey and Abrams (Chapters 4 and 3) provide insights into the role of fire in maintaining oak ecosystems and an understanding of the oak regeneration problem that developed during the latter half of the twentieth century. Van Lear and Brose (Chapter 18) provide further insights into regeneration problems and describe a promising new regeneration method that employs shelterwood cutting and prescribed fire. In this chapter, I examine silvicultural approaches to managing oaks and then review methods for managing stands and landscapes to provide three important elements of wildlife habitat; mast, deadwood and cavity trees, and old growth.

BACKGROUND

We know more about managing stands for wildlife habitat than we do about managing forest landscapes to sustain wildlife diversity. Silvicultural guidelines allow us to manipulate stand structure to create specific habitat conditions. The effects on wildlife of landscape mosaics of different stand age-class and cover-type combinations are not well understood. It has been difficult to identify habitat features to which individual species respond (DeGraaf et al. 1998) and to measure species responses to landscape patterns created by forest management (Welsh and Healy 1993).

Despite these uncertainties, several generalizations can be made about vertebrate wildlife species in eastern forests. The majority of forest-dwelling species are habitat generalists. Forest cover type is a poor predictor of wildlife species distributions, because most species occur in many forest cover types. Most species also use stands of more than one age or tree size (DeGraaf et al. 1992). Thus, maintaining a variety of age classes and cover types is important for maintaining wildlife diversity in forested landscapes.

Scanlon (1992) reviewed the habitat requirements of forest-dwelling amphibians, reptiles, birds, and mammals in Massachusetts and concluded that optimally the bulk of the forest landscape should be in a mature stage (sawtimber). Seedling stands, however, were essential for maintaining wildlife diversity, because 79% of all forest-associated bird species and 88% of the mammals used seedling forests to meet part or all of their habitat requirements. Scanlon (1992) and DeGraaf et al. (1992) suggested maintaining a sawtimber-to-seedling ratio of about 5:1 on the New England landscape, and that stands should be composed of northern hardwood and transition hardwood forest associations. Both authors emphasized the critical role of forested wetlands and nonforest habitats in maintaining regional faunal diversity.

The structure of the landscape determines how we approach wildlife management and select silvicultural and habitat management systems. In fragmented landscapes of the Midwest and Mid-Atlantic, forests occur in isolated patches in agricultural and suburban landscapes. In these landscapes, wildlife management emphasizes increasing patch sizes and connectivity among patches. Management is designed to protect, restore, and enhance the limited habitat base. In the extensively forested landscapes of New England and Appalachia, wildlife management emphasizes maintaining a diversity of stand conditions within the forest matrix and protecting rare and uncommon habitats. Forest commodity production is also important in these landscapes. In the following section I consider predominantly forested landscapes where timber management is appropriate.

SILVICULTURAL SYSTEMS

Maintaining oaks and other nut-producing trees as dominant components of oak forests represents the greatest technical challenge to managers. Meeting this challenge will require using all the silvicultural tools available to us and a number of other vegetation-manipulation techniques (see, e.g., Vose et al. 1999). The loss of oak dominance often goes unnoticed because regeneration of some sort usually develops rapidly when oak forests are harvested. Landowners are often unaware that a significant ecological change has occurred following "selective cuts" that remove 20–30% of the basal area and most of the high-quality oaks. As destructive as high grading is to wildlife habitat in oak forests, it is acceptable to many landowners because structural integrity is equated with ecological integrity. Suggestions that more aggressive silviculture would be appropriate are usually rejected, often emphatically. Yet, clearcutting, patch cutting, and controlled burning are usually acceptable when used to improve wildlife habitat or to restore an ecosystem, even though the timber is also harvested. Therefore, public perceptions of timber management and of the motivations for management are important elements in selecting and implementing silvicultural systems. If we expect to be able to use the tools that are most effective at sustaining oaks, we must be able to explain the ecological values of oak ecosystems and the role of disturbance in maintaining this system.

Early silvicultural guidelines for upland oak forest recommended even-aged management systems and the clearcutting method of regeneration (Roach and Gingrich 1968). Even-aged systems and area regulation of harvest are conceptually simple, and wildlife managers generally embraced even-aged management as appropriate for providing wildlife habitat in oak forests (Shaw 1971). Even-aged management produced a landscape dominated by stands that were of mast-producing age, yet stands of all age classes were available to meet the needs of various wildlife species (Scanlon 1992). Over time it became apparent that the clearcutting method often failed to regenerate oaks, especially on mesic sites.

The clearcutting method of regeneration worked when large advance oak regeneration dominated the understory, but in many cases when oak stands were harvested by clearcutting, rapidly growing shade-intolerant species or well-established shade-tolerant species replaced oaks (Chapters 5 and 18). Shelterwood regeneration methods, which removed the overstory in two or more stages, were designed to ensure adequate oak regeneration. The first cut, or seed cut, was made within three years of a bumper acorn crop and was often preceded by a herbicide treatment that removed understory stems that would compete with oak regeneration. The final overstory removal occurred only after adequate oak regeneration was present. Shelterwood regeneration has been successful on many, but not all, sites (Schuler and Miller 1995). The method does require five to ten years to implement, careful evaluation of regeneration, and flexibility in scheduling the harvests (Wolf 1988). In Chapter 18 Van Lear and Brose describe a recently developed method of regeneration that employs shelterwood cutting and controlled burning and that has great potential. The method has been successful on high-quality sites in the Piedmont Region, and I encourage experimenting with shelterwood-burn techniques in other regions. The shelterwood-burn method may be more economically attractive than other shelterwood systems, but the use of fire will be difficult on many properties. I am comfortable with even-aged management and clearcutting, shelterwood, and shelterwood-burn methods of regeneration for providing wildlife habitat in oak-dominated landscapes. The shelterwood-burn regeneration method is appealing to me because it returns fire to the landscape and may provide a number of benefits to wildlife. The method can easily be modified to retain trees with high wildlife value in the next stand.

To be effective for wildlife management, even-aged systems need to meet several conditions. First, rotations need to be relatively long. I recommend rotations long enough to grow large-diameter, high-quality sawlogs (e.g., 80–120 years). Second, rare and fragile habitats, such as wetlands, steep slopes, eaves, and rock outcrops, need to be protected. Rare plants and animals tend to be concentrated in uncommon habitats. Third, some areas should be allowed to develop old-growth characteristics. This is particularly important on public lands, where ecologic, aesthetic, and spiritual values outweigh timber values. Maintaining old-growth oak forest will require the use of fire and active management of the deer population. Finally, silvicultural treatments should provide for reasonable numbers of snags, den trees, nest trees, and other structures with high value for wildlife within stands.

I must emphasize, however, that I would be comfortable with any silvicultural system that provided the features described above and regenerated oaks. Uneven-aged management has generally been ineffective for oaks. The selection method of regeneration eventually produces stands dominated by the most shade-tolerant species that are capable of growing on the site (Marquis et al. 1992). Single-tree selection in oak stands usually results in the conversion of these stands to dominance by more shade-tolerant species, such as sugar maple and beech on better sites and red maple, sugar maple, and beech on poorer sites (Gottschalk 1983). The xeric to dry-mesic oak-hickory forests of the southern Missouri Ozarks provide a notable exception to this general rule. Oak reproduction tends to accumulate in these stands, and the single-tree selection method has maintained oak dominance in these forests (Loewenstein et al. 1995, Dey et al. 1996). Even-aged management is also effective in these ecosystems.

Eastern oak forests are characterized by great species diversity, and no single silvicultural system will be universally applicable. The compositional diversity among stands within the same forest association, and often within those growing in close proximity, is evident in the descriptions of old-growth stands provided by Braun (1950). The striking feature of most stands described by Braun is that they were composed of mixtures of species that were tolerant, intolerant, and intermediately tolerant of shade. It is clear that natural disturbances were variable in intensity and frequency. Braun (p. 32) noted that the eastern deciduous forest had been called the *Quercus-Fagus* formation by earlier ecologists; oaks and beeches were widespread, co-occurring dominants, a condition that would be difficult to duplicate by using either even-aged or uneven-aged silvicultural techniques exclusively. The great diversity of site conditions within eastern oak-dominated landscapes also requires more than one approach to silviculture. For example, an 8,100 ha oak-dominated forest in the Ridge and Valley region of West Virginia included 14 distinct forest cover types, ranging from extremely xeric table-mountain pine (*Pinus pungens*) stands to mesic cove hardwoods (Gill et al. 1975). The most xeric types were fire dependent, while various combinations of even- and uneven-aged techniques would have been appropriate on the more productive sites.

Franklin (1997) has proposed replacing traditional harvesting systems with "variable retention" harvesting systems. Harvesting prescriptions would be designed to meet specific management objectives for individual stands, and treatments could range from full canopy removal to single-tree selection. Variable retention has great appeal for wildlife management, because it allows flexible disturbance regimes within and among stands.

Franklin (1997) emphasizes the importance of retaining and restoring structural complexity within stands. Western coniferous types that require centuries to reach old-growth conditions have been structurally simplified by management that includes clearcutting, site preparation, planting, and short rotations. Structural complexity has been less compromised in eastern deciduous forest landscapes that have been less intensively managed and where rotations are proportionally much longer. In oak forests, I see species diversity as the more challenging issue. In fact, fire and regeneration techniques tend to simplify structure in oak stands by reducing the understory and midstory layers. Shelterwood systems remove the midstory and create a single layer overstory of uniformly spaced trees. Repeated fire has a similar effect, producing parklike stands that are less than fully stocked. These structural conditions concentrate energy in grasses, forbs, fruits, and tree seeds, where it is available for wildlife. Structural complexity is provided at the landscape level by providing stands of different age classes and retaining structural features such as dens, cavities, and coarse woody debris within stands.

Oaks thrive in environments characterized by disturbance and stress, and regeneration is a process that requires time (Chapter 5). Successful regeneration requires a series of treatments that establish seedlings, grow them to large size in the understory, and promote their development after overstory removal. We have the tools to accomplish these ends. The difficulty lies in explaining to landowners and stakeholders the values of oak ecosystems and the role of disturbance and silviculture in sustaining these systems.

MANAGING FOR MAST

Wildlife managers have always stressed maintaining the mast-producing capability of forest land, because acorns and other tree seeds are important to so many species. Recommendations to promote seed production for wildlife generally include (1) periodic thinnings to promote vigorous crowns and rapid growth of mast producers; (2) managing for a diversity of mast-producing species, with a mixture of oaks consisting of one-third in the white oak group and two-thirds in the red oak group; and (3) maintaining about half of the management unit in mast-producing stands (Shaw 1971, Dellinger 1973). Mast-producing stands include oak types greater than 40 years old, sawtimber-size hardwood types with 50% of the basal area in oak, and any cover type with more than 30 square feet of basal area in oak and sawtimber.

These recommendations were developed in the late 1960s, when even-aged management and the clearcutting method of regeneration were considered the most effective treatment for oaks (Roach and Gingrich 1968). The goals for acorn production were based on estimates of the dietary needs of acorn-consuming wildlife during the dormant season (Goodrum et al. 1971, Shaw 1971). These management recommendations were intended for extensive tracts of forest land managed for multiple uses that included the sustained yield of high-quality sawtimber, even-aged silvicultural systems, and relatively long (80–120 years) rotations. With the exception of adding guidelines for cavity and den trees, little refinement has occurred in recent years. In my experience, this approach has sustained wildlife populations and habitat on extensive tracts. Scientists now recognize the difficulty of retaining oaks and hickories as stands are harvested and the need for improved regeneration methods (Chapters 5 and 18).

Managing Stands

The goal of treating stands to enhance acorn production has been difficult to achieve, because of variation in productivity among individuals, variation among species, and the weak relationship between tree diameter (or basal area) and acorn production per unit of crown area (Chapter 10). Open-grown trees flower and fruit more abundantly than trees in closed canopy, and it is well established that full crown release will promote profuse and regular flowering and seed production by the seed trees (Matthews 1963). Releasing the crown of an individual oak will increase acorn production by that tree (Healy 1997b), but thinning oak stands has produced mixed results, decreasing per hectare acorn production about as often as increasing it (Beck 1993).

Part of the difficulty lies in the extreme variation in acorn production among individual trees of the same species. Most stands consist of a mixture of good and poor acorn producers. For example, among 120 red oaks observed for 11 years in Massachusetts, 39% were better than average producers, and they provided 61% of the acorns collected. About 25% of the trees were poor producers and they provided only 12% of the total acorn collection (Healy et al. 1999). If superior producers could be identified and released from competition with poor producers, thinning immature stands would increase acorn production, but practical methods for identifying good seed producers in well-stocked stands have not been developed. The only time thinning can be guaranteed to increase acorn production per hectare is when other species are removed to enhance crown development in oaks.

Sharp (1958) described the best acorn producers as uncrowded, dominant trees of good form, but red oaks meeting these criteria have been found to include both good and poor producers (Healy et al. 1999). We have been unable to identify any physical characteristic that separates good from poor producers among the dominant trees in a stand. At present, the only sure way to identify good producers is by monitoring an individual tree's acorn production for several years (Johnson 1994a, Healy et al. 1999).

One approach was developed by recording acorn production of 120 northern red oaks over 11 years (Healy et al. 1999). Trees whose mean 11-year production exceeded the 11-year population mean were considered good producers. The analysis simulated how successfully, on average, a forester could identify the best acorn producers in the population by annually marking all trees that produced more acorns than the year's stand mean. The most efficient procedure was to select trees that were identified as better than average in at least one year during periods of two to five consecutive years. Requiring trees to produce more acorns than the stand average for two or more consecutive years was found to be too restrictive a criterion to be useful.

Three consecutive years of observation generally resulted in the inclusion of most (72–100%) good producers in the marked group. On average, after three consecutive years, 60% of the total population and 87% of the good producers were marked. Monitoring for five years did identify more of the good producers, but it also resulted in marking so many trees that there was little room for silvicultural manipulation. The system was most efficient when the period of observation included one or more years of fair to good production. The years of observation need not be consecutive. Field procedure includes inspecting tree crowns with binoculars prior to acorn fall, classifying trees as above or below average producers, and marking good producers. This method of identifying superior acorn producers is effective and conceptually simple, but it requires at least three years to implement (Healy et al. 1999).

At least one additional factor must be considered when manipulating stand structure to promote acorn production. The stand density giving maximum seed production per tree is usually much lower than that giving maximum seed production per hectare (Harlow and Eikum 1963, Matthews 1963). The weak correlation between tree diameter at breast height (dbh) (or basal area [BA]) and acorn production per unit of crown area described by Greenberg and Parresol (Chapter 10) is an important factor. Acorn production per tree is positively correlated with crown area, which is positively correlated with tree diameter and basal area. Tree diameter, however, explains little of the variation in acorn production among individuals. Large trees tend to produce more acorns than small ones because they have larger crowns, but they are not necessarily more productive per unit of crown area. Greenberg and Parresol found that black, red, and white oaks < 25 cm dbh produced fewer acorns per m^2 of BA than larger trees, and that acorn production per m^2 BA appeared to decline in red, scarlet, and white oaks > 76 cm dbh. Thus, over a wide range of mean stand diameter, the potential for acorn production is primarily a function of total crown area, regardless of whether the crown area comes from a few large trees or many smaller ones. Because of the ease of measurement, basal area of dominant and codominant trees provides the best indicator of per hectare acorn production potential.

Providing Mast across the Landscape

The traditional approach for providing mast with even-aged management systems has been to extend the rotation length. For example, with a 100-year rotation, 50% of the area would be at least 50 years old, and 1% would be regenerated annually. For eastern oak forests, substantial mast production usually occurs by age 40, so about 60% of the area would be producing mast. The forest would be a matrix of 40+ year-old stands with scattered patches of regeneration.

Rotation lengths of 80 to 120 years are used to grow quality hardwood sawtimber, and management usually includes thinnings that favor mast production. Mature, even-aged oak stands are quite productive. In Massachusetts, the mean dry weight of the acorn crop ranged from 77 to 549 kg/ha (30,000–220,000 sound acorns/ha) over three successive years (Healy 1997b). In the best year, several stands produced more than 900 kg/ha. Similar values have been reported for mature, mixed-oak stands in Missouri (Christisen and Kearby 1984), while acorn production was less over a four-year period in the Shenandoah Valley of Virginia (3–396 kg/ha) (McShea and Schwede 1993). In sum, even-aged management approaches with extended rotations are effective at maintaining mast production potential across the landscape.

A number of alternatives to even-aged management are being proposed for eastern oak forests. Franklin (1997) proposed variable-retention harvesting systems that could be used to meet any number of specific management objectives for a stand. Modified shelterwood cuttings and treatments designed to produce two or three age class stands are being tried as alternatives to the clearcutting method of regeneration in eastern oak forests. Modified shelterwood cuts retain the residual overstory as the next cohort develops. These treatments leave 10–30 ft.^{2} of ba/acre in mast-producing trees, so the next stand will contain a small cohort of older trees (1–2 × rotation length) throughout its life.

"Variable-retention regeneration methods" will retain some mast production in a stand at all times. It is not clear if these methods will increase mast production at the landscape scale, because overall production will depend on the total basal area in mast-producing trees. Variable retention methods certainly should be tried, especially where they are used in conjunction with fire to restore oak savannas and oak-pine mixtures.

DEADWOOD

Dead trees have important ecological and physical roles in forested ecosystems (Spies et al. 1988). Larger dead material, called coarse woody debris (CWD) includes any standing dead or fallen tree stems greater than some minimum diameter, usually 7–20 cm in eastern forests (Muller and Liu 1991, Van Lear 1993). CWD provides habitat for numerous plants and animals and plays a significant role in nutrient cycling, soil development, and sediment transport (Tyrell and Crow 1994). Traditional silviculture has been directed almost exclusively at living trees, while ecosystem management places strong emphasis on incorporating deadwood into forest management (Franklin 1997). Recognizing the importance of deadwood, and also live cavity trees, will require revising our basic silvicultural premises (Franklin 1997).

The challenge of providing CWD differs between eastern deciduous forests and western coniferous forests, and because much of the literature comes from western forests it is worth considering these differences. Coarse woody debris dynamics in eastern deciduous forests are characterized by rapid rates of decay. Twenty-three years after all trees were felled and left in place in a mature northern hardwood stand in New Hampshire, the mass of dead boles had declined by approximately 90% (Arthur et al. 1998). A comparable loss of the CWD from a western old-growth Douglas-fir (*Pseudotsuga menziesii*) forest following a stand-replacing fire would have taken about 100 years (Spies et al. 1988).

Because of the rapid decay rates in eastern forests, CWD reaches only modest accumulations under old-growth conditions (Muller and Liu 1991). The apparent upper limit for CWD biomass was 30–35 mg/ha in warm temperate deciduous forests and 50 mg/ha for cool temperate forests (Muller and Liu 1991). In contrast, CWD ranged from 72 mg/ha to 174 mg/ha among Douglas-fir forests aged 40 to 900 years old in western Oregon and Washington (Spies et al. 1988). The length of time required to reach old-growth conditions is considerably longer for western coniferous forests (400–500 years) than for eastern deciduous forests (150–200 years).

The total amount of CWD is often greatest early in succession, lowest in mature forests, and somewhat greater in older forests (Gore and Patterson 1986, Spies et al. 1988). CWD early in succession is derived from the previous stand, while CWD in older stands comes from mortality within the current stand. For managed eastern forests, the greatest mass of CWD is found immediately after stand regeneration, and this material decays rapidly within 20–30 years after logging (Gore and Patterson 1986). In midwestern deciduous forests, the volume of down wood decreased with increasing stand age between 10 to 70 years. Beyond age 70, down wood volume begins to increase and old-growth forest may have more than three times the volume of deadwood than 70–90-year-old upland forests (Spetich et al. 1999).

In addition to stand age, CWD is strongly influenced by site productivity and disturbance history. The amount of deadwood increases along a site's productivity gradient, so young, productive sites can have more CWD than older sites with low productivity. In older stands, the time since a disturbance is often more closely related to the density and volume of CWD than is stand age, because discrete disturbance events can add large amounts of CWD (Goebel and Hix 1996, Spetich et al. 1999).

For the wildlife manager, the diameter distribution of snags and logs is of more interest than the total volume, because larger vertebrates such as pileated woodpeckers and black bears require large diameter snags and den trees. Snag density may be greatest during the stem exclusion phase of stand development and even young even-aged stands may have snag densities equal to those of old-growth stands (McComb and Muller 1983). In contrast to density, the diameter distribution of snags and logs increases with stand age. Older stands have greater mean stand diameters and larger diameter snags than younger ones (McComb and Muller 1983). In fact, the diameter distribution of live trees provides a good indicator of the diameter distribution of standing dead trees. In midwestern old-growth forests, there was a relatively constant ratio of dead to live trees of about 0.1 across all diameter classes $\leq$ 65 cm (Spetich et al. 1999). Thus, the size distribution of snags was similar to that of live trees, but snags were only about 10% as abundant. Obviously it takes time to grow large-diameter trees, and thus to have available large-diameter CWD. It will take about 50 years to grow large-diameter trees ($\geq$ 60 cm dbh) on excellent oak sites (red oak site index 80), and longer on lower quality sites.

One method for providing adequate CWD and large-diameter snags and logs is to extend the rotation length. Maximum tree size and dead-wood volume should increase to about age 200, and perhaps beyond (Spetich et al. 1999). Two-hundred-year rotations are probably impractical for timber production, and they would be less than optimum for mast production. because seed production declines in larger-diameter trees (Chapter 10). Such long rotations would be possible only on better sites and with long-lived species.

An alternative is to retain individual trees beyond rotation age. Even-aged management with the shelterwood method of regeneration lends itself to retention of healthy individuals until they reach advanced age and size. Many other approaches could be developed for leaving trees on site through their natural life and decay cycle.

LIVING CAVITY AND DEN TREES

Retaining live cavity trees is a separate issue from providing dead trees. Cavity formation is a process that involves the action of wood-decay fungi in living trees (DeGraaf and Shigo 1985). Killing sound trees will eventually provide CWD, but it will not lead to cavity formation.

Live cavity trees are used by mammals ranging in size from white-footed mice (*Peromyscus leucopus*) to black bears (*Ursus americanus*), for shelter, denning, and rearing young (DeGraaf and Shigo 1985). Many woodpecker species also prefer to nest in a column of decay surrounded by sound wood (Conner et al. 1975, Evans and Conner 1979). In comparison with snags, cavity trees have long life spans, persisting for many decades. Dead trees are usually ignored in silvicultural treatments, but cavity trees are traditionally viewed as defective and targeted for removal. Specific marking guidelines must be developed to ensure that cavity trees are identified and retained during silvicultural treatments.

The general philosophy of thinning has been to save trees with best growth form and potential and to remove trees of poor form and those unlikely to survive to the next cutting cycle. This "save the best" philosophy is clearly effective for timber and mast production, and we also believe it is appropriate for providing cavity trees. There is a general perception that providing enough cavities for wildlife will lead to the development of excessive amounts of cull, but this has not been the case in our experience. In mature oak forests in Massachusetts, cavities occurred in 3% of the live trees, and cavity trees accounted for 3.8% of the live basal area (Healy et al. 1989). We believe these figures are representative of the opportunity costs for providing cavities for wildlife at a landscape scale. We emphasize, however, the variation among stands; cavity trees accounted for 0.0 to 15.8% of the live basal area among 13 stands.

Cavities were found in trees of all quality classes in the Massachusetts study. The proportion of trees with cavities in each quality class was: rotten cull, 38%; rough cull, 11%; acceptable, 4%; preferred, 3%; and live sapling, 1%. Although relatively few preferred and acceptable trees had cavities, 20% of all cavity trees were in these quality classes. Only 9% of the cavity trees were rotten culls. Live saplings (2.5–12.6 cm dbh) had the lowest proportion of cavities, but 20% of the cavity trees were live saplings, and these cavities were extensively used by white-footed mice.

Clearly, thinnings can be conducted in these stands that will retain most cavity trees and have little impact on timber yield. Overall, 96% of the acceptable trees, 89% of the rough cull, and 63% of the rotten cull trees did not contain cavities. Thus, there is considerable room for improving timber growing stock quality, promoting mast production, and retaining cavity trees.

I recommend retaining all cavity trees during cultural operations. Per hectare guidelines for cavity trees are of limited use when treating stands, because cavity trees are not uniformly distributed within or among stands. I emphasize retaining vigorous and otherwise healthy cavity trees. Preferred and acceptable trees are characterized as being mature and free from serious damage, having less than 20% total board foot cull, and having a life expectancy of 10 or more years. Many of these trees have the potential to persist well beyond rotation age and ultimately produce very large cavities. We were able to identify about 80% of cavity trees in fully stocked, mature oak stands by examining trees from the ground during the dormant season (Healy et al. 1989). Binoculars are particularly helpful for confirming the presence of cavities. Examining trees for cavities adds little time to inventory and marking procedures, because trees are already being examined for growth form and quality.

Finally, it is important to recognize that random processes play a major role in cavity tree development in hardwood forests. The abundance of cavity trees is variable, even among similar forests, and much of the variability is unrelated to stand age, diameter, site index, or other stand and topographic features (Carey 1983).

SUSTAINING OLD GROWTH AND OLD-GROWTH VALUES

Not much old growth remains in eastern oak forests, and most that exists occurs in small patches (White and Lloyd 1995). In the Northeast, Lake States, and southern mountains, old-growth stands are often embedded in a matrix of second-growth forest. In the Midwest, old-growth stands are usually small and isolated in a matrix of agricultural land (Parker 1989), and opportunities for enhancing old growth are limited. In extensively forested landscapes, managers have unprecedented opportunities for enhancing old growth and managing second growth for old-growth attributes because of the age, history, and extent of forest land. Much of today's eastern oak forest originated in the late 1800s and early 1900s following extensive cutting and the abandonment of agricultural lands. Many stands are now approaching a century in age and have been free from major disturbance for 50 or more years. Within this maturing landscape, primary forests or stands that have never been plowed or converted to pasture hold the most potential for developing old-growth characteristics. These stands have usually been cut over several times, but they have always remained forested. Primary forest stands are most likely to contain the original flora and least likely to have been invaded by exotic plant species.

Old-growth and old second-growth eastern hardwood forests are structurally similar. Stem density, basal area, and mean stand diameter fall within similar ranges (Parker 1989, Goebel and Hix 1996). On productive sites, the most striking features of old-growth stands are the large trees and the large size of the tip-up mounds, cavities, and down woody material associated with them. Our experience with primary and second-growth stands on good sites in West Virginia suggests that these large structures begin to emerge around 100 years of age. Mean stand diameter and maximum tree size were about the same in 100- to 120-year-old stands and a 220-year-old stand, but the oldest trees had distinctive bark and crown characteristics (see Healy and Brooks 1988 for stand description).

Large, old trees and the stands that contain them are highly valued for spiritual inspiration, aesthetics, and wildlife habitat (White and Lloyd 1995). The ecological role of old growth in the eastern landscape is not well understood, because old growth remnants are scattered and small. We think primary second-growth forest offers the best opportunities for buffering existing old growth and developing larger patches with some old-growth attributes. An example of the potential to develop old-growth attributes can be found on the 22,663 ha Quabbin Reservation, managed by the Metropolitan District Commission to provide Boston's water supply. About 80% of the arable land in this region of the Connecticut Valley had been cleared for agriculture by the mid-1800s. White pine stands developed naturally as pastures and fields were abandoned during the latter half of the nineteenth century. Commercial harvest of old-field white pine led to the development of the modern forest dominated by red oak, red maple, and white pine (Hosley and Ziebarth 1935, Foster 1992). The Quabbin Reservation is 93% forested; 99% of the forest is older than 50 years, and 59% is older than 90 years (Metropolitan District Commission 1995 p. 13). About 13% of the area has been identified as primary forest.

Mladenoff et al. (1994) provide a model that could be used for enhancing old-growth or primary forest characteristics within a working forest landscape. The model was developed in north-central Wisconsin, where scattered remnants of the original old-growth hemlock and northern hardwood forest occurred in a matrix of managed second-growth forest. Old growth patches were identified and each was surrounded by a 100-m restoration zone, and then a 300-m secondary zone. The restoration zone served as an unmanipulated buffer, moderate harvest was allowed in the secondary zone, and the remainder of the landscape was used for general commodity management. This zoning system greatly increased the connectivity of old-growth patches and decreased the edge-to-interior ratio. With their definition of 100-m edge effect, all of the old growth became interior forest. Approximately 75% of the area remained in general commodity production, while about 12% was reserved as old growth and another 12% was managed conservatively with even-aged systems. Establishing large reserves was not an option in these commercial woodlands of mixed ownership. The shade-tolerant hemlock and northern hardwood cover types are well suited to uneven-aged management with the single-tree selection system, and fire was not part of the natural disturbance regime.

Restoring natural fire regimes will be a major challenge to sustaining old growth and mature, primary oak forest. Some of the best examples of mature, primary oak stands on the Quabbin have little or no advance oak regeneration and have well-developed understories of shade tolerant species. Similar conditions exist on mesic and dry-mesic sites throughout the east (Chapter 4), and complete protection of old-growth oaks will result in their gradual replacement by shade tolerant species (Fralish et al. 1991). Shelterwood cutting and prescribed fire can be used to regenerate stands managed for timber (Chapter 18), but the role of fire in maintaining old growth and primary forest needs further investigation.

Chapter 22

Goals and Guidelines for Managing Oak Ecosystems for Wildlife

WILLIAM M. HEALY AND WILLIAM J. MCSHEA

This volume is a compilation of what we know about oak forests and woodlands and selected species that use these habitats. Much of the information deals with eastern oak forests because those forests have a longer history of economic value and wildlife research. In this final chapter we provide goals for the management of oak forests and woodlands and prescriptions for reaching these goals. We do this with full knowledge that our guidelines will not fit all forests, but with hope that they are malleable enough to be used by most managers.

ESTABLISHING GOALS

Establishing goals is the first and most difficult step in instituting ecosystem management (Lawton 1997). There is general agreement that ecosystem management should be goal driven and that goals should be explicit, operational, and based on our best understanding of ecosystem function. The primary obstacle to implementing ecosystem management on national forest land has been the inability to reach consensus on management goals, rather than a lack of ecological or technical knowledge (Kotar 1997). Despite intensive planning efforts since the passage of the National Forest Management Act in 1976, the U.S. Forest Service has been unable to implement ecosystem management on a broad scale. Management decisions can be blocked by administrative appeal or by litigation at almost any stage in the process. Moving forward with management will require developing a decision-making process that is viewed as open and fair by all parties (Rauscher 1999).

Most management deals with individual forests or districts. Establishing regional management goals will be even more challenging, as there is no mechanism for reaching consensus and coordinating efforts across landscapes composed of diverse ownerships. Regional-scale ecosystem management would obviously require unprecedented cooperation among agencies and landowners. The closest we come to this type of management is the effort in California to preserve oak woodlands (see Chapter 19). The only comparable effort in the eastern United States that we are aware of is in the shortleaf pine (*Pinus echinata*) hardwood forests of the Quachita and Ozark National Forests, in the Quachita Mountain Region of Arkansas (Baker 1994). Oaks are important in these ecosystems and the goal of ecosystem management is to maintain mixed pine-hardwood stands that reflect indigenous vegetation and historical stand structures. The results of these efforts continue to be evaluated (Baker 1994). The social issues involved in establishing ecosystem management goals across broad regions are complex, and helpful reviews are provided by Endter-Wada et al. (1998) and Rauscher (1999).

The overall goal of ecosystem management is often stated as sustaining ecological integrity, the connotation of *integrity* being complete, unimpaired systems. More specific goals of ecosystem management include maintaining viable populations of native species, sustaining productive capacity, genetic diversity, and evolutionary and ecological processes, and accommodating human use (Grumbine 1994, Franklin 1997). Although there is general agreement among ecologists concerning these philosophical goals, the difficulty for managers arises in describing an oak forest that would best meet these goals.

A review of postglacial history illustrates the dynamic nature of North American forests (see Chapter 3). There is no natural baseline that can be used as a standard for decision making, because forest communities have been continuously reorganized, as individual plant species have responded to climate change over the past 10,000 years (Foster and Motzkin 1998). The rate of change has accelerated during the past three centuries, as human activity has altered the landscape (see Chapter 4). Wildlife communities have evolved with these changes, and at continental scale, most vertebrate species have survived the recent period of forest clearing and exploitation. We do not recommend setting goals based on some past or "virgin" condition, as those conditions were dynamic and, because of the loss of species, are not reproducible. However, we do recommend trying to maintain ecological functions, and with oaks that means attending to disturbance patterns and sources.

DISTURBANCE AND OAKS

Natural disturbance patterns should guide the development of management techniques for forests ecosystem management (Seymour and Hunter 1999). The techniques used to meet management goals should reflect the ecological pattern of the system. Patterns of natural disturbance vary among the physiographic regions of North America, reflecting patterns for climate, topography, and soils (Runkle 1990). Large-scale disturbances, such as blowdowns and stand-replacing fires, occurred throughout the eastern deciduous forest but were more important around the edges of the biome. In the central portions of the eastern forest, small-scale disturbances associated with the loss of individual trees were most common. Oaks are difficult to fit into this continuum of canopy disturbance. Shade-tolerant species clearly dominate in forests affected primarily by single tree gaps, while shade-intolerant species dominate in forests affected by catastrophic disturbance (Runkle 1990). Oaks require a combination of understory and overstory disturbance, which was generally provided by repeated surface fires. The shelterwood-burn regeneration method described by Van Lear and Brose in Chapter 18 emulates this disturbance pattern.

From an ecosystem management perspective, fire is the most important and difficult disturbance factor to replicate. Fire clearly played a critical role in the developent of oak forests and woodland, and human activity was a major source of fire ignition. The suite of natural canopy disturbances associated with weather events and individual tree mortality can be simulated with various cutting treatments, but the use of fire is limited by safety and air quality concerns. Despite the difficulties, we believe it is important to reintroduce fire into oak landscapes, and we encourage further development of controlled burning as an ecosystem management tool. Controlled burning will be particularly useful for three activities: restoration of fire-dependent ecosystems, such as oak savannas and pine-oak types; maintenance of old-growth oak reserves; and regeneration of managed oak forests.

Pathogens frequently serve as natural disturbance events within oak forests (see Chapter 6). Outbreaks that result in widespread canopy opening, and even mortality, are commonly caused by both natural and introduced agents. Whereas other disturbance events are presently managed for, pathogen outbreaks are frequently managed against. There is a real danger to oak forests from introduced pathogens, as the loss of American chestnut (*Castanea dentata*) and elm (*Ulmus americana*) attest;

but with native pathogens it is difficult, if not impossible, to determine when active management is appropriate. Most symptoms measured by managers are not due to single pathogens or vectors. Oak decline results from a combination of pathogens with environmental and demographic factors (Chapter 6), and managers should not expect to be able to isolate a single factor for action. We can offer no concrete advice regarding management of pathogen outbreaks, except that these should be considered an important part of disturbance regimes within oak forests and not ignored or treated as unexpected.

There is an additive nature to disturbance events. Management designed to mimic disturbance regimes cannot focus on the historical pattern of a single factor, such as fire, when determining the frequency or scope of planned activities. Charcoal records may indicate how often a region has experienced fire but not the rate or scope of storm damage or pathogen outbreaks. Because advanced oak regeneration depends partially on the degree of canopy opening, these unrecorded events cannot be ignored. Only detailed monitoring of current forests will provide a baseline of current disturbance regimes.

GOALS FOR OAK FORESTS

We believe that the quantity and quality of tree seed production distinguishes oak communities from other temperate forest associations. Oaks were dominant over much of the eastern deciduous forest, but this biome contained a rich mixture of nut-producing trees, including American beech, chestnut, hickories (*Carya* spp.), and walnuts (*Juglans nigra, J. cinera*) (Braun 1950). Seeds from these species are relatively high in energy and are available during the dormant season, when supplies of forage and invertebrate foods for wildlife are at a minimum. Seed production has undoubtedly diminished with the loss of chestnut and the decline of beech during this century. Yet, seed production continues to play a fundamental role in the organization and dynamics of eastern wildlife communities (see Chapters 7 and 13).

We believe that the philosophical goal for oak ecosystems should be maintaining the dominance of oaks and associated nut-producing trees. Developing operational goals has proven more difficult, and we can offer only broad goals based on our experience. We recommend maintaining a minimum of half the forest in mast-producing stands on landscape units of $\geq$ 5,000 ha. We have worked in forests where oak cover types ranged from 48% (central Massachusetts, 22,663 ha: Metropolitan District Commission 1995) to 86% (West Virginia, 8,100 ha: Gill et al. 1975). Both areas exhibited oak ecosystem dynamics, with annual fluctuations in acorn production determining wildlife abundance, movements, and distribution. In Massachusetts, oaks were common associates in pine and northern hardwood cover types, so oaks were more widespread than cover type alone would indicate. Mast-producing stands should have 30% to 80% of the basal area in nut-producing species (*Fagus, Castanea, Quercus, Juglans,* and *Carya*). Mast-producing stands should be maintained at stocking levels of 80% to 100%, with dominant trees having diameters of $\geq$ 30 cm. Overstory removal should not be attempted unless there is enough oak reproduction present to satisfy the regeneration-stocking guidelines developed for the region (Gottschalk 1983, Loftis 1990a, Sander et al. 1992).

Herbivory and Oaks

Herbivory has received little attention in the ecosystem management literature, yet herbivory is an important factor in oak ecosystems and one that we believe will become increasingly important for ecosystem management. Grazing by wildlife and domestic livestock has always been recognized as important in western oak woodlands (Chapter 19). Livestock grazing has largely been eliminated from eastern forests, leaving white-tailed deer (*Odocoileus virginianus*) as the most influential herbivore. Much of the existing eastern forest regenerated in the early 1900s when deer numbers were at historically low levels. Today, deer numbers are at historic highs in many areas, and some evidence suggests that deer may be imposing unprecedented browsing pressure on native plant species everywhere that large predators such as wolves (*Canis lupus*) were extirpated in eastern North America (Crête 1999).

Selective foraging by deer can affect plant species composition, structure, diversity, and successional pathways (Healy 1997a). Deer have significant impacts on endangered and threatened plants, especially herbaceous species (Miller et al. 1992). The negative effects of deer browsing on forest regeneration have frequently been noted, and deer seem to be responsible for the absence of adequate tree regeneration over large areas and in many forest cover types (McWilliams et al. 1995, Frederickson et al. 1998). Despite an extensive literature, deer effects on forest vegetation dynamics are poorly understood, because deer density has rarely been measured with precision (Gill 1992). Consequently, it is difficult to make comparisons among studies or to identify deer densities associated with particular plant community responses. The absolute density of deer is less important than the density of deer relative to forage resources (deCalesta and Stout 1997). The appropriate density of deer for an oak forest is one that permits the development of sufficient oak regeneration to ensure the dominance of oak in future stands.

There are many administrative and social challenges to integrating deer management and ecosystem management. Wildlife and forests are managed by different agencies, usually with different goals and values. Deer population has been manipulated largely through public hunting. Hunting participation rates are declining, the hunting population is aging, and access to land for hunting is declining. Deer hunters and advocates of rare plants and other aspects of forest ecosystem diversity often have different interests and goals for deer management (Sinclair 1997). Despite these difficulties, public hunting can be an effective tool for meeting regional and local ecosystem management goals. We encourage continued dialogue about deer management goals among all stakeholders and management agencies. Forest managers should encourage responsible hunting practices and access policies that permit adequate harvest, especially of female deer.

Herbivory by gypsy moth larva (*Lymantria dispar*) has had severe impacts on eastern oak forests (see Chapter 7). Whereas herbivory by deer affects primarily the seedlings and saplings, herbivory by gypsy moth larva attacks adult trees and affects both adult survival and acorn production. Low levels of gypsy moth population are maintained by predation by small mammals, but moth outbreaks do occur. Since our goal in managing oak ecosystems is to maintain seed production, and gypsy moth larva are an introduced pest, we recommend management against severe outbreaks that would result in widespread canopy loss. The replacement of oak canopies by red maple following widespread mortality also supports the active prevention of widespread mortality (Chapter 7).

MONITORING OF OAK FORESTS

A manager's ability to meet or maintain specific goals can be measured only by monitoring of forest resources. Monitoring serves two functions: to provide baseline data on factors that influence forest condition and to record progress toward management goals. As mentioned previously, several forest features, such as disturbance and herbivory, have not been measured adequately to provide managers with sufficient relevant information. Factors that affect regeneration, through either opening the canopy or altering the understory, must be monitored before sustainable management of oak forests can be achieved.

In an ecosystem management context, monitoring should be designed to measure progress toward goals and test hypotheses about management treatments. Defining clear operational goals is the first step in developing a focused, scientifically designed monitoring program; but even with clear operational goals, monitoring, especially of wildlife, presents an immense challenge (Franklin 1997). For many areas, we lack basic inventories or the ability to predict if species known to exist in the region actually will be found on a particular patch. Estimating animal populations, even of common and abundant species, on large areas (10^5–10^6 ha) is technically possible but generally impractical. For example, mark-recapture methods can be used to estimate the abundance of wild turkeys (*Meleagris gallopavo*), but obtaining an adequate sample has proven economically and logistically impractical (Weinstein et al. 1995). Similar difficulties are encountered with all wide-ranging species, especially rare ones. In addition, many species that consume or disperse acorns exhibit large annual fluctuations in abundance (Elkinton et al. 1996, Brooks et al. 1998). We do not understand the relationships between ecosystem function and abundance for many of these species, so the results of annual surveys are difficult to interpret.

Because of the difficulties outlined above, we recommend focusing monitoring efforts on habitat attributes such as dominant tree species, stand age, and regeneration characteristics. In contrast with methods for estimating animal abundance, methods for monitoring forest resources are well developed and applicable at local and regional scales (He et al. 1998, Scott 1998). McWilliams et al. (Chapter 2) demonstrate how the U.S. Forest Service's Forest Inventory and Analysis (FIA) data can be used to describe the abundance and distribution of oaks and to estimate changes over time. Field inventory data can be integrated with classified satellite imagery to assess forest composition across large, heterogeneous landscapes (He et al. 1998). Comparable technology does not exist for wildlife assessment.

There are several additional reasons why we emphasize monitoring habitat conditions. First, operational goals for wildlife management are more likely to be expressed in terms of habitat condition and distribution than in terms of animal abundance. Secondly, vegetation and physiographic features form the basic map data for management and planning. Finally, vegetation data that can be applied to forest landscape modeling has much more utility for projecting future ecosystem conditions than does data on wildlife abundance.

CONCLUSION: WORKING AT THE RIGHT SCALE

Ecosystem management of oak forests should operate over a broad range of spatial and temporal scales (Society of American Foresters 1991, Franklin 1997). Ideally, time frames should encompass the natural forest seres for the region and the spatial scales should meet the needs of the widest-ranging wildlife species. With regards to a time frame for management, eastern forest seres span from one to four centuries from initiation to demise. Forty years is a common span between seedling regeneration and acorn production for individual trees. With regards to spatial scale, perhaps the widest-ranging eastern species was the passenger pigeon, flocks of which traversed the length and breadth of the eastern deciduous forest (Bucher 1992). Today, black bears *(Ursus americanus)* hold that distinction.Adult females occupy home ranges of from 15 km^2 to 50 km^2, and those of adult males are several times larger (Kolenosky and Strathearn 1987a). These figures are for individual animals; a viable population of black bears would occupy well over 1000 km^2.

Managers, especially those working with private landowners, may despair of having any influence at these scales. Ownership and economic trends are clearly moving toward smaller properties, shorter landowner tenure, and short-term economic planning. Continuity of ownership is more assured on public lands, but even on national forests the amount of land may be less than the optimum for ecosystem management (e.g., Society of American Foresters 1991). The ecosystem management of oak forests needs the leadership of an agency whose vision is long-term and whose scope is large. We believe that state agencies are best situated to take this lead, as both federal and private landholders are generally responsive to state agencies. Regulatory guidelines, like those outlined for California (see Chapter 19), are best tailored at the state level. Whether this leadership is best taken by the state wildlife or forest agency depends on the priority wildlife receives within each agency.

Despite the present limitations, we encourage managers to plan and assess the effects of management activities at a variety of spatial scales and over long time periods. In our experience, many private landowners are willing to forgo some economic return to provide ecological benefits or meet the needs of wildlife, if they are made aware of these values (Schaberg et al. 1999). As previous chapters demonstrate, our knowledge of oak and wildlife dynamics is quite detailed, and we have the technical knowledge to regenerate and sustain oaks at the stand level. That knowledge makes it possible to apply at least some aspects of ecosystem management across the landscapes that we influence. The challenge will be convincing landowners and stakeholders to do what is necessary to sustain and restore oak ecosystems over large spatial and temporal scales. We hope this volume provides the information needed to start the process.

References

Abrams, M. D. 1985. Fire history of oak gallery forests in a northeast Kansas tallgrass prairie. American Midland Naturalist 114:188–191.

Abrams, M. D. 1986. Historical development of gallery forests in northeast Kansas. Vegetatio 65:29–37.

Abrams, M. D. 1990. Adaptations and responses to drought in *Quercus* species of North America. Tree Physiology 7:227.

Abrams, M. D. 1992. Fire and the development of oak forests. BioScience 42:346–353.

Abrams, M. D. 1996. Distribution, historical development and ecophysiological attributes of oak species in the eastern United States. Annales des Sciences Forestieres 53:487–512.

Abrams, M. D. 1998. The red maple paradox. BioScience 48:355–364.

Abrams, M. D., and J. A. Downs. 1990. Successional replacement of old-growth white oak by mixed mesophytic hardwoods in southwestern Pennsylvania. Canadian Journal of Forest Research 20:1864–1870.

Abrams, M. D., and D. M. McCay. 1996. Vegetation-site relationships of witness trees (1780–1856) in the presettlement forests of eastern West Virginia. Canadian Journal of Forest Research 26:217–224.

Abrams, M. D., and G. J. Nowacki. 1992. Historical variation in fire, oak recruitment, and post-logging accelerated succession in central Pennsylvania. Bulletin of the Torrey Botanical Club 119:19–28.

Abrams, M. D., and D. A. Orwig. 1994. Temperate hardwoods. Pages 289–302 in: Encyclopedia of Agricultural Sciences (C. J. Arntzen, ed.). Academic Press, New York.

Abrams, M. D., D. A. Orwig, and T. E. Demeo. 1995. Dendrochronological analysis of successional dynamics for presettlement-origin white pine–mixed oak forest in southern Appalachians, USA. Journal of Ecology 83:123–133.

Abrams, M. D., and C. M. Ruffner. 1995. Physiographic analysis of witness-tree distribution (1765–1798) and present forest cover through north central Pennsylvania. Canadian Journal of Forest Research 25:659–668.

Abrams, M. D., and F. K. Seischab. 1997. Does the absence of sediment charcoal provide substantial evidence against the fire and oak hypothesis? Journal of Ecology 85:373–375.

Adams, T. E., P. B. Sands, W. H. Weitkamp, and N. K. McDougald. 1992. Oak seedling establishment on California rangelands. Journal of Range Management 45:93–98.

Agee, J. K., and D. R. Johnson. 1988. Ecosystem Management for Parks and Wilderness. University of Washington Press, Seattle, Washington.

Aizen, M. A., and W. A. Patterson, III. 1990. Acorn size and geographical range in the North American oaks (*Quercus* L.). Journal of Biogeography 17:327–332.

Alban, D. H., and E. C. Berry. 1994. Effects of earthworm invasion on morphology, carbon, and nitrogen of a forest soil. Applied Soil Ecology 1:243–249.

Albon, S. D., B. Mitchell, and B. W. Staines. 1983. Fertility and body weight in female red deer: A density dependent relationship. Journal of Animal Ecology 52:969–980.

Alerich, C. A. 1993. Forest statistics for Pennsylvania—1978 and 1989. U.S. Forest Service Resource Bulletin NE-126.

Alig, R. J., W. G. Hohenstein, B. C. Murphy, and R. G. Haight. 1990. Changes in area of timberland in the United States 1952–2040 by ownership, forest type, and region and state. U.S. Forest Service General Technical Report SE-64.

Allen, B. H., B. A. Holzman, and R. R. Evett. 1991. A classification system for California's hardwood rangelands. Hilgardia 59:1–45.

Allen, D., and T. Bowersox. 1989. Regeneration in oak stands following gypsy moth defoliations. Pages 67–73 in: Proceedings of the Seventh Central Hardwood Conference (G. Rink and C. A. Budelsky, eds.). U.S. Forest Service General Technical Report NC-132.

Allen, L. S. 1992. Livestock-wildlife coordination in the encinal oak woodlands: Coronado National Forest. Pages 109–110 in: Ecology and Management of Oak and Associated Woodlands: Perspectives in the southwestern United States and northern Mexico (P. F. Ffolliott, G. J. Gottfried, D. A. Bennett, V. M. Hernandez C., A. Ortega-Rubio, and R. H. Hamre, technical coordinators.). U.S. Forest Service General Technical Report RM-218.

Allen-Diaz, B. H., and J. W. Bartolome. 1992. Survival of *Quercus douglasii* (Fagaceae) seedlings under the influence of fire and grazing. Madrono 39:47–53.

Alt, G. L. 1980. Hunting vulnerability of bears. Pennsylvania Game News 51:7–10.

Anderson, R. C. 1997. Native pests: The impact of deer in highly fragmented habitats. Pages 117–134 in: Conservation in Highly Fragmented Landscapes (M. W. Schwartz, ed.). Chapman and Hall, New York.

Anderson, R. C., and A. J. Katz. 1993. Recovery of browse-sensitive tree species following release from white-tailed deer (*Odocoileus virginianus* Zimmerman) browsing pressure. Biological Conservation 63:203–208.

Anderson R. M., and R. M. May. 1980. Infectious diseases and population cycles of forest insects. Science 210:658–661.

Anderson, R. M., and R. M. May. 1981. The population dynamics of microparasites and their invertebrate hosts. Philosophical Transactions of the Royal Society Bulletin 291:451–524.

Andreadis, T. G., and Weseloh, R. M. 1990. Discovery of *Entomophaga maimaiga* in North American gypsy moth, *Lymantria dispar*. Proceedings of the National Academy of Science 87:2461–2465.

Anthony, R. G., and N. S. Smith. 1977. Ecological relationships between mule deer and white-tailed deer in southeastern Arizona. Ecological Monograph 47:255–277.

Arends, E., and J. F. McCormick 1987. Replacement of oak-chestnut forests in the Great Smoky Mountains. Pages 305–316 in: Proceedings of the Sixth Central Hardwood Forest Conference (R. L. Hay, F. W. Woods, and H. DeSelm, eds.). Knoxville, TN.

Arthur, M. A., R. D. Parately, and B. A. Blankenship. 1998. Single and repeated fires affect survival and regeneration of woody and herbaceous species in an oak-pine forest. Journal of the Torrey Botanical Club 125:225–236.

Ashe, W. W. 1911. Chestnut in Tennessee. In: Forest Studies. Tennessee Geological Survey Series Bulletin 10-B.

Ashton, P. M. S., and G. P. Berlyn. 1994. A comparison of leaf physiology and anatomy of *Quercus* (section Erythrobalanus-Fagaceae) species in different light environments. American Journal of Botany 81:589–597.

Atkeson, T. D., and A. S. Johnson. 1979. Succession of small mammals on pine plantations in the Georgia Piedmont. American Midland Naturalist 101:385–392.

Auchmoody, L. R., and H. C. Smith. 1993. Survival of northern red oak acorns after fall burning. U.S. Forest Service Research Paper NE-678.

Auchmoody, L. R., H. C. Smith and R. S. Walters. 1994. Planting northern red oak acorns: Is size and planting depth important? U.S. Forest Service Research Paper NE-693.

Augspurger, M. K., D. H. Van Lear, S. K. Cox, and D. R. Phillips. 1987. Regeneration of hardwood coppice following clearcutting with and without prescribed fire. Pages 89–92 in: Proceedings of the Fourth Biennial Southern Silvicultural Research Conference. U.S. Forest Service General Technical Report WO-3.

Augustine, D. J., and L. E. Frelich. 1998. Effects of white-tailed deer on populations of an understory forb in fragmented deciduous forests. Conservation Biology 12:995–1004.

Augustine, D. J., and P. A. Jordon. 1998. Predictors of white-tailed deer grazing intensity in fragmented deciduous forests. Journal of Wildlife Management 62:1076–1085.

Augustine, D. J., and S. J. McNaughton. 1998. Ungulate effects on the functional species composition of plant communities: Herbivore selectivity and plant tolerance. Journal of Wildlife Management 62:1165–1183.

Axelrod, D. I. 1958. Evolution of the Madro-Tertiary geoflora. Botanical Review 24:433–509.

Bahari, Z. A., S. G. Pallardy, and W. C. Parker. 1985. Photosynthesis, water relations and drought adaptation of woody species of oak-hickory forests in central Missouri. Forest Science 31:557–569.

Bailey, R. W., and K. T. Rinell. 1967. Management of the eastern wild turkey in the northern hardwoods. Pages 261–302 in: The Wild Turkey and Its Management (O. H. Hewitt, ed.). Wildlife Society, Washington, DC.

Bailey, R. W., H. G. Uhlig, and G. Breiding. 1951. Wild turkey management in West Virginia. Conservation Commission of West Virginia Bulletin 2.

Baker, J. B. 1994. An overview of stand-level ecosystem management research in the Ouachita/Ozark National Forests. Pages 18–28 in: Proceedings of the Symposium on Ecosystem Management Research in the Ouachita Mountains: Pretreatment Conditions and Preliminary Findings (J. Baker, compiler). U.S. Forest Service General Technical Report SO-112.

Baker, M. B., Jr., L. F. DeBano, and P. F. Ffolliott. 1995. Hydrology and watershed management in the Madrean archipelago. Pages 329–337 in: Biodiversity and management of the Madrean archipelago: The sky islands of southwestern United States and northern Mexico (L. F. DeBano, P. F. Ffolliott, A. Ortega-Rubio, G. J. Gottfried, R. H. Hamre, and C. B. Edminister, technical coordinators). U.S. Forest Service General Technical Report RM-264.

Baker, R. G., E. A. Bettis III, D. P. Schwert, D. G. Horton, C. A. Chumbley, L. A. Gonzalez, and M. K. Reagan. 1996. Holocene paleoenvironments of northeast Iowa. Ecological Monographs 66:203–234.

Baker, R. H. 1984. Origin, classification and distribution. Pages 1–18 in: White-tailed Deer: Ecology and Management (L. K. Halls, ed.). Stackpole Books, Harrisburg, PA.

Balch, R. E. 1927. Frost kills oak. Journal of Forestry 24:949–950.

Balda, R. P. 1969. Foliage use by birds of the oak-juniper woodland and ponderosa pine forests in southeastern Arizona. Condor 71:399–412.

Balda, R. P., and A. C. Kamil. 1989. A comparative study of cache recovery by three corvid species. Animal Behavior 38:761–769.

Barbour, A. G., and D. Fish. 1993. The biological and social phenomenon of Lyme disease. Science 260:1610–1616.

Barden, L. S., and F. W. Woods. 1976. Effects of fire on pine and pine-hardwood forests in the southern Appalachians. Forest Science 22:399–403.

Barnes, B. V. 1991. Deciduous forests of North America. Pages 219–344 in: Ecosystems of the World 7: Temperate Deciduous Forests (E. Rohrig and B. Ulrich, eds.). Elsevier, Amsterdam.

Barnes, T. A., and D. H. Van Lear. 1998. Prescribed fire effects on hardwood advance regeneration in mixed hardwood stands. Southern Journal of Applied Forestry 22:138–142.

Barnett, R. J. 1977. The effect of burial by squirrels on germination and survival of oak and hickory nuts. American Midland Naturalist 98:319–330.

Barrett, J. W., ed. 1995. Regional silviculture of the United States. 3rd ed. John Wiley and Sons, New York.

Barrett, L. I. 1931. Influence of forest litter on the germination and early survival of oak chestnut, *Quercus montana*. Wildlife Ecology 12:476–484.

Barrett, R. H. 1980. Mammals of California oak habitats—management implications. Pages 275–291 in: Proceedings of the Symposium on the Ecology, Management, and Utilization of California Oaks. U.S. Forest Service General Technical Report PSW-44.

Bartolome, J. W. 1987. California annual grassland and oak savannah. Rangelands 9:122–125.

Bartolome. J. W., P. C. Muick, and M. P. McClaren. 1987. Natural regeneration of Californian hardwoods. Pages 26–31 in: Proceedings of a Symposium on Multiple-use Management of California's Hardwood Resources. U.S. Forest Service General Technical Report PSW-100.

Barton, A. M. 1991. Factors controlling the elevational positions of pines in the Chiricahua Mountains, Arizona: Drought, competition, and fire. Ph.D. diss., University of Michigan, Ann Arbor.

Bartram, W. [1791] 1955. Travels through North and South Carolina, Georgia, east and west Florida, the Cherokee country, the extensive territories of the Muscogulges, or Creek confederacy, and the country of the Choctaws; containing an account of the soil and natural productions of those regions, together with observations on the manners of the Indians. In: Travels of William Bartram (M. Van Doren, ed.). Dover Publications, New York.

Barwick, L. H., W. M. Hetrick, and L. E. Williams. 1973. Foods of young Florida wild turkeys. Proceedings of the Annual Conference of the Southeast Association of Fish and Game Commission 27:92–102.

Barwick, L. H., and D. W. Speake. 1973. Seasonal movements and activities of wild turkey gobblers in Alabama. Pages 125–133 in: Wild Turkey Management: Current Problems and Programs (G. C. Sanderson and H. C. Schultz, eds.). Missouri Chapter of the Wildlife Society and University of Missouri Press, Columbia.

Bate-Smith, E. C. 1972. Attractants and repellents in higher animals. Pages 45–56 in: Phytochemical Ecology (J. B. Harborne, ed.). Academic Press, New York.

Batek, M. J., A. J. Rebertus, W. A. Schroeder, T. L. Haithcoat, E. Compas, and R. P. Guyette. 1999. Reconstruction of early nineteenth-century vegetation and fire regimes in the Missouri Ozarks. Journal of Biogeography 26:397–412.

Batzli, G. O. 1977. Population dynamics of the white-footed mouse in floodplain and upland forests. American Midland Naturalist 97:18–32.

Baughman, W. M., and D. C. Guynn, Jr. 1993. Wild turkey food habits in pine plantations in South Carolina. Proceedings of the Annual Conference of the Southeast Association of Fish and Wildlife Agencies 47:163–169.

Baumgras, P. 1944. Experimental feeding of captive fox squirrels. Journal of Wildlife Management 8:296–300.

Bazzaz, F. A. 1979. The physiological ecology of succession. Annual Review of Ecology and Systematics 10:351–371.

Beasom, S. L., and D. Wilson. 1992. Rio Grande turkey. Pages 306–330 in: The Wild Turkey: Biology and Management (J. G. Dickson, ed.). Stackpole Books, Harrisburg, PA.

Beck, D. E. 1977. Twelve-year acorn yield in southern Appalachian oaks. U.S. Forest Service Research Note SE-244.

Beck, D. E. 1993. Acorns and oak regeneration. Pages 96–104 in: Oak regeneration: Serious problems, practical recommendations. U.S. Forest Service General Technical Report SE-84.

Beck, D. E., and Olson, D. F., Jr. 1968. Seed production in southern Appalachian oak stands. U.S. Forest Service Research Note SE-91.

Beck, J. R., and D. O. Beck. 1955. A method for nutritional evaluation of wildlife foods. Journal of Wildlife Management 19:198–205.

Beecham, J. 1980. Some population characteristics of two black bear populations in Idaho. International Conference on Bear Research and Management 4:201–204.

Beecham, J., D. G. Reynolds, and M. J. Hornocker. 1983. Black bear denning activities and den characteristics in west-central Idaho. International Conference on Bear Research and Management 5:79–86.

Beeman, L. E., and M. R. Pelton. 1980. Seasonal foods and feeding ecology of black bears in the Smoky Mountains. International Conference on Bear Research and Management 4:141–147.

Beilmann, A. P., and L. G. Brenner. 1951. The recent intrusion of forests in the Ozarks. Annals of the Missouri Botanical Garden 38:261–282.

Bennett, A. T. D. 1993. Spatial memory in a food storing corvid: I. Near tall landmarks are primarily used. Journal of Comparative Physiology, A. Sensory, Neural, and Behavioral Physiology 173:193–207.

Bennett, D. A. 1992. Fuelwood extraction in southeastern Arizona. Pages 96–97 in: Ecology and management of oak and associated woodlands: Perspectives in the southwestern United States and northern Mexico (P. F. Ffolliott, G. J. Gottfried, D. A. Bennett, V. M. Hernandez, C. A. Ortega-Rubio, and R. H. Hamre, technical coordinators). U.S. Forest Service General Technical Report RM-218.

Bennett, D. A. 1995. Fuelwood harvesting in the sky islands of southeastern Arizona. Pages 519–523 in: Biodiversity and management of the Madrean archipelago: The sky islands of southwestern United States and northwestern Mexico (L. F. DeBano, P. F. Ffolliott, A. Ortega-Rubio, G. J. Gottfried, R. H. Hamre, and C. B. Edminster, technical coordinators). U.S. Forest Service General Technical Report RM-264.

Bennett, L. J., and P. F. English. 1941. November foods of the wild turkey. Pennsylvania Game News 11:8.

Bennett, L. J., P. F. English, and R. L. Watts. 1943. The food habits of black bears in Pennsylvania. Journal of Mammalogy 24:25–31.

Berg, E. E., and J. L. Hamrick. 1994. Spatial and genetic structure of two sandhill oaks: *Quercus laevis* and (Fagaceae). American Journal of Botany 8:7–14.

Bernhardt, E. A., and T. J. Swiecki. 1991. Minimum input techniques for valley oak restocking. U.S. Forest Service General Technical Report PSW-126:2–8.

Bess, H. A. 1961. Population ecology of the gypsy moth *Porthetria dispar* L. (Lepidoptera: Lymantridae). Connecticut Agriculture Experimental Station Bulletin 646.

Bess, H. A., S. H. Spurr, and E. W. Littlefield. 1947. Forest site conditions and the gypsy moth. Harvard Forest Bulletin 22.

Billingsley, B. B., Jr., and D. H. Arner. 1970. The nutritive value and digestibility of some winter foods of the eastern wild turkey. Journal of Wildlife Management 34:176–182.

Binkley, D., D. Richter, M. B. David, and B. Caldwell. 1992. Soil chemistry in a loblolly/longleaf pine forest with interval burning. Ecological Applications 2:157–164.

Birch, T. W. 1996. Private forest-land owners of the United States, 1994. U.S. Forest Service Resource Bulletin NE-134.

Bjorkbom, J. C. 1979. Seed production and advance regeneration in Allegheny hardwood forests. U.S. Forest Service Research Paper NE-435.

Black Bear Conservation Committee. 1996. Black bear management handbook for Louisiana, Mississippi, southern Arkansas, and east Texas. Black Bear Conservation Committee, Baton Rouge, LA.

Black Bear Conservation Committee. 1997. Black bear restoration plan. Black Bear Conservation Committee, Baton Rouge, LA.

Blankenship, L. H. 1992. Physiology. Pages 84–100 in: The Wild Turkey: Biology and Management (J. G. Dickson, ed.). Stackpole Books, Harrisburg, PA.

Block, W. M. 1990. Geographic variation in foraging ecologies of breeding and nonbreeeding birds in oak woodlands. Studies in Avian Biology 13:264–269.

Block, W. M., J. L. Ganey, K. E. Severson, and M. L. Morrison. 1992. Use of oaks by neotropical migratory birds in the Southwest. Pages 65–70 in: Ecology and management of oak and associated woodlands: Perspectives in the southwestern United States and northern Mexico (P. F. Ffolliott, G. J. Gott-fried, D. A. Bennett, V. M. Hernandez C., A. Ortega-Rubio, and R. H. Hamre, technical coordinators). U.S. Forest Service General Technical Report RM-218.

Block, W. M., and M. L. Morrison, 1990. Wildlife diversity of the central Sierra foothills. California Agriculture 44:19–22.

Block, W. M., M. L. Morrison, and J. Verner. 1990. Wildlife and oak woodland interdependence. Fremontia 18:72–76.

Bock, C. E., and J. H. Bock. 1974. Geographical ecology of the acorn woodpecker: Diversity versus abundance of resources. American Naturalist 208:694–698.

Boerner, R. E. J., and J. A. Brinkman. 1996. Ten years of tree seedling establishment and mortality in an Ohio deciduous forest complex. Bulletin of the Torrey Botanical Club 123:309–317.

Boettner, G. H., J. S. Elkinton, and C. J. Boettner. In press. Effects of a biological control introduction on three non-target native species of Saturniid moths. Conservation Biology.

Bolsinger, C. L. 1988. The hardwoods of California's timberlands, woodlands, and savannas. U.S. Forest Service Resource Bulletin PNW-RB-148.

Bond, W. J., and B. W. van Wilgen. 1996. Fire and Plants. Chapman and Hall, London.

Borelli, S., P. F. Ffolliott, and G. J. Gottfried. 1994. Natural regeneration in the encinal woodlands of southeastern Arizona. Southwestern Naturalist 39:179–183.

Bormann, F. H., and G. E. Likens. 1979. Pattern and Process in a Forested Ecosystem. Springer-Verlag, New York.

Bossema, I. 1979. Jays and oaks: An eco-ethological study of a symbiosis. Behavior 70:1–117.

Boyce, M. S., and A. Haney. 1997. Ecosystem Management. Yale University Press, New Haven.

Boyce, S. G., and W. H. Martin. 1993. The future of terrestrial communities in the Southeastern Coastal Plain. Pages 339–366 in: Biodiversity of the Southeastern United States: Upland Terrestrial Communities. (W. H. Martin, S. G. Boyce, and A. C. Echternacht, eds.). John Wiley and Sons, New York.

Boyd, C. E., and R. Oglesby. 1960. Eastern wildlife development: Mast production study. Texas Game and Fish Commission W-27-D-14, Job 2, Plan 2.

Boyer, W. D. 1990. Growing-season burns for control of hardwoods in longleaf pine stands. U.S. Forest Service Research Paper SO-256.

Boyer, W. D. 1993. Season of burn and hardwood development in young longleaf pine stands. Pages 511–515 in: Proceedings of the 7th Biennial Southern Silviculture Research Conference, Mobile, AL. U.S. Forest Service General Technical Report SO-93.

Braun, E. L. 1950. Deciduous forests of eastern North America. The Blakiston Co., Philadelphia.

Breda, N., A. Granier, and G. Aussenac. 1995. Effects of thinning on soil and tree water relations, transpiration and growth in an oak forest (*Quercus petraea* [Matt.] Liebl.). Tree Physiology 15:295–306.

Brodbeck, D. R. 1994. Memory for spatial and local cues: A comparison of a storing and a nonstoring species. Animal Learning and Behavior 22:119–133.

Brooks, R. T., D. B. Kittredge, and C. L. Alerich. 1993. Forest resources of southern New England. U.S. Forest Service Resource Bulletin NE-127.

Brooks, R. T., H. R. Smith, and W. M. Healy. 1998. Small-mammal abundance at three elevations on a mountain in central Vermont, USA: A sixteen-year record. Forest Ecology and Management 110:181–193.

Brose, P. H., and D. H. Van Lear. 1998. Responses of hardwood advance regeneration to seasonal prescribed fires in oak-dominated shelterwood stands. Canadian Journal of Forest Research 28:331–339.

Brose, P. H., and D. H. Van Lear. 1999. Effects of seasonal prescribed fires on residual overstory trees in oak-dominated shelterwood stands. Southern Journal of Applied Forestry 23:88–93.

Brose, P. H., D. H. Van Lear, and R. Cooper. 1999. Using shelterwood harvests and prescribed fire to regenerate oak stands on productive upland sites. Forest Ecology and Management 113:125–141.

Brose, P. H., D. H. Van Lear, and P. D. Keyser. 1999. A shelterwood-burn technique for regenerating oak stands on productive upland sites in the Piedmont region. Southern Journal of Applied Forestry 23:158–163.

Brown, D. E. 1982. Madrean evergreen woodland. Pages 59–65 in: Biotic communities of the American Southwest—United States and Mexico (D. E. Brown, ed.). Desert Plants 4.

Brown, D. E. 1984. The effect of drought on white-tailed deer recruitment in the arid Southwest. Pages 7–12 in: Deer in the Southwest: A Workshop (P. R. Krausman, and N. S. Smith, eds.). School of Renewable Natural Resources, University of Arizona, Tucson.

Brown, D. E., and C. H. Lowe. 1980. Biotic communities of the Southwest. U.S. Forest Service General Technical Report RM-78.

Brown, J. H., and E. J. Heske. 1990. Temporal changes in a Chihuahuan Desert rodent community. Oikos 59:290–302.

Bucher, E. H. 1992. The causes of extinction of the passenger pigeon. Current Ornithology 9:1–36.

Buckner, E. 1983. Archaeological and historical basis for forest succession in eastern North America. Pages 182–187 in: Proceedings of 1982 Society of American Forestry National Convention, SAF Publication 83-104.

Buell, M. F., H. F. Buell, and J. A. Small. 1954. Fire in the history of Mettler's Woods. Bulletin of the Torrey Botanical Club 81:253–255.

Bunnell, F. L., and D. E. N. Tait. 1981. Population dynamics of bears: Implications. Pages 75–98 in: Dynamics of Large Mammal Populations (C. W. Fowler and T. D. Smith, eds.). John Wiley and Sons, New York.

Burcham, L. T. 1970. Ecological significance of alien plants in California grasslands. Pages 36–39 in: Proceedings of the Association of American Geographers 2.

Burget, M. L. 1957. The wild turkey in Colorado. Colorado Game and Fish Department, P-R Project W-39-R.

Burns, P. Y., D. M. Christisen, and J. M. Nichols. 1954. Acorn production in the Missouri Ozarks. University of Missouri, College of Agriculture, Agriculture Experimental Station Bulletin 611.

Burns, R. M., and B. H. Honkala, eds. 1990. Silvics of North America. Vol. 2, Hardwoods. U.S. Department of Agriculture Handbook 654.

Buttrick, P. L. 1925. Chestnut in North Carolina. In: Chestnut and the chestnut blight in North Carolina. North Carolina Geological and Economic Survey Economic Paper 56.

Byrne, R., E. Edlund, and S. Mensing. 1991. Holocene changes in the distribution and abundance of oaks. Pages 182–188 in: Proceedings of a Symposium on Oak Woodlands and Hardwood Rangeland Management. U.S. Forest Service General Technical Report PSW-126.

Cahalane, V. H. 1942. Caching and recovery of food by the western fox squirrel. Journal of Wildlife Management. 6:338–352.

California Department of Forestry and Fire Protection. 1988. California's forests and rangelands: Growing conflicts over changing uses. Forest and Rangeland Resource Assessment Program, Sacramento, CA.

Campbell, R. W. 1967. The analysis of numerical changes in gypsy moth populations. Forest Science Monographs 15:1–33.

Campbell, R. W. 1973. Numerical behavior of a gypsy moth population system. Forest Science 19:162–167.

Campbell, R. W. 1975. The gypsy moth and its natural enemies. United States Department of Agriculture Information Bulletin 381.

Campbell, R. W., and R. J. Sloan. 1977. Natural regulation of innocuous gypsy moth populations. Environmental Entomology 6:315–322.

Campbell, R. W., and R. J. Sloan. 1978. Natural maintenance and decline of gypsy moth outbreaks. Environmental Entomology 7:389–395.

Canham, C. D. 1985. Suppression and release during canopy recruitment in *Acer saccharum*. Bulletin of the Torrey Botanical Club 112:134–145.

Canham, C. D. 1988. Growth and canopy architecture of shade-tolerant trees: Response to canopy gaps. Ecology 69:786–795.

Canham, C. D. 1989. Different responses to gaps among shade-tolerant tree species. Ecology 70:548–550.

Canham, C. D., J. S. Denslow, W. J. Platt, J. R. Runkle, T. A. Spies, and P. S. White. 1990. Light regimes beneath closed canopies and tree-fall gaps in temperate and tropical forests. Canadian Journal of Forest Resources 20:620–631.

Canham, C. D., and O. L. Loucks. 1984. Catastrophic windthrow in the pre-settlement forests of Wisconsin. Ecology 65:803–809.

Carey, A. B. 1983. Cavities in trees in hardwood forests. Pages 167–184 in: Proceedings of a Symposium on Snag Habitat Management (J. W. Davis, G. A. Goodwin, and R. A. Ockenfels, technical coordinators). U.S. Forest Service General Technical Report RM-99.

Carey, A. B., and R. O. Curtis. 1996. Conservation of biodiversity: A useful paradigm for forest ecosystem management. Wildlife Society Bulletin 24:610–620.

Carlock, D. M., R. H. Conley, J. H. Collins, P. E. Hale, K. G. Johnson, and M. R. Pelton. 1983. The tri-state black bear study. Technical Report 83-9, University of Tennessee, Knoxville.

Carney, D. W. 1985. Population dynamics and denning ecology of black bears in Shenandoah National Park, Virginia. Master's thesis, Virginia Polytechnic Institute and State University, Blacksburg.

Carpenter, S. R., and J. R. Kitchell. 1988. Consumer control of lake productivity. BioScience 38:764–769.

Carpenter, S. R., J. R. Kitchell, and J. R. Hodgson. 1985. Cascading trophic interactions and lake productivity. BioScience 35:634–639.

Carvell, K. L., and W. R. Maxey. 1969. Wildfire destroys! West Virginia Agriculture and Forestry Bulletin 2:4–5, 12.

Catlin, G. [1844] 1993. Letters and notes on the manners, customs, and conditions of the North American Indians written during eight years' travel (1832–1839) amongst the wildest tribes of Indians in North America. Vols. I and II. Dover Publications, New York.

Cecich, R. A. 1997. Influence of weather on pollination and acorn production in two species of Missouri oaks. Pages 252–261 in: Proceedings of the 11th Central Hardwood Forest Conference, U.S. Forest Service General Technical Report NC-188.

Childers, E. L., T. L. Sharik, and C. S. Adkisson. 1986. Effects of loblolly pine plantations on songbird dynamics in the Virginia piedmont. Journal of Wildlife Management 50:406–413.

Christisen, D. M., and W. H. Kearby. 1984. Mast measurement and production in Missouri (with special reference to acorns). Missouri Department of Conservation Terrestrial Series 13, Jefferson City.

Christisen, D. M., and L. J. Korschgen. 1955. Acorn yield and wildlife usage in Missouri. North American Wildlife Conference 20:337–357.

Chung-MacCoubrey, A. L. 1993. Effects of tannins on protein digestibility and detoxification activity in gray squirrels (*Sciurus carolinensis*). Master's thesis, Virginia Polytechnic Institute and State University, Blacksburg.

Chung-MacCoubrey, A. L., E. E. Hagerman, and R. L. Kirkpatrick. 1997. Effects of tannin on digestion and detoxification activity in gray squirrels. Physiological Zoology 70:270–277.

Clapper, R. B., and G. F. Gravatt. 1943. The American chestnut: Its past, present, and future. Southern Lumberman 65:227–229.

Clark, J. D, D. L. Clapp, K. G. Smith, and B. Ederington. 1994. Black bear habitat use in relation to food availability in the interior highlands of Arkansas. International Conference on Bear Research and Management 9:309–318.

Clark, J. D., W. R. Guthrie, and W. B. Owen. 1987. Fall foods of black bears in Arkansas. Proceedings of the Annual Conference of the Southeastern Association of Fish and Wildlife Agencies 41:432–437.

Clark, J. S. 1986. Coastal forest tree populations in a changing environment, southeastern Long Island, New York. Ecological Monographs 56:259–277.

Clark, J. S. 1997. Facing short-term extrapolation with long-term evidence: Holocene fire in the north-eastern U.S. forests. Journal of Ecology 85:377–380.

Clark, J. S., and P. D. Royall. 1995. Transformation of a northern hardwood forest by aboriginal (Iroquois) fire: Charcoal evidence from Crawford Lake, Ontario, Canada. The Holocene 5:1–9.

Clark, J. S., and P. D. Royall. 1996. Local and regional sediment charcoal evidence for fire regimes in presettlement north-eastern North America. Journal of Ecology 84:365–382.

Clarke, M. F., and D. L. Kramer. 1994. Scatter-hoarding by a larder-hoarding rodent: Intraspecific variation in the hoarding behavior of the eastern chipmunk, *Tamias striatus*. Animal Behavior 48:299–308.

Clary, W. P., and A. R. Tiedemann. 1992. Ecology and values of gambel oak woodlands. Pages 78–95 in: Ecology and management of oak and associated woodlands: Perspectives in the southwestern United States and northern Mexico. U.S. Forest Service General Technical Report RM-218.

Clatterbuck, W. K. 1991. Forest development following disturbances by fire and by timber cutting for charcoal production. Pages 60–65 in: Proceedings from Fire and the Environment: Ecological and Cultural Perspectives (S. C. Nodvin and T. A. Waldrop, eds.). U.S. Forest Service General Technical Report SE-69.

Clatterbuck, W. K., and F. T. Bonner 1985. Utilization of seed reserves in *Quercus* during storage. Seed Science and Technical 13:121–128.

Clayton, N. S., and J. R. Krebs. 1994. Memory for spatial and object-specific cues in food-storing and nonstoring birds. Journal of Comparative Psychology 174:371–379.

Cobb, S. W., A. E. Miller, and R. Zahner. 1985. Recurrent shoot flushes in scarlet oak stump sprouts. Forest Science 31:725–730.

Connell, J. H. 1971. On the role of natural enemies in preventing competitive exclusion in some marine animals and in rain forest trees. Pages 298–312 in: Dynamics of Populations (P. J. Den Boer and G. Gradwell, eds.). PUDOC, Oosterbeck, Netherlands.

Conner, R. C., J. D. Born, A. W. Green, and R. A. O'Brien. 1990. Forest resources of Arizona. U.S. Forest Service Resource Bulletin INT-69.

Conner, R. N., R. G. Hooper, H. S. Crawford, and H. S. Mosby. 1975. Woodpecker nesting habitat in cut and uncut woodlands in Virginia. Journal of Wildlife Management 39:144–150.

Cooper, W. S. 1922. The broad-sclerophyll vegetation of California. Carnegie Institution, Washington, DC. Publication 319.

Cottam, C., A. L. Nelson, and T. E. Clarke. 1939. Notes on early winter food habits of the black bear in George Washington National Forest. Journal of Mammalogy 20:310–314.

Cottam, G. 1949. The phytosociology of an oak woods in southwestern Wisconsin. Ecology 30:271–287.

Crawley, M. J., and C. R. Long. 1995. Alternate bearing, predator satiation and seedling recruitment in *Quercus robur.* Linnean Journal of Ecology 83:683–696.

Crête, M. 1999. The distribution of deer biomass in North America supports the hypothesis of exploitation ecosystems. Ecology Letters 2:223–227.

Crockett, B. C. 1973. Quantitative evaluation of winter roost sites of the Rio Grande turkey in north-central Oklahoma. Pages 211–218 in: Wild Turkey Management: Current Problems and Programs (G. C. Sanderson and H. C. Schultz, eds.). Missouri Chapter of the Wildlife Society and University of Missouri Press, Columbia.

Cronon, W. 1983. Changes in the land Indians, colonists, and the ecology of New England. Hill and Wang, New York.

Crouch, G. L. 1981. Effects of deer on forest vegetation. Part 3. Pages 449–457 in: Mule and Black-tailed Deer of North America (O. C. Wallmo, ed.). University of Nebraska Press, Lincoln.

Crow, T. R. 1988. Reproductive mode and mechanisms for self-replacement of northern red oak (*Quercus rubra*)—a review. Forest Science 34:19–40.

Crow, T. R. 1992. Population dynamics and growth patterns for a cohort of northern red oak (*Quercus rubra*) seedlings. Oecologia 91:192–200.

Crowley, M. 1985. Reduction of oak fecundity by low density herbivore populations. Nature. 314:163–164.

Crunkilton, D. D., S. G. Pallardy, and H. E. Garrett. 1992. Water relations and gas exchange of northern red oak seedlings planted in a central Missouri clearcut and shelterwood. Forest Ecology Management 53:117–129.

Culbertson, A. B. 1948. Annual variation of winter foods taken by wild turkeys on the Virginia state forests. Virginia Wildlife 9:14–16.

Cummings, J. R., and S. H. Vessey. 1994. Agricultural influences of movement patterns of white-footed mice (*Peromyscus leucopus*). American Midland Naturalist 132:209–218.

Cunningham, R. J., and C. Hauser. 1989. The decline of the Ozark forest between 1880 and 1920. Pages 34–37 in: Pine-hardwood mixtures: A symposium on the management and ecology of the type (T. A. Waldrop, ed.). U.S. Forest Service General Technical Report SE-58.

Curtis, J. T. 1959. The Vegetation of Wisconsin. University Wisconsin Press, Madison.

Cushing, E. J. 1965. Problems in the Quartenary phytogeography of the Great Lakes region. Pages 403–419 in: The Quartenary of the United States (H. E. Wright, Jr. and D. G. Frey, eds.). Princeton University Press, Princeton, NJ.

Cutter, B. E., and R. P. Guyette. 1994. Fire frequency on an oak-hickory ridgetop in the Missouri Ozarks. American Midland Naturalist 132:393–398.

Cwynar, L. C. 1977. The recent fire history of Barron Township, Algonquin Park. Canadian Journal of Botany 55:1524–1538.

Dalke, P. D., W. K. Clark, Jr., and L. J. Korschgen. 1942. Food habit trends of the wild turkey in Missouri as determined by dropping analysis. Journal of Wildlife Management 6:237–243.

Danielsen, K. C., and W. L. Halvorson, 1991. Valley oak seedling growth associated with selected grass species. Pages 9–11 in: U.S. Forest Service General Technical Report PSW-126.

Darley-Hill, S., and W. C. Johnson. 1981. Acorn dispersal by the blue jay (*Cyanocitta cristata*). Oecologia 50:231–232.

Daubenmire, R. 1943. Vegetation zonation in the Rocky Mountains. Botanical Review 9:325–393.

Davis, F. W. 1995. Vegetation change in blue oak and blue oak–foothill pine woodland. Report to California Department of Forestry and Fire Protection, Sacramento, CA.

Davis, M. A., K. J. Wrage, and P. B. Reich. 1998. Competition intensity between herbaceous vegetation and tree seedlings along a water-light gradient supports the resource supply-demand theory. Journal of Ecology 86:652–661.

Davis, M. B. 1976. Pleistocene biogeography of temperate deciduous forests. Geoscience and Man 13:13–26.

Davis, M. B. 1985. Historical consideration. 1. History of the vegetation of the Mirror Lake Watershed. Pages 42–65 in: An Ecosystem Approach to Aquatic Ecology: Mirror Lake and Its Environment (G. E. Likens, ed.). Springer-Verlag, New York.

Davis, M. B. 1996. Extent and location. Pages 18–32 in: Eastern Old-growth Forests: Prospects for Rediscovery and Recovery (M. B. Davis, ed.). Island Press, Washington, DC..

DeBano, L. F., D. G. Neary, and P. F. Ffolliott. 1998. Fire's Effects on Ecosystems. John Wiley and Sons, New York.

deCalesta, D. S. 1994. Effect of white-tailed deer on songbirds within managed forests in Pennsylvania. Journal of Wildlife Management 58:711–717.

deCalesta, D. S. 1997. Deer and ecosystem management. Pages 267–279 in: The

Science of Overabundance: Deer Ecology and Population Management (W. J. McShea, H. B. Underwood, and J. H. Rappole, eds.). Smithsonian Institution Press, Washington, DC.

deCalesta, D. S., and S. L. Stout. 1997. Relative deer density and sustainability: A conceptual framework for integrating deer management with ecosystem management. Wildlife Society Bulletin 25:252–258.

Decker, S. R. 1988. Nutritive quality and metabolizable energy of eight wild turkey foods in New Hampshire. Master's thesis, University of New Hampshire, Durham.

Decker, S. R., P. J. Pekins, and W. W. Mautz. 1991. Nutritional evaluation of winter foods of wild turkeys. Canadian Journal of Zoology 69:2128–2132.

DeGange, A. R., J. W. Fitzpatrick, J. N. Layne, and G. E. Woolfenden. 1989. Acorn harvesting by Florida scrub jays. Ecology 70:348–356.

DeGraaf, R. M., and A. L. Shigo. 1985. Managing cavity trees for wildlife in the Northeast. U.S. Forest Service General Technical Report NE-101.

DeGraaf, R. M., J. B. Hestbeck, and M. Yamasaki. 1998. Associations between breeding bird abundance and stand structure in the White Mountains, New Hampshire and Maine, USA. Forest Ecology and Management 103:217–233.

DeGraaf, R. M., M. Yamasaki, W. B. Leak, and J. W. Lanier. 1992. New England Wildlife: Management of Forested Habitats. U.S. Forest Service, Northeastern Forest Experiment Station, Radnor, PA.

Delcourt, H. R. 1979. Late Quarternary vegetation history of the western highland rim and adjacent Cumberland Plateau of Tennessee. Ecological Monographs 49:255–280.

Delcourt, H. R. 1987. The impacts of prehistoric agriculture and land occupation on natural vegetation. Trends in Ecology and Evolution 2:39–44.

Delcourt, H. R., and P. A. Delcourt. 1997. Pre-Columbiam Native American use of fire on southern Appalachian landscapes. Conservation Biology 11:1010–1014.

Delcourt, P. A., and H. R. Delcourt. 1987. Long-term forest dynamics of the temperate zone. Ecological Studies 63, Springer, New York.

Delcourt, P. A. , H. R. Delcourt, P. A. Cridlebaugh, and J. Chapman. 1986. Holocene ethnobotanical and paleoecological record of human impact on vegetation in the Little Tennessee River Valley, Tennessee. Quarternary Research 25:330–349.

Delcourt, P. A., H. R. Delcourt, D. F. Morse, and P. A. Morse. 1993. History, evolution, and organization of vegetation and human culture. Pages 47–79 in: Biodiversity of the Southeastern United States Lowland Terrestrial Communities (W. H. Martin, S. G. Boyce, and A. C. Echternacht, eds.). John Wiley and Sons, New York.

Dellinger, G. P. 1973. Habitat management for turkeys in the oak-hickory forests of Missouri. Pages 235–244 in: Wild Turkey Management: Current Problems and Programs. Missouri Chapter of the Wildlife Society and University of Missouri Press, Columbia.

Denevan, W. M. 1992. The pristine myth: The landscape of the Americas in 1492. Annals of the Association of American Geographers 82:369–385.

Dey, D. C. 1993. Predicting quantity and quality of reproduction in the uplands. Pages 138–145 in: Oak regeneration: Serious problems, practical recommendations (D. L. Loftis and C. E. McGee, eds.). U.S. Forest Service General Technical Report SE-84.

Dey, D. C., and R. P. Guyette. 2000. Anthropogenic fire history and red oak forests in south-central Ontario. Forestry Chronicle 76:339–347.

Dey, D. C., and W. C. Parker. 1996. Regeneration of red oak (*Quercus rubra* L.) using shelterwood systems: Ecophysiology, silviculture and management recommendations. Ontario Ministry for Natural Resources, Ontario Forestry Research Institute, Forestry Research Information Paper 126.

Dey, D. C., and W. C. Parker. 1997. Morphological indicators of stock quality and field performance of red oak (*Quercus rubra* L.) seedlings underplanted in a central Ontario shelterwood. New Forestry 14:145–156.

Dey, D. C., P. S. Johnson, and H. E. Garrett. 1996. Modeling the regeneration of oak stands in the Missouri Ozark Highlands. Canadian Journal of Forest Resources 26:573–583.

Diamond, S. J. 1989. Vegetation, wildlife, and human foraging in prehistoric western Virginia. Master's thesis, Virginia Polytechnic Institute and State University, Blacksburg.

Dickson, J. G. 1990. Oak and flowering dogwood production for eastern wild turkeys. Proceedings of the National Wild Turkey Symposium 6:90–95.

Dickson, J. G., C. D. Adams, and S. H. Hanley. 1978. Response of turkey populations to habitat variables in Louisiana. Wildlife Society Bulletin 6:163–166.

Dixon, M. D., W. C. Johnson, and C. S. Adkisson. 1997a. Effects of caching on acorn tannin levels and blue jay dietary performance. Condor 99:756–764.

Dixon, M. D., W. C. Johnson, and C. S. Adkisson. 1997b. Effects of weevil larvae on acorn use by blue jays. Oecologia (Berlin) 111:201–208.

Doak, S. C. 1989. Modeling patterns of land use and ownership. Final report to California Department of Forestry and Fire Protection, Sacramento, CA.

Doak, S. C., and W. Stewart. 1986. A model of economic forces affecting California's hardwood resource: Monitoring and policy implications. Report submitted to the Forest and Rangeland Assessment Program, California Department of Forestry and Fires Protection, Sacramento, CA.

Doane, C. C. 1970. Primary pathogens and their role in the development of an epizootic in the gypsy moth. Journal of Invertebrate Pathololology 15:21–33.

Doane, C. C. 1976. Ecology of pathogens of the gypsy moth. Pages 285–293 in:

Perspectives in Forest Entomology (J. Anderson and H. Kaya, eds.). Academic Press, New York.

Doane, C. C., and M. L. McManus, eds. 1981. The gypsy moth: Research towards integrated pest management. U.S. Department of Agriculture, Washington, DC.

Dodge, S. L., and J. R. Harman. 1985. Woodlot composition and successional trends in south-central lower Michigan. Michigan Botanist 24:43–54.

Dooley, J. L., Jr., and M. A. Bowers. 1996. Influences of patch size and microhabitat on the demography of two old-field rodents. Oikos 75:453–462.

Dorney, C. H., and J. R. Dorney. 1989. An unusual oak savanna in northeastern Wisconsin: The effect of Indian-caused fire. American Midland Naturalist 122:103–113.

Downing, T. E., and P. F. Ffolliott. 1983. The social dimension of rangeland management. Pages 19–23 in: Wildlife and range research needs in northern Mexico and southwestern United States (D. R. Patton, J. M. de la Puente E., P. F. Ffolliott, S. Gallina, E. T. Bartlett, technical coordinators). U.S. Forest Service General Technical Report WO-36.

Downs, A. A. 1944. Estimating acorn crops for wildlife in the southern Appalachians. Journal of Wildlife Management 8:339–340.

Downs, A. A., and W. E. McQuilken. 1944. Seed production of southern Appalachian oaks. Journal of Forestry 42:913–920.

Drake, W. E. 1991. Evaluation of an approach to improve acorn production during thinning. Pages 429–441 in: Proceedings of the Eighth Central Hardwood Forest Conference (L. H. McCormick and K. W. Gottschalk, eds.). U.S. Forest Service General Technical Report NE-148.

Drooz, A. T. 1980. A review of the biology of the elm spanworm (Lepidoptera: Geometridae). Great Lakes Entomologist 13:49–53.

Drooz, A. T., ed. 1985. Insects of eastern forests. U.S. Forest Service Miscellaneous Publication 1426.

Dunbar, D. M., and G. R. Stephens. 1976. The bionomics of the two-lined chestnut borer. Pages 73–83 in: Perspectives in Forest Entomology (J. F. Anderson and H. K. Kaya, eds.). Academic Press, New York.

Durkee, L. H. 1971. A pollen profile from Woden Bog in north-central Iowa. Ecology 52:837–844.

Dutrow, G. F., and H. A. Devine. 1981. Economics of forest management for multiple outputs: timber and turkey. Pages 114–121 in: Proceedings of the Symposium on Habitat Requirements and Habitat Management for the Wild Turkey in the Southeast (P. T. Bromley and R. L. Carlton, eds.). Virginia Wild Turkey Foundation, Elliston.

Duvendeck, J. P. 1962. The value of acorns in the diet of Michigan deer. Journal of Wildlife Management 26:371–379.

Dwyer, G., and J. S. Elkinton. 1993. Using simple models to predict virus epizootics in gypsy moth populations. Journal of Animal Ecology 61:1–11.

Eagle, T. C., and M. R. Pelton. 1983. Seasonal nutrition of black bears in the Great Smoky Mountains National Park. International Conference on Bear Research and Management 6:94–101.

Eaton, S. W., F. M. Evans, J. W. Glidden, and B. D. Penrod. 1976. Annual range of wild turkeys in southwestern New York. New York Fish and Game Journal 23:20–33.

Eaton, S. W., T. W. Moore, and E. N. Saylor. 1970. A ten-year study of the food habits of a northern population of wild turkeys. Science Studies 26:43–64.

Ehrenfeld, J. G. 1982. The history of the vegetation and the land of Morristown National Historical Park, New Jersey, since 1700. Bulletin New Jersey Academy of Science 27:1–19.

Eiler, J. H., Wathen, W. G., and Pelton, M. R. 1989. Reproduction in black bears in the southern Appalachian mountains. Journal of Wildlife Management 53:353–360.

Eis, S., E. H. Garman, and L. F. Ebell. 1965. Relation between cone production and diameter increment of douglas fir (*Pseudotsuga menziesii* (Mirb.) Franco), grand fir (*Pinus grandis* (Dougl.) Lindl.) and western white pine (*Pinus monticola* Dougl.). Canadian Journal of Botany 43:1553–1559.

Eisenberg, J. F. 1981. The Mammalian Radiations. University of Chicago Press, Chicago.

Elkinton, J. S., J. R. Gould, A. M. Liebhold, H. R. Smith, and W. E. Wallner. 1989. Are gypsy moth populations in North America regulated at low density? Pages 233–249 in: The Lymantriidae: Comparisons of Features of New and Old World Tussock Moths. Northeastern Forest Service Experimental Station Publication, Broomall, PA.

Elkinton, J. S., A. E. Hajek, G. H. Boettner, and E. E. Simons. 1991. Distribution and apparent spread of *Entomophaga maimaiga* (Zygomycetes: Entomophthorales) in gypsy moth (Lepidoptera: Lymantriidae) populations in North America. Environmental Entomology 20:1601–1605.

Elkinton, J. S., W. M. Healy, J. P. Buonaccorsi, G. H. Boettner, A. M. Hazzard, H. R. Smith, and A. M. Leobhold. 1996. Interactions among gypsy moths, white-footed mice, and acorns. Ecology 77:2332–2334.

Elkinton, J. S., and A. M. Liebhold. 1990. Population dynamics of gypsy moth in North America. Annual Review of Entomology 35:571–596.

Elliott, P. F. 1974. Social behavior and foraging ecology of the eastern chipmunk (*Tamias striatus*) in the Adirondack Mountains. Smithsonian Contributions in Zoology 265:1–107.

Elliott, P. F. 1988. Foraging behavior of a central-place forager: Field tests of theoretical predictions. American Naturalist 131:159–174.

Ellis, J. E., and J. B. Lewis. 1967. Mobility and annual range of wild turkeys in Missouri. Journal of Wildlife Management 31:568–581.

Elowe, K. D., and W. E. Dodge. 1989. Factors affecting black bear reproductive success and cub survival. Journal of Wildlife Management 53:962–968.

Endter-Wada, J., D. Blahna, R. Krannich, and M. A. Brunson. 1998. A framework for understanding social science contributions to ecosystem management. Ecological Applications 8:891–904.

Engle, D. M., T. G. Bidwell and R. E. Masters. 1996. Restoring Cross Timbers ecosystems with fire. Pages 190–199 in: Transactions of the 61st North American Wildlife and Naural Resource Conference (K. G. Wadsworth and R. E. McCabe, eds.). Wildlife Management Institute, Washington, DC.

Ensminger, M. E., J. E. Oldfield, W. W. Heinemann. 1990. Feeds and nutrition. 2nd ed. Ensminger Publishing, Clovis, CA.

Epifanio, C. R., M. J. Singer, and X. Huang. 1991. Hydrologic impacts of oak harvesting and evaluation of the modified universal soil loss equation. Pages 189–193 in: Proceedings of the Symposium on Oak Woodlands and Hardwood Rangeland Management. U.S. Forest Service General Technical Report PSW-126.

Evans, K. E., and R. N. Conner. 1979. Snag management. Pages 214–225 in: Workshop Proceedings: Management of North Central and Northeastern Forests for Nongame Birds (R. M. DeGraaf and K. E. Evans, compilers). U.S. Forest Service General Technical Report NC-51.

Everett, D. D., Jr., D. W. Speake, and W. K. Maddox. 1985. Habitat use by wild turkeys in northwest Alabama. Proceedings of the Annual Conference of the Southeastern Association of Fish and Wildlife Agencies 39:479–488.

Exum, J. H., J. A. McGlincy, D. W. Speake, J. L. Buckner, and F. M. Stanley. 1987. Ecology of the eastern wild turkey in an intensively managed pine forest in southern Alabama. Bulletin of Tall Timbers Research Station, Tallahassee, FL.

Eyre, F. H., ed. 1980. Forest Cover Types of the United States and Canada. Society of American Foresters, Washington, DC.

Faber-Langendoen, D., and M. A. Davis. 1995. Effects of fire frequency on tree cover at Allison savanna, east central Minnesota, USA. Natural Areas Journal 15:319–328.

Fahey, T. J., and W. A. Reiners. 1981. Fire in the forests of Maine and New Hampshire. Bulletin of the Torrey Botanical Club 108:363–373.

Fairweather, M. L., and R. L. Gilbertson. 1992. *Inonotus andersonii:* A wood decay fungus of oak trees in Arizona. Pages 195–198 in: Ecology and management of oak and associated woodlands: Perspectives in the southwestern United States and northern Mexico (P. F. Ffolliott, G. J. Gottfried, D. A. Bennett, V. M. Hernandez C., A. Ortega-Rubio, and R. H. Hamre, technical coordinators). U.S. Forest Service General Technical Report RM-218.

Farentinos, R. C. 1972. Observations on the ecology of the tassel-eared squirrel. Journal of Wildlife Management 36:1234–1239.

Feicht, D. L., S. L. C. Fosbroke, and M. J. Twery. 1993. Forest stand condition after 13 years of gypsy moth infestation. Pages 130–144 in: Proceedings of the North Central Hardwood Forest Conference (A. R. Gillespie, G. R. Parker,

P. E. Pope, and G. Rink, eds.). U.S. Forest Service General Technical Report NC-161.

Feldhamer, G. A., T. P. Kilbane, and D. W. Sharp. 1989. Cumulative effect of winter on acorn yield and deer body weight. Journal of Wildlife Management 53:292–295.

Feldhamer, G. A., D. W. Sharp, and T. Davin. 1992. Acorn yield and yearling white-tailed deer on Land Between The Lakes, Tennessee. Journal of the Tennessee Academy of Science 67:46–48.

Felix, A. C., III, T. L. Sharik, and B. S. McGinnes. 1986. Effects of pine conversion on food plants of northern bobwhite quail, eastern wild turkey, and white-tailed deer in the Virginia Piedmont. Southern Journal of Applied Forestry. 10:47–52.

Ffolliott, P. F. 1980. West live oak. Page 118 in: Forest Cover Types of the United States and Canada (F. H. Eyre, ed.). Society of American Foresters, Washington, DC.

Ffolliott, P. F. 1989. Arid zone forestry program: State of knowledge and experience in North America. Arizona Agricultural Experiment Station, Technical Bulletin 264.

Ffolliott, P. F. 1992. Multiple values of woodlands in the southwestern United States and northern Mexico. Pages 17–23 in: Ecology and management of oak and associated woodlands: Perspectives in the southwestern United States and northern Mexico (P. F. Ffolliott, G. J. Gottfried, D. A. Bennett, V. M. Hernandez C., A. Ortega-Rubio, and R. H. Hamre, technical coordinators). U.S. Forest Service General Technical Report RM-218.

Ffolliott, P. F., and G. J. Gottfried. 1992. Growth, yield, and utilization of oak woodlands in the southwestern United States. Pages 34–38 in: Ecology and management of oak and associated woodlands: Perspectives in the southwestern United States and northern Mexico (P. F. Ffolliott, G. J. Gottfried, D. A. Bennett, V. M. Hernandez C., A. Ortega-Rubio, and R. H. Hamre, technical coordinators). U.S. Forest Service General Technical Report RM-218.

Ffolliott, P. F., and G. J. Gottfried. 1999. Water use by Emory oak in southeastern Arizona. Hydrology and Water Resources in Arizona and the Southwest 29:43–48.

Ffolliott, P. F., and D. P. Guertin. 1987. Opportunities for multiple use values in the encinal oak woodlands of North America. Pages 182–189 in: Strategies for classification of natural vegetation for food production in arid zones (E. F. Aldon, C. E. Gonzales Vicente, and W. H. Moir, technical coordinators). U.S. Forest Service, General Technical Report RM-150.

Ffolliott, P. F., V. L. Lopes, C. Esquivel, and I. Sanchez Cohen. 1993. Conservation and sustainable development of encinal woodlands: A watershed management approach. Pages 61–66 in: Making sustainability operational:

Fourth Mexico–U.S. symposium (H. Manzanilla, D. Shaw, C. Aguirre-Bravo, L. Iglesias Gutierrez, and R. H. Hamre, technical coordinators). U.S. Forest Service General Technical Report RM-240.

Ffolliott, P. F., L. F. DeBano, M. B. Baker, Jr., G. J. Gottfried, G. Solis-Garza, C. B. Edminster, D. G. Neary, L. S. Allen, and R. H. Hamre (technical coordinators). 1996. Effects of fire on Madrean Province ecosystems: A symposium proceedings. U.S. Forest Service General Technical Report RM-289.

Firestone, M. K. 1995. Nutrient cycling in managed oak woodland-grass ecosystem. Final report to the Integrated Hardwood Range Management Program.

Fish, D. 1993. Population ecology of *Ixodes dammini*. Pages 25–42 in: Ecology and Management of Lyme disease (H. Ginsberg, ed.). Rutgers University Press, New Brunswick, NJ.

Flake, L. D., R. A. Craft, and W. L. Tucker. 1995. Vegetation characteristics of wild turkey roost sites during summer in south-central South Dakota. Proceedings of the National Wild Turkey Symposium 7:159–164.

Fleck, D. C. 1994. Chemical mediation of vertebrate-aided seed dispersal. Ph.D. diss., University of Colorado, Boulder.

Flowerdew, J. R. 1972. The effect of supplementary food on a population of wood mice (*Apodemus sylvaticus*). Journal of Animal Ecology 41:553–566.

Forbes, E. B., L. F. Marcy, A. L. Voris, and C. E. French. 1941. The digestive capacities of the white-tailed deer. Journal of Wildlife Management 5:108–114.

Forbush, E. H., and C. H. Fernald. 1896. The gypsy moth. Wright and Porter, Boston.

Ford, W. M., A. S. Johnson, P. E. Hale, and J. M. Wentworth. 1997. Influences of forest type, stand age, and weather on deer weights and antler size in the southern Appalachians. Southern Journal of Applied Forestry 21:11–18.

Fordham, R. A. 1971. Field populations of deermice with supplemental food. Ecology 52:138–146.

Fosbroke, D. E., and R. R. Hicks, Jr. 1989. Tree mortality following gypsy moth defoliation in southwestern Pennsylvania. Pages 74–80 in: Proceedings of the Seventh Central Hardwood Conference (G. Rink and C. A. Budelsky, eds.). U.S. Forest Service General Technical Report NC-132.

Foster, D. R. 1992. Land-use history (1730–1990) and vegetation dynamics in central New England, USA. Journal of Ecology 80:753–772.

Foster, D. R., J. D. Aber, J. M. Melillo, R. D. Bowden, and F. A. Bazzaz. 1997. Forest response to disturbance and anthropogenic stress. BioScience 47:437–445.

Foster, D. R., D. H. Knight, and J. F. Franklin. 1998. Landscape patterns and legacies resulting from large, infrequent forest disturbances. Ecosystems 1:497–510.

Foster, D. R., and G. Motzkin. 1998. Ecology and conservation in the cultural landscape of New England: Lessons from nature's history. Northeastern Naturalist 5:111–126.

Foster, D. R., G. Motzkin, and B. Slater. 1998. Land-use history as long-term broad-scale disturbance: Regional forest dynamics in central New England. Ecosystems 1:96–119.

Foster, D. R., and T. M. Zebryk. 1993. Long-term vegetation dynamics and disturbance history of a *Tsuga*-dominated forest in New England. Ecology 74:982–998.

Fox, J. F. 1974. Coevolution of white oak and its seed predators. Ph.D. diss., University of Chicago, Chicago.

Fox, J. F. 1982. Adaptation of gray squirrel behavior to autumn germination by white oak acorns. Evolution 36:800–809.

Fralish, J. S., F. B. Crooks, J. L. Chambers, and F. M. Harty. 1991. Comparison of presettlement, second-growth and old-growth forest on six site types in the Illinois Shawnee Hills. American Midland Naturalist 125:294–309.

Franklin, J. F. 1997. Ecosystem management: An overview. Pages 21–53 in: Ecosystem Management (M. S. Boyce and A. Haney, eds.). Yale University Press, New Haven, CT.

Frederickson, T. S., B. Ross, W. Hoffman, M. Lester, J. Beyea, M. L. Morrison, and B. N. Johnson. 1998. Adequacy of natural hardwood regeneration on forestlands in northeastern Pennsylvania. Northern Journal of Applied Forestry 15:130–134.

Frelich, L. E. 1995. Old forest in the Lake States today and before European settlement. Natural Areas Journal 15:157–167.

Frelich, L. E., and C. G. Lorimer. 1985. Current and predicted long-term effects of deer browsing in hemlock forests in Michigan, USA. Biological Conservation 34:99–120.

Frelich, L. E., and C. G. Lorimer. 1991. Natural disturbance regimes in hemlock-hardwood forests of the upper Great Lakes region. Ecological Monograph 61:145–164.

Frost, W. E., and S. B. Edinger. 1991. Effects of tree canopies on soil characteristics of annual rangeland. Journal of Range Management 44:286–288.

Frost, W. E., J. W. Bartolome, and J. M. Connor. 1997. Understory-canopy relationships in oak woodlands and savannas. Pages 183–190 in: Proceedings of a Symposium on Oak Woodlands: Ecology, Management, and Urban Interface Issues. U.S. Forest Service General Technical Report PSW-GTR-160.

Frost, W. E., and N. K. McDougald. 1989. Tree canopy effects on herbaceous production of annual rangeland during drought. Journal of Range Management 42:281–283.

Fuller, J. L. 1997. Holocene forest dynamics in southern Ontario, Canada: Fine-resolution pollen data. Canadian Journal of Botany 75:1714–1727.

Fuller, J. L., D. R. Foster, J. S. Mclachlan, and N. Drake. 1998. Impact of human activity on regional forest composition and dynamics in central New England. Ecosystems 1:76–95.

Galford, J. R. 1986. Primary infestation of sprouting chestnut, red, and white oak acorns by *Valentinia glandella* (Lepidoptera: Blastrobasidae). Entomological News 97:109–112.

Galford, J. R., J. W. Peacock, and S. L. Wright. 1988. Insects and other pests affecting oak regeneration. Pages 219–225 in: Guidelines for regenerating Appalachian hardwood stands (H. C. Smith, A. W. Perkey, and W. E. Kidd Jr., eds.). Society of American Foresters Publication 88-03.

Ganey, J. L., and R. Balda. 1989. Distribution and habitat use of Mexican spotted owl in Arizona. Condor 91:355–361.

Ganey, J. L., R. B. Duncan, and W. M. Block. 1992. Use of oak and associated woodlands by Mexican spotted owls in Arizona. Pages 125–128 in: Ecology and management of oak and associated woodlands: Perspectives in the southwestern United States and northern Mexico (P. F. Ffolliott, G. J. Gottfried, D. A. Bennett, V. M. Hernandez C., A. Ortega-Rubio, and R. H. Hamre, technical coordinators). U.S. Forest Service General Technical Report RM-218.

Gansner, D. A., S. L. Arner, and R. H. Widmann. 1993. After two decades of gypsy moth, is there any oak left? Northern Journal of Applied Forestry 10:184–186.

Gardner, D. T., and D. H. Arner. 1968. Food supplements and wild turkey reproduction. Transactions of the North American Wildlife and Natural Resources Conference 33:250–258.

Garner, M. E. 1999. Risk mitigation of wildfire hazards at the wildland urban interface of northwest Arkansas. Master's thesis, University of Arkansas, Center for Advanced Spatial Technologies.

Garner, N. P. 1986. Seasonal movements, habitat selection, and food habits of black bears (*Ursus americanus*) in Shenandoah National Park, Virginia. Master's thesis, Virginia Polytechnic Institute and State University, Blacksburg.

Garren, K. H. 1943. Effects of fire on vegetation of the southeastern United States. Botanical Review 9:617–654.

Garrison, B. 1996. Vertebrate wildlife species and habitat associations. In: Guidelines for managing California's hardwood rangelands (R. B. Standiford and P. Tinnin, eds.). University of California Division of Agriculture and Natural Resources Leaflet 3368.

Garrison, B., and R. B. Standiford. 1996. Oaks and habitats of the hardwood rangeland. In: Guidelines for managing California's hardwood rangelands (R. B. Standiford and P. Tinnin, eds.). University of California Division of Agriculture and Natural Resources Leaflet 3368.

Garshelis, D. L. 1978. Movements ecology and activity behavior of black bears in the Great Smoky Mountains National Park. Master's thesis, University of Tennessee, Knoxville.

Garshelis, D. L, and M. R. Pelton. 1980. Activity of black bears in the Great Smoky Mountains National Park. Journal of Mammalogy 61:8–19.

Garshelis, D. L, and M. R. Pelton. 1981. Movements of black bears in the Great Smoky Mountains National Park. Journal of Wildlife Management 45:912–925.

Gashwiler, J. S. 1979. Deer mouse reproduction and its relationship to the tree seed crop. American Midland Naturalist 102:95–104.

George, M. 1987. Management of hardwood range: A historical review. Agronomy and Range Science, Range Science Report 12, University of California, Davis.

Gibbs, J. N., and D. W. French. 1980. The transmission of oak wilt. U.S. Forest Service Research Paper NC-185.

Gibson, L. P. 1982. Insects that damage northern red oak acorns. U.S. Forest Service, Northeastern Forest Experiment Station, Research Paper NE-220.

Gilbert, B. S., and C. J. Krebs. 1981. Effects of extra food on *Peromyscus* and *Clethrionomys* populations in the southern Yukon. Oecologia 51:326–331.

Gilbert, G. S., and S. P. Hubbell. 1996. Plant diseases and the conservation of tropical forests. BioScience 46:98–106.

Gilbert, J. R., W. S. Kordek, J. Collins, and R. Conley. 1978. Interpreting sex and age data from legal kills of bears. Proceedings of the Eastern Black Bear Workshop 4:253–263.

Gill, J. D., J. W. Thomas, W. M. Healy, J. C. Pack, and H. R. Sanderson. 1975. Comparison of seven forest types for game in West Virginia. Journal of Wildlife Management 39:662–768.

Gill, R. M. A. 1992. A review of damage by mammals in north temperate forests: 1. Deer. Forestry 65:145–169.

Giusti, G. A, and P. J. Tinnin. 1993. A planner's guide for oak woodlands. Publication of the Integrated Hardwood Range Management Program, University of California Division of Agriculture and Natural Resources Leaflet 3369.

Glaser, R. W. 1915. Wilt of gypsy moth caterpillars. Journal of Agricultural Research 4:101–128.

Gleason, H. A. 1913. The relation of forest distribution and prairie fires in the Middle West. Torreya 13:173–181.

Glitzenstein, J. S., C. D. Canham, M. J. McDonnell, and D. R. Streng. 1990. Effects of environment and land-use history on upland forests of the Cary Arboretum, Hudson Valley, New York. Bulletin of the Torrey Botanical Club 117:106–122.

Glover, F. A. 1948. Winter activities of wild turkey in West Virginia. Journal of Wildlife Management 12:416–427.

Godfrey, C. L., K. Needham, M. R. Vaughan, J. C. Vashon, D. D. Martin, and G. T. Blank, Jr. 2000. A technique for and risk associated with entering tree dens used by black bears. Wildlife Society Bulletin 28:131–140.

Godwin, K. D., G. A. Hurst, and B. D. Leopold. 1994. Movements of wild turkey gobblers in central Mississippi. Proceedings of the Annual Conference of the Southeast Association of Fish and Wildlife Agencies 48:117–122.

Goebel, P. C., and D. M. Hix. 1996. Development of mixed-oak forests in southeastern Ohio: A comparison of second-growth and old-growth forests. Forest Ecology and Management 84:1–21.

Goldberg, D. E. 1982. The distribution of evergreen and deciduous trees relative to soil types: An example from the Sierra Madre, Mexico, and a general model. Ecology 63:942–951.

Goldsmith, A., M. E. Walraven, D. Graber, and M. White. 1981. Ecology of the black bear in Sequoia National Park. National Park Service Final Report, Contract CY-8000-4-0022.

Good, H. G., and L. C. Webb. 1940. Spring foods of the wild turkey in Alabama. Alabama Game and Fish News 12:3–4.

Goodrum, P. D. 1959. Acorns in the diet of wildlife. Proceedings of the Southeastern Association of Game and Fish Commissioners 13:54–57.

Goodrum, P. D., V. H. Reid, and C. E. Boyd. 1971. Acorn yields, characteristics, and management criteria of oaks for wildlife. Journal of Wildlife Management 35:520–532.

Gordon, D. R., J. M. Welker, J. W. Menke, and K. J. Rice. 1989. Competition for soil water between annual plants and blue oak seedlings. Oecologia 79:533–541.

Gore, J. A., and W. A. Patterson, III. 1986. Mass of downed wood in northern hardwood forests in New Hampshire: Potential effect of forest management. Canadian Journal of Forest Research 16:335–339.

Gotelli, N. J. 1995. A Primer of Ecology. Sinauer Associates, Sunderland, MA.

Gottfried, G. J., and P. F. Ffolliott. 1993. Silvicultural prescriptions for sustained productivity of the pinyon-juniper and encinal oak woodlands. Pages 185–192 in: Making sustainability operational: Fourth Mexico–U.S. symposium (H. Manzanilla, D. Shaw, C. Aguirre-Bravo, L. Iglesias Gutierrez, and R. H. Hamre, technical coordinators). U.S. Forest Service General Technical Report RM-240.

Gottfried, G. J., P. F. Ffolliott, and L. F. DeBano. 1995. Forests and woodlands of the sky islands: Stand characteristics and silvicultural prescriptions. Pages 152–164 in: Biodiversity and management of the Madrean archipelago: The sky islands of southwestern United States and northern Mexico (L. F. DeBano, P. F. Ffolliott, A. Ortega-Rubio, G. J. Gottfried, R. H. Hamre, and C. B. Edminister, technical coordinators). U.S. Forest Service General Technical Report RM-264.

Gottschalk, K. W. 1983. Management strategies for successful regeneration: oak-hickory. Pages 190–213 in: Proceedings of a Symposium on Regenerating Hardwood Stands (J. Finley, R. S. Cochran, and J. R. Grace, coordinators). Pennsylvania State Forestry Issues Conference, School of Forest Resources, University Park.

Gottschalk, K. W. 1987. Effects of shading on growth and development of northern red oak, black oak, black cherry, and red maple seedlings. Vol. 2, Biomass partitioning and prediction. Pages 99–110 in: Proceedings of the 6th Central Hardwood Forest Conference (R. L. Hay, F. W. Woods, and H. DeSelm, eds.). University of Tennessee, Knoxville.

Gottschalk, K. W. 1989. Gypsy moth effects on mast production. Pages 42–50 in: Proceedings of Southern Appalachian Mast Management Workshop (C. E. McGee, ed.). University of Tennessee, Knoxville.

Gottschalk, K. W. 1993. Silvicultural guidelines for forest stands threatened by the gypsy moth. U.S. Forest Service General Technical Report NE-171.

Gottschalk, K. W. 1994. Shade, leaf growth and crown development of *Quercus rubra, Quercus velutina, Prunus serotina* and *Acer rubrum* seedlings. Tree Physiolology 14:735–749.

Gould, J. R., J. S. Elkinton, and W. E. Wallner. 1990. Density-dependent suppression of experimentally created gypsy moth, *Lymantria dispar* (Lepidoptera: Lymantriidae) populations by natural enemies. Journal of Animal Ecology 59:213–233.

Graber, D. M., and M. White. 1983. Black bear food habits in Yosemite National Park. International Conference on Bear Research and Management 5:1–10.

Gravatt, G. F. 1925. The chestnut blight in North Carolina. Pages 13–17 in: Chestnut and the chestnut blight in North Carolina. North Carolina Geological and Economic Survey Economic Paper 56.

Graves, W. C. 1980. Annual oak mast yields from visual estimates. Pages 279–274 in: Proceedings of the Symposium on the Ecology, Management, and Utilization of California Oaks. U.S. Forest Service General Technical Report PSW-44.

Greenberg, C. H. 2000. Individual variation in acorn production by five species of southern Appalachian oaks. Forest Ecology and Management 132:199–210.

Greenberg, C. H. 1998. Summary: Acorn production by southern Appalachian oaks, 1993–97. U.S. Forest Service In-house Report, Bent Creek Experimental Forest, Asheville, NC.

Greenberg, C. H., and W. H. McNab. 1998. Forest disturbance in hurricane-related downbursts in the Appalachian mountains of North Carolina. Forest Ecology and Management 104:179–191.

Greenberg, C. H., and B. R. Parresol. 2000. Acorn production characteristics of southern Appalachian oaks: A simple method to predict within-year acorn crop size. U.S. Forest Service, Southern Research Station, Research Paper SRS-19.

Greenfell, W. E., Jr., and A. J. Brody. 1983. Seasonal foods of black bears in Tahoe National Forest, California. California Fish and Game 69:132–150.

Greenwood, G. B., R. K. Marose, and J. M. Stenback. 1993. Extent and ownership of California's hardwood rangelands. Unpublished report for Strategic Planning Program, California Department of Forestry and Fire Protection, Sacramento.

Griffin, J. R. 1971. Oak regeneration in the upper Carmel Valley, California. Ecology 52:862–868.

Griffin, J. R. 1973. Valley oaks: The end of an era? Fremontia 1:5–9.

Griffin, J. R. 1977. Oak woodland. Pages 382–415 in: Terrestrial Vegetation of California (M. G. Barbour and J. Major, eds.). John Wiley and Sons, New York.

Griffin, J. R., and W. B. Critchfield. 1972. The distribution of forest trees in California. U.S. Forest Service Research Paper PSW-820.

Grimm, E. C. 1983. Chronology and dynamics of vegetation change in the prairie-woodland region of southern Minnesota, USA. New Phytologist 93:311–350.

Grimm, E. C. 1984. Fire and other factors controlling the Big Woods vegetation of Minnesota in the mid-nineteenth century. Ecological Monographs 54:291–311.

Grinnell, J. 1936. Up-hill planters. Condor 38:80–82.

Grumbine, R. E. 1992. Ghost Bears: Exploring the Biodiversity Crisis. Island Press, Washington, DC.

Grumbine, R. E. 1994. What is ecosystem management? Conservation Biology 8:27–38.

Grumbine, R. E. 1997. Reflections on "What is ecosystem management?" Conservation Biology 11:41–47.

Guyette, R. P., and B. E. Cutter. 1991. Tree-ring analysis of fire history of a post oak savanna in the Missouri Ozarks. Natural Areas Journal 11:93–99.

Guyette, R. P., and B. E. Cutter. 1997. Fire history, population, and calcium cycling in the Current River watershed. Pages 354–372 in: Proceedings of the 11th Central Hardwood Forest Conference (S. G. Pallardy, R. A. Cecich, E. H. Garrett, and P. S. Johnson, eds.). U.S. Forest Service General Technical Report NC-188.

Guyette, R. P., and D. C. Dey. 1995. A presettlement fire history in an oak-pine forest near Basin Lake, Algonquin Park, Ontario. Ontario Ministry of Natural Resources, Ontario Forest Research Institute, Forest Research Report 132.

Guyette, R. P., and D. C. Dey. 1997. Historic shortleaf pine (*Pinus echinata* Mill.) abundance and fire frequency in a mixed oak-pine forest (MOFEP, site 8). Pages 136–149 in: Proceedings of the Missouri Ozark forest ecosystem project: An experimental approach to landscape research (B. L. Brookshire and S. R. Shifley, eds.). U.S. Forest Service General Technical Report NC-193.

Guyette, R., M. Dey, and D. C. Dey. 1999. An Ozark fire history. Missouri Conservationist 60:5–7.

Guyette, R. P., D. C. Dey, and C. McDonell. 1995. Determining fire history from old white pine stumps in an oak-pine forest in Bracebridge, Ontario. Ontario Ministry Natural Resources, Ontario Forest Research Institute, Forest Research Report 133.

Guyette, R. P., and E. A. McGinnes. 1982. Fire history of an Ozark glade in Missouri. Transactions of the Missouri Academy of Science 16:85–93.

Gysel, L. W. 1956. Measurement of acorn crops. Forestry Science 2:305–313.

Gysel, L. W. 1958. Prediction of acorn crops. Forest Science 4:239–245.

Haack, R. A., and J. W. Byler. 1993. Insects and pathogens: Regulators of forest ecosystems. Journal of Forestry 91:32–37.

Hadj-Chikh, L. Z., M. A. Steele, and P. D. Smallwood. 1996. Caching decisions by grey squirrels: A test of the handling-time and perishability hypotheses. Animal Behaviour 52:941–948.

Haines, D. A., V. J. Johnson, and W. A. Main. 1975. Wildfire atlas of the northeastern and north central states. U.S. Forest Service General Technical Report NC-16.

Hajek, A. E., J. S. Elkinton, and J. J. Witcosky. 1996. Introduction and spread of the fungal pathogen *Entomophaga maimaiga* (Zygomycetes: Entomophthorales) along the leading edge of gypsy moth (Lepidoptera: Lymantriidae) spread. Environmental Entomology 25:1235–1247.

Hajek, A. E., R. A. Humber, and J. S. Elkinton. 1995. Mysterious origin of *Entomophaga maimaiga* in North America. American Entomology 41:31–42.

Hajek, A. E., R. A. Humber, J. S. Elkinton, B. May, S. R. A. Walsh, and J. C. Silver. 1990. Allozyme and RFLP analyses confirm *Entomophaga maimaiga* responsible for 1989 epizootics in North American gypsy moth populations. Proceedings of the National Academy of Science 87:6979–6982.

Hall, L. M., M. R. George, D. D. McCreary, and T. E. Adams. 1992. Effects of cattle grazing on blue oak seedling damage and survival. Journal of Range Management 45:503–506.

Hallwachs, W. 1994. The clumsy dance between agoutis and plants: Scatterhoarding by Costa Rican dry forest agoutis (*Dasyprocta punctata:* Dasyproctidae: Rodentia). Ph.D. diss., Cornell University, Ithaca, NY.

Halvorson, C. H. 1982. Rodent occurrence, habitat disturbance, and seed fall in a larch-fir forest. Ecology 63:423–433.

Hamilton, W. J., Jr. 1941. The food of small forest mammals in eastern United States. Journal of Mammalogy 22:250–263.

Hamilton, R. J., and R. L. Marchinton. 1980. Denning and related activities of black bears in the coastal plain of North Carolina. International Conference on Bear Research and Management 4:121–126.

Hane, J. 1999. Developmental responses of oaks and yellow-poplar following shelterwood and clearcut treatments. Master's thesis, Department of Forest Resources, Clemson University, Clemson, SC.

Hannon, S. J., R. L. Mumme, W. D. Koenig, S. Spon, and F. A. Pitelka. 1987. Poor acorn crop, dominance, and decline in numbers of acorn woodpeckers. Journal of Animal Ecology 56:197–207.

Hansen, L., and G. O. Batzli. 1978. The influence of food availability on the white-footed mouse: Populations in isolated woodlots. Canadian Journal of Zoology 56:2530–2541.

Hansen, L., and G. O. Batzli. 1979. Influence of supplemental food on local populations of *Peromyscus leucopus*. Journal of Mammalogy 60:335–342.

Hansen, M. H., T. Frieswyk, and J. F. Glover, and J. F. Kelly. 1992. The eastwide forest inventory data base: User's manual. U.S. Forest Service General Technical Report NC-151.

Hanson, P. J., J. G. Isebrands, and R. E. Dickson. 1987. Carbon budgets of *Quercus rubra* L. seedlings at selected stages of growth: Influence of light. Pages 269–276 in: Proceedings of the 6th Central Hardwood Forest Conference (R. L. Hay, F. W. Woods, and H. DeSelm, eds.). University of Tennessee, Knoxville.

Hare, R. C. 1965. The contribution of bark to fire resistance of southern trees. Journal of Forestry 63:248–251.

Harlow, R. F. 1961. Characteristics and status of Florida black bears. Transactions of the North American Wildlife Conference 26:481–495.

Harlow, R. F., and R. L. Eikum. 1963. The effect of stand density on the acorn production of turkey oaks. Proceedings of the Annual Conference of the Southeastern Association of Game and Fish Commission 17:126–133.

Harlow, R. F., and J. K. Jones, Jr., eds. 1965. The white-tailed deer in Florida. Technical Bulletin, Florida Game and Freshwater Fish Commission 9:1–240.

Harlow, R. F., J. B. Whelan, H. S. Crawford, and J. E. Skeen. 1975. Deer foods during years of oak mast and abundance. Journal of Wildlife Management 39:330–336.

Harmon, M. E. 1982. Fire history of the westernmost portion of Great Smoky Mountains National Park. Bulletin of the Torrey Botanical Club 109:74–79.

Harmon, M. E. 1984. Survival of trees after low-intensity surface fires in the Great Smoky Mountains National Park. Ecology 65:796–802.

Haroldson, K. J. 1996. Energy requirements for winter survival of wild turkeys. Proceedings of the National Wild Turkey Symposium 7:6–14.

Harper, K. T., F. J. Wagstaff, and L. M. Kunzler. 1985. Biology and management of the gambel oak vegetative type: A literature review. U.S. Forest Service General Technical Report INT-179.

Harrington, R. A., B. J. Brown, and P. B. Reich. 1989. Ecophysiology of exotic and native shrubs in southern Wisconsin. I. Relationship of leaf characteristics, resource availability, and phenology to seasonal patterns of carbon gain. Oecologia 80:356–367.

Harrison, J. S., and P. A. Werner. 1984. Colonization by oak seedlings into a heterogeneous successional habitat. Canadian Journal of Botany 62:559–563.

Havera, S. P., and K. E. Smith. 1979. A nutritional comparison of selected fox squirrel foods. Journal of Wildlife Management 43:691–704.

Hawley, R. C., and A. F. Hawes. 1912. Forestry in New England. John Wiley and Sons, New York.

Haworth, K., and G. R. McPherson. 1994. Effects of *Quercus emoryi* on herbaceous vegetation in a semi-arid savanna. Vegetatio 112:153–159.

Hayden, A. H. 1979. Wild turkey food habits and nutritional studies. Pennsylvania Game Commission, Harrisburg. Final Report, Project 04030, Job 7.

Hayden, A. H., and E. Nelson. 1963. The effects of starvation and limited rations on reproduction of game farm wild turkeys. Transactions of the Northeastern Section, Wildlife Society 20:1–11.

Hayes, S. G., and M. R. Pelton. 1994. Habitat characteristics of female black bear dens in northwestern Arkansas. International Conference on Bear Research and Management 9:411–418.

He, H. S., D. J. Mladenoff, V. C. Radeloff, and T. R. Crow. 1998. Integration of GIS data and classified satellite imagery for regional forest assessment. Ecological Applications 8:1072–1083.

Healy, W. M. 1981. Habitat requirements of the wild turkey in the southeastern mountains. Pages 24–34 in: Proceedings of the Symposium on Habitat Requirements and Habitat Management for the Wild Turkey in the Southeast (P. T. Bromley and R. L. Carlton, eds.). Virginia Wild Turkey Foundation, Elliston, VA.

Healy, W. M. 1992a. Behavior. Pages 46–65 in: The Wild Turkey: Biology and Management (J. G. Dickson, ed.). Stackpole Books, Harrisburg, PA.

Healy, W. M. 1992b. Population influences: Environment. Pages 129–143 in: The Wild Turkey: Biology and Management (J. G. Dickson, ed.). Stackpole Books, Harrisburg, PA.

Healy, W. M. 1997a. Influence of deer on the structure and composition of oak forests in Central Massachusetts. Pages 249–266 in: The Science of Overabundance: Deer Ecology and Population Management. (W. J. McShea, H. B. Underwood, and J. H. Rappole, eds.). Smithsonian Institution Press, Washington, DC.

Healy, W. M. 1997b. Thinning New England oak stands to enhance acorn production. Northern Journal of Applied Forestry 14:152–156.

Healy, W. M., and R. T. Brooks. 1988. Small mammal abundance in northern hardwood stands in West Virginia. Journal of Wildlife Management 52:491–496.

Healy, W. M., R. T. Brooks, and R. M. DeGraaf. 1989. Cavity trees in sawtimber-size oak stands in central Massachusetts. Northern Journal of Applied Forest 6:61–65.

Healy, W. M., K. W. Gottschalk, R. P. Long, and P. M. Wargo. 1997. Changes in eastern forests: Chestnut is gone. Are oaks far behind? Transactions of North American Wildlife and Natural Resources Conference 62:249–263.

Healy, W. M., A. M. Lewis, and E. F. Boose. 1999. Variation of red oak acorn production. Forest Ecology and Management 116:1–11.

Hecklau, J. D., W. F. Porter, and W. M. Shields. 1982. Feasibility of transplanting wild turkeys into areas of restricted forest cover and high human density. Transactions of the Northeastern Section, Wildlife Society 39:96–104.

Heim, S. 1987. Spring food habits of New Hampshire deer. Master's thesis, University of New Hampshire, Durham.

Heinselman, M. L. 1973. Fire in the virgin forests of the Boundary Water Canoe Area, Minnesota. Quarterly Research 3:329–382.

Hellgren, E. C. 1993. Status, distribution, and summer food habits of black bears in Big Bend National Park. Southwest Naturalist 38:77–80.

Hellgren, E. C., and M. R. Vaughan. 1988. Seasonal food habits of black bears in Great Dismal Swamp, Virginia–North Carolina. Proceedings of the Annual Conference of the Southeastern Association of Fish and Wildlife Agencies 42:295–305.

Hellgren, E. C., and M. R. Vaughan. 1989. Denning ecology of black bears in a southeastern wetland. Journal of Wildlife Management 53:347–353.

Hendricks, D. M. 1985. Arizona Soils. University of Arizona Press, Tucson.

Hengst, G. E., and J. O. Dawson. 1994. Bark properties and fire resistance of selected tree species from the central hardwood region of North America. Canadian Journal of Forest Resources 24:688–696.

Henning, S. J., and D. I. Dickmann. 1996. Vegetative responses to prescribed burning in a mature red pine stand. Northern Journal of Applied Forestry 13:140–146.

Henry, J. D., and J. M. S. Swan. 1974. Reconstructing forest history from live and dead plant material: An approach to the study of forest succession in southwest New Hampshire. Ecology 55:772–783.

Herrera, C., P. Jordano, J. Guitian, and A. Traveset. 1998. Annual variability in seed production by woody plants and the masting concept: Reassessment of principles and relationship to pollination and seed dispersal. American Naturalist 152:576–594.

Herrick, D. W., and D. A. Gansner. 1987. Gypsy moth on a new frontier: Forest tree defoliation and mortality. Northern Journal of Applied Forestry 4:128–133.

Hicks, R. R., Jr. 1998. Ecology and Management of Central Hardwood Forests. John Wiley and Sons, New York.

Higgins, J. C. 1997. Survival, home range, and spatial relationships of Virginia's exploited black bear population. Master's thesis, Virginia Polytechnic Institute and State University, Blacksburg.

Hildebrand, D. C., and M. N. Schroth. 1967. A new species of *Erwinia* causing drippy nut disease of live oaks. Phytopathology 57:250–253.

Hinkle, C. R., W. C. McComb, J. M. Safley, Jr., and P. A. Schmalzer. 1993. Mixed mesophytic forests. Pages 203–253 in: Biodiversity of the southeastern United States upland terrestrial communities (W. H. Martin, S. G. Boyce, and A. C. Echternacht, eds.). John Wiley and Sons, New York.

Hix, D. M., and C. G. Lorimer. 1991. Early stand development on former oak sites in southwestern Wisconsin. Forest Ecology and Management 42:169–193.

Hix, D. M., D. F. Fosbroke, R. R. Hicks, Jr., and K. W. Gottschalk. 1991. Development of regeneration following gypsy moth defoliation of Appalachian Plateau and Ridge and Valley hardwood stands. Pages 347–359 in: Proceedings of the Eighth Central Hardwood Forest Conference. U.S. Forest Service General Technical Report NE-148.

Hodges, J. D., and E. S. Gardiner. 1993. Ecology and physiology of oak regeneration. Pages 54–65 in: Oak Regeneration: Serious Problems, Practical Recommendations (D. Loftis and C. E. McGee, eds.). U.S. Forest Service General Technical Report SE-84.

Hoffard, W. H., D. H. Marx, and D. H. Brown. 1995. The health of southern forests. U.S. Forest Service Protection Report R-8 PR 27.

Hoffman, D. M. 1962. The wild turkey in eastern Colorado. Colorado Game and Fish Department Technical Bulletin 12, Denver.

Hoffman, D. M. 1968. Roosting sites and habits of Merriam's turkeys in Colorado. Journal of Wildlife Management 32:859–866.

Hoffman, R. W., M. P. Luttrell, and W. R. Davidson. 1995. Reproductive performance of Merriam's wild turkeys with suspected *Mycoplasma* infection. Proceedings of the National Wild Turkey Symposium 7:145–151.

Hoffman, P. W., H. G. Shaw, M. A. Rumble, B. F. Wakeling, S. D. Schmnitz, C. M. Mollohan, R. Engel-Wilson, and D. A. Hengel. 1993. Guidelines for managing Merriam's wild turkey. U.S. Forest Service, Rocky Mountain Forest and Range Experiment Station, Fort Collins, Colorado.

Holland, V. L. 1980. Effect of blue oak on rangeland forage production in central California. Pages 314–318 in: Proceedings of a Symposium on the Ecology, Management and Utilization of California Oaks. U.S. Forest Service General Technical Report PSW-44.

Holland, V. L., and J. Morton. 1980. Effect of blue oak on nutritional quality of rangeland forage in central California. Pages 319–322 in: Proceedings of a Symposium on the Ecology, Management and Utilization of California Oaks. U.S. Forest Service General Technical Report PSW-44.

Holmes, T. H. 1990. Botanical trends in Northern California Oak Woodland. Rangelands 12:3–7.

Holt, R. D. 1984. Spatial heterogeneity, indirect interactions, and the coexistence of prey species. American Naturalist 124:377–406.

Holter, J. B., and H. H. Hayes. 1977. Growth in white-tailed deer fawns fed varying energy and constant protein. Journal of Wildlife Management 41:506–510.

Holzman, B. A. 1993. Vegetation change in California's blue oak woodlands 1932–1992. Ph.D. diss., University of California, Berkeley.

Holzman, B. A., and B. H. Allen-Diaz. 1991. Vegetation change in blue oak woodlands in California. Pages 189–193 in: Proceedings of a Symposium on Oak Woodlands and Hardwood Rangeland Management. U.S. Forest Service General Technical Report PSW-126.

Horsley, S. B., and D. A. Marquis. 1983. Interference by weeds and deer with Allegheny hardwood reproduction. Canadian Journal of Forest Research 13:61–69.

Hosley, N. W., and R. K. Ziebarth. 1935. Some winter relations of the white-tailed deer to the forests in north central Massachusetts. Ecology 16:535–553.

Hough, A. F. 1965. A twenty-year record of understory vegetational change in a virgin Pennsylvania forest. Ecology 46:370–373.

Hough, A. F., and R. D. Forbes. 1943. The ecology and silvics of forests in the high plateaus of Pennsylvania. Ecological Monographs 13:301–320.

Houston, D. R. 1981. Effects of defoliation on trees and stands. Pages 217–297 in: The Gypsy Moth: Research Toward Integrated Pest Management (C. G. Doane and M. L. McManus, eds.). U.S. Department of Agriculture Technical Bulletin 1584.

Howe, C. D., and J. H. White. 1913. Trent Watershed Survey. Commission of Conservation, Canada, Committee on Forests. Bryant Press, Toronto.

Howe, H. F. 1986. Seed dispersal by fruit-eating birds and mammals. Pages 123–190 in: Seed Dispersal (D. R. Murray, ed.). Academic Press, Sydney, Australia.

Howe, H. F. 1989. Scatter- and clump-dispersal and seedling demography: Hypothesis and implications. Oecologia 79:417–426.

Howe, H. F., and G. F. Estabrook. 1977. On intraspecific competition for avian dispersers in tropical trees. American Naturalist 111:817–832.

Howe, H. F., and J. Smallwood. 1982. Ecology of seed dispersal. Annual Review of Ecology and Systematics 13:201–228.

Howe, H. F., and L. C. Westley. 1988. Ecological Relationships between Plants and Animals. Oxford University Press, Oxford.

Howell, D. L., and C. L. Kucera. 1956. Composition of pre-settlement forests in three counties of Missouri. Bulletin of the Torrey Botanical Club 83:207–217.

Hubbard, J. A., and G. R. McPherson. 1997. Acorn selection by Mexican jays: A test of a tri-trophic symbiotic relationship hypothesis. Oecologia 110:143–146.

Huddle, J. A., and S. G. Pallardy. 1996. Effects of soil and stem base heating on survival, resprouting and gas exchange of *Acer* and *Quercus* seedlings. Tree Physiology 16:583–589.

Humiston, G. 1995. California rangeland water quality management plan. State Water Resources Control Board, Sacramento, CA.

Hunter, A. F., and G. Dwyer. 1998. Outbreaks and interacting factors: Insect population explosions synthesized and dissected. Integrative Biology 1:166–177.

Hunter, M. L., Jr. 1999. Maintaining Biodiversity in Forest Ecosystems. Cambridge University Press, New York.

Huntsinger, L. 1997. California's oak woodlands revisited: Changes in owners, use, and management, 1985 to 1992. Pages 626–630 in: Proceedings of the Symposium on Oak Woodlands: Ecology, Management, and Urban Interface Issues. U.S. Forest Service General Technical Report PSW-162.

Hurly, T. A., and S. A. Lourie. 1997. Scatterhoarding and larderhoarding by red squirrels: Size, dispersion, and allocation of hoards. Journal of Mammalogy, 78:528–537.

Hurst, G. A. 1992. Foods and feeding. Pages 66–83 in: The Wild Turkey: Biology and Management (J. G. Dickson, ed.). Stackpole Books, Harrisburg, PA.

Hutto, R. L. 1985. Seasonal changes in the habitat distribution of transient insectivorous birds in southeastern Arizona: Competition mediated? The Auk 102:120–132.

Hyink, D. M., and S. M. Zedaker. 1987. Stand dynamics and the evaluation of forest decline. Tree Physiology 3:17–26.

Ignatoski, F. J. 1973. Status of wild turkeys in Michigan. Pages 49–53 in: Wild Turkey Management: Current Problems and Programs (G. C. Sanderson and H. C. Schultz, eds.). Missouri Chapter of the Wildlife Society and University of Missouri Press, Columbia.

Iida, S. 1996. Quantitative analysis of acorn transportation by rodents using magnetic locator. Vegetatio 124:39–43.

Inman, R. M., and M. R. Pelton. In press. Energetic production by black bear foods in the Smoky Mountains. Ursus.

Irwin, L. L., and F. M. Hammond. 1985. Managing black bear habitat for food items in Wyoming. Wildlife Society Bulletin 13:477–483.

Jackson, L. E., R. B. Strauss, M. K. Firestone, and J. W. Bartolome, 1990. Influence of tree canopies on grassland productivity and nitrogen dynamics in deciduous oak savanna. Agriculture, Ecosystems and Environment 32:89–105.

Jacobs, L. F., and E. R. Liman. 1991. Grey squirrels remember the locations of buried nuts. Animal Behaviour 41:103–110.

Jansen, H. C. 1987. The effect of blue oak removal on herbaceous production on a foothill site in the northern Sierra Nevada. Pages 343–350 in: Proceedings of a Symposium on Multiple-Use Management of California's Hardwood Resources. U.S. Forest Service General Technical Report PSW-100.

Janzen, D. H. 1970. Herbivores and the number of tree species in tropical forests. American Naturalist 104:501–528.

Janzen, D. H. 1971. Seed predation by animals. Annual Review of Ecology and Systematics 2:465–492.

Jędrzejewska, B., and W. Jędrzejewski. 1998. Predation in Vertebrate Communities: The Biaalowieza Primeval Forest as a Case Study. Springer-Verlag, Berlin.

Jenkins, M. A., and G. R. Parker. 1998. Composition and diversity of woody vegetation in silvicultural openings of southern Indiana forests. Forest Ecology and Management 109:57–74.

Jenkins, S. H., and R. A. Peters. 1992. Spatial patterns of food storage by Merriam's kangaroo rats. Behavioral Ecology 3:60–65.

Jenkins, S. H., A. Rothstein, and W. C. H. Green. 1996. Food hoarding by kangaroo rats: A test of alternative hypotheses. Ecology 76:2470–2481.

Jensen, T. S., and O. F. Nielsen. 1986. Rodents as seed dispersers in a heath-oak wood succession. Oecologia 70:214–221.

Jepson, W. L. 1910. The Silva of California. Vol. 2. University of California, Berkeley.

Johnson, K. G., and M. R. Pelton. 1980. Environmental relationships and the denning period of black bears in Tennessee. Journal of Mammalogy 61:653–660.

Johnson, K. G., and M. R. Pelton. 1981. Selection and availability of dens for black bears in Tennessee. Journal of Wildlife Management 45:111–119.

Johnson, K. G., D. O. Johnson, and M. R. Pelton. 1978. Simulation of winter heat loss for a black bear in a closed tree den. Proceedings of the Eastern Black Bear Workshop 4:155–166.

Johnson, P. S. 1974. Survival and growth of northern red oak seedlings following a prescribed burn. U.S. Forest Service Research Note NC-177.

Johnson, P. S. 1977. Predicting oak stump sprouting and sprout development in the Missouri Ozarks. U.S. Forest Service Research Paper NC-149.

Johnson, P. S. 1979. Shoot elongation of black oak and white oak sprouts. Canadian Journal of Forest Resources 9:489–494.

Johnson, P. S. 1992. Oak overstory/reproduction relations in two xeric ecosystems in Michigan. Forest Ecology and Management 48:233–248.

Johnson, P. S. 1993a. Perspectives on ecology and silviculture of oak-dominated forests in the central and eastern states. U.S. Forest Service General Technical Report NC-153.

Johnson, P. S. 1993b. Sources of oak reproduction. Pages 112–131 in: Oak regeneration: Serious problems, practical recommendations (D. L. Loftis and C. E. McGee, eds.). U.S. Forest Service General Technical Report SE-84.

Johnson, P. S. 1994a. How to manage oak forests for acorn production. U.S. Forest Service Technical Brief TB-NC-1.

Johnson, P. S. 1994b. The silviculture of northern red oak. Pages 33–68 in: U.S. Forest Service General Technical Report NC-173.

Johnson, P. S., C. D. Dale, K. R. Davidson, and J. R. Law. 1986. Planting northern red oak in the Missouri Ozarks: A prescription. Northern Journal of Applied Forestry 3:66–68.

Johnson, P. S., and R. D. Jacobs. 1981. Northern red oak regeneration after preherbicided clearcutting and shelterwood removal cutting. U.S. Forest Service Research Paper NC-202.

Johnson, P. S., R. D. Jacobs, A. J. Martin, and E. D. Godel. 1989. Regenerating northern red oak: Three successful case studies. Northern Journal of Applied Forestry 6:174–178.

Johnson, R. L. 1975. Natural regeneration and development of Nuttall oak and associated species. U.S. Forest Service Research Paper SO-104.

Johnson, W. C., and C. S. Adkisson. 1986. Airlifting the oaks. Natural History 95:40–47.

Johnson, W. C., C. S. Adkisson, T. R. Crow, and M. D. Dixon. 1997. Nut caching by blue jays (*Cyanocitta cristata* L.): Implications for tree demography. American Midland Naturalist 138:357–370.

Johnson, W. C., L. Thomas, and C. S. Adkisson. 1993. Dietary circumvention of acorn tannins by blue jays: Implications for oak demography. Oecologia 94:159–164.

Johnson, W. C., and T. Webb. 1989. The role of blue jays (*Cyanocitta cristata*) in the postglacial dispersal of Fagaceous trees in eastern North America. Journal of Biogeography 16:561–571.

Johnson, W. T., and H. H. Lyon. 1976. Insects that Feed on Trees and Shrubs: An Illustrated Practical Guide. Cornell University Press, Ithaca, NY.

Jones C. G., R. S. Ostfeld, M. P. Richard, E. M. Schauber, and J. O. Wolff. 1998. Chain reactions linking acorns to gypsy moth outbreaks and Lyme disease risk. Science 279:1023–1026.

Jonkel, J. C., and I. McT. Cowan. 1971. The black bear in the spruce-fir forest. Wildlife Monograph 27:1–57.

Kane, D. M. 1989. Factors influencing the vulnerability of black bears to hunters in northern New Hampshire. Master's thesis, University of New Hampshire, Durham.

Kartesz, J. T. 1994. A Synonymized Checklist of the Vascular Flora of the United States, Canada, and Greenland. Vol. 1. Timber Press, Portland, OR.

Kasbohm, J. W. 1994. Response of black bears to gypsy moth infestation in Shenandoah National Park, Virginia. Ph.D. diss., Virginia Polytechnic Institute and State University, Blacksburg.

Kasbohm, J. W., M. R. Vaughan, and J. G. Kraus. 1994. Behavioral response of black bears to gypsy moth infestation in Shenandoah National Park, Virginia. International Conference on Bear Research and Management 9:461–470.

Kasbohm, J. W., M. R. Vaughan, and J. G. Kraus. 1995. Food habits and nutrition of black bears during a gypsy moth infestation. Canadian Journal of Zoology 73:1771–1775.

Kasbohm, J. W., M. R. Vaughan, and J. G. Kraus. 1996a. Black bear denning during a gypsy moth infestation. Wildlife Society Bulletin 24:62–70.

Kasbohm, J. W., M. R. Vaughan, and J. G. Kraus. 1996b. Effects of gypsy moth infestation on black bear reproduction and survival. Journal of Wildlife Management 60:408–416.

Kasbohm, J. W., M. R. Vaughn, and J. G. Kraus. 1998. Black bear home range dynamics and movement patterns during a gypsy moth infestation. Ursus 10:259–267.

Kaufman, D. W., and G. A. Kaufman. 1989. Population biology. Pages 233–270 in: Advances in the Study of *Peromyscus* (Rodentia) (G. L. Kirkland, Jr., and J. N. Layne, eds.). Texas Tech University Press, Lubock.

Kaufman, D. W., G. A. Kaufman, and E. J. Finick. 1995. Temporal variation in abundance of *Peromyscus leucopus* in wooded habitats of eastern Kansas. American Midland Naturalist 133:7–17.

Kaul, R. B. 1985. The reproductive morphology of *Quercus* (Fagaceae). American Journal of Botany 72:1962–1977.

Kawamichi, M. 1980. Food, food hoarding and seasonal changes of Siberian chipmunks. Japanese Journal of Ecology 30:211–220.

Kay, B. L. 1987. Long-term effects of blue oak removal on forage production, forage quality, soil and oak regeneration. Pages 351–357 in: Proceedings of a Symposium on Multiple-use Management of California's Hardwood Resources. U.S. Forest Service General Technical Report PSW-100.

Kays, J. S., D. W. Smith, and S. M. Zedaker. 1985. Season of harvest and site quality effects on hardwood regeneration in the Virginia Piedmont. Pages 137–154 in: U.S. Forest Service General Technical Report SO-54.

Kearny, T. H., and R. H. Peebles. 1960. Arizona Flora. University of California Press, Berkeley.

Keever, C. 1953. Present composition of some stands of the former oak-chestnut forest in the southern Blue Ridge Mountains. Ecology 34:44–54.

Kelley, R. L., G. A. Hurst, and D. E. Steffen. 1988. Home ranges of wild turkey gobblers in central Mississippi. Proceedings of the Annual Conference of the Southeastern Association of Fish and Wildlife Agencies 42:470–475.

Kelly, D. 1994. The evolutionary ecology of mast seeding. Trends in Ecological Evolution 9:465–470.

Kelty, M. J. 1988. Sources of hardwood regeneration and factors that influence these sources. Pages 17–30 in: Guidelines for regenerating Appalachian hardwood stands. Morgantown, WV. Society of American Foresters Publication 88-03.

Kendall, M. G., and Stuart, A. 1979. The advanced theory of statistics. 4th ed. Vol. 2, Inference and relationship. Griffin and Co., London.

Kennamer, J. E., and D. H. Arner. 1967. Winter food available to the wild turkey in a hardwood forest. Proceedings Annual Conference Southeastern Association Fish and Game Commission 21:123–129.

Ketterson, E. D., V. Nolan, Jr., M. J. Cawthorn, P. G. Parker, and C. Ziegenfus. 1996. Phenotypic engineering: Using hormones to explore the mechanistic and functional bases of phenotypic variation in nature. Ibis 138:1–17.

Keyser, P. D., P. H. Brose, D. H. Van Lear, and K. M. Burtner. 1996. Enhancing oak regeneration with fire in shelterwood stands: Preliminary Trials. Pages 215–219 in: Transactions of the 61st North American Wildlife and Natural Resources Conference (K. Wadsworth and R. McCabe, eds.). Wildlife Management Institute, Washington, DC.

Kilburn, P. D. 1960. Effects of logging and fire on xerophytic forests in northern Michigan. Bulletin of the Torrey Botanical Club 87:42–45.

Kilpatrick, H. J., T. P. Husband, and C. A. Pringle. 1988. Winter roost site characteristics of eastern wild turkeys. Journal of Wildlife Management 52:461–463.

Kilpatrick, H. J., P. J. Pekins, and W. W. Mautz. 1991. Nutritional evaluation of simulated spring diets of white-tailed deer. Transactions of the Northeastern Section of the Wildlife Society 48:63–69.

King, C. M. 1983. The relationship between beech (*Nothofagus* sp.) seedfall and populations of mice (*Mus musculus*), and the demographic and dietary responses of stoats (*Mustela erminea*) in three New Zealand forests. Journal of Animal Ecology 52:141–166.

King, T. R., and H. E. McClure. 1944. Chemical composition of some American wild feedstuffs. Journal of Agricultural Research 69:33–46.

Kirkpatrick, R. L., J. P. Fontenot, and R. F. Harlow. 1969. Seasonal changes in rumen chemical components as related to forages consumed by white-tailed deer in the southeast. Transactions of the North American Wildlife and Natural Resources Conference 34:229–238.

Kittredge, J., and A. K. Chittenden. 1929. Oak forests of northern Michigan. Michigan State College Agricultural Experimental Station Special Bulletin 190.

Kleiner, K. W., M. D. Abrams, and J. C. Schultz. 1992. The impact of water and nutrient deficiencies on the growth, gas exchange and water relations of red oak and chestnut oak. Tree Physiology 11:271–287.

Kline, V. M., and G. Cottam. 1979. Vegetation response to climate and fire in the Driftless Area of Wisconsin. Ecology 60:261–268.

Knight, R. S. 1985. A model of episodic, abiotic dispersal for oaks (*Quercus robur*). South African Journal of Botany 51:265–269.

Knox, W. M. 1997. Historical changes in the abundance and distribution of deer in Virginia. Pages 27–36 in: The Science of Overabundance: Deer Ecology and Population Management (W. J. McShea, H. B. Underwood, and J. H. Rappole, eds.). Smithsonian Institution Press, Washington, DC.

Koenig, W. D. 1991. The effects of tannins and lipids on digestion of acorns by acorn woodpeckers. The Auk 108:79–88.

Koenig, W. D. 1999. Spatial autocorrelation of ecological phenomena. Trends in Ecology and Evolution 14:22–26.

Koenig, W. D., W. J. Carmen, W. T. Stanback, and R. L. Mumme. 1991. Determinants of acorn productivity among five species of oaks in central coastal California. Pages 136–142 in: Proceedings of the Symposium on Oak Woodlands and Hardwood Rangeland Management (R. B. Standiford, technical coordinator). U.S. Forest Service General Technical Report PSW-126.

Koenig, W. D., and J. Haydock. 1999. Oaks, acorns, and the geographical ecology of acorn woodpeckers. Journal of Biogeography 26:159–165.

Koenig, W. D., and J. M. H. Knops. 1998a. Scale of mast-seeding and tree-ring growth. Nature 396:225–226.

Koenig, W. D., and J. M. H. Knops. 1998b. Testing for spatial autocorrelation in ecological studies. Ecography 21:423–429.

Koenig, W. D., and J. M. H. Knops. 2000. Patterns of annual seed production by Northern Hemisphere trees: A global perspective. American Naturalist 155:59–69.

Koenig, W. D., J. M. H. Knops, W. J. Carmen, M. T. Stanback, and R. L. Mumme. 1994. Estimating acorn crops using visual surveys. Canadian Journal of Forest Research 24:2105–2112.

Koenig, W. D., J. M. H. Knops, W. J. Carmen, M. T. Stanback, and R. L. Mumme. 1996. Acorn production by oaks in central coastal California: Influence of weather at three levels. Canadian Journal of Forest Research 26:1677–1683.

Koenig, W. D., J. M. H. Knops, W. J. Carmen, and M. T. Stanback. 1999. Spatial dynamics in the absence of dispersal: Acorn production by oaks in central coastal California. Ecography 22: 499–506.

Koenig, W. D., D. R. McCullough, C. E. Vaughn, J. M. H. Knops, and W. J. Carmen. 1999. Synchrony and asynchrony of acorn production at two coastal California sites. Madroño 46:20–24.

Koenig, W. D., and R. L. Mumme. 1987. Population ecology of the cooperatively breeding acorn woodpecker. Monograph of Population Biology 24, Princeton University Press, Princeton, NJ.

Koenig, W. D., R. L. Mumme, W. J. Carmen, and M. T. Stanback. 1994. Acorn production by oaks in central coastal California: Variation within and among years. Ecology 75:99–109.

Kolb, T. E., and K. C. Steiner. 1990. Growth and biomass partitioning of northern red oak and yellow-poplar seedlings: Effects of shading and grass root competition. Forest Science 36:34–44.

Kolb, T. E., K. C. Steiner, L. H. McCormick, and T. W. Bowersox. 1990. Growth response of northern red oak and yellow poplar seedlings to light, soil moisture and nutrients in relation to ecological strategy. Forest Ecology and Management 38:65–78.

Kolenosky, G. B., and S. M. Strathearn. 1987a. Black bear. Pages 442–454 in: Wild Furbearer Management and Conservation in North America (M. Novak, J. A. Baker, M. E. Obbard, and B. Malloch, eds.). Ontario Ministry of Natural Resources and Ontario Trappers Association, Toronto.

Kolenosky, G. B., and S. M. Strathearn. 1987b. Winter denning of black bears in east-central Ontario. International Conference on Bear Research and Management 7:305–316.

Kollmann, J., and H. P. Schill. 1996. Spatial patterns of dispersal, seed predation and germination during colonization of abandoned grassland by *Quercus petrea* and *Corylus avellana.* Vegetatio 125:193–205.

Komarek, E. V. 1974. Appalachian mountains, mesophytic forest, and grasslands. Pages 269–272 in: Fire and Ecosystems (T. T. Kozlowski and C. E. Ahlgren, eds.). Academic Press, New York.

Korschgen, L. J. 1962. Foods of Missouri deer, with some management implications. Journal of Wildlife Management 26:164–172.

Korschgen, L. J. 1967. Feeding habits and food. Pages 137–198 in: The Wild Turkey and Its Management (O. H. Hewitt, ed.). Wildlife Society, Washington, DC.

Korschgen, L. J. 1973. April foods of wild turkeys in Missouri. Pages 143–150 in: Wild Turkey Management: Current Problems and Programs (G. C. Sanderson and H. C. Schultz, eds.). Missouri Chapter of the Wildlife Society and University of Missouri Press, Columbia.

Korstian, C. F. 1927. Factors controlling germination and early survival in oaks. Yale University School of Forestry Bulletin 19:1–122.

Korstian, C. F., and P. W. Stickel. 1927. The natural replacement of blight-killed chestnut. U.S. Department of Agriculture Miscellaneous Circular 100.

Kotar, J. 1997. Silviculture and ecosystem management. Pages 265–275 in: Ecosystem Management (M. S. Boyce and A. Haney, eds.). Yale University Press, New Haven.

Kothmann, H. G., and G. W. Litton. 1975. Utilization of man-made roosts by turkey in west Texas. Proceedings of the National Wild Turkey Symposium 3:159–163.

Kozicky, E. L. 1942. Pennsylvania wild turkey food habits based on dropping analysis. Pennsylvannia Game News 13:10–11, 28–29, 31.

Kozlowski, T. T., P. J. Kramer and S. G. Pallardy. 1991. The Physiological Ecology of Woody Plants. Academic Press, San Diego.

Kramer, P. J., and J. S. Boyer. 1995. Water Relations of Plants and Soils. Academic Press, San Diego.

Krausman, P. R., and N. S. Smith, eds. 1984. Deer in the Southwest: A workshop. School of Renewable Natural Resources, University of Arizona, Tucson.

Krausman, P. R., and N. S. Smith, eds. 1990. Managing wildlife in the Southwest. Arizona Chapter of the Wildlife Society, Phoenix.

Kruger, E. 1992. Survival, growth, and ecophsiology of northern red oak (*Quercus rubra* L.) and competing tree regeneration in response to fire and related disturbances. Ph.D. diss., Department of Forestry, University of Wisconsin.

Kruger, E. L., and P. B. Reich. 1993a. Coppicing affects growth, root: Shoot relations and ecophysiology of potted *Quercus rubra* seedlings. Physiologia Plantarum 89:751–760.

Kruger, E. L., and P. B. Reich. 1993b. Coppicing alters ecophysiology of *Quercus rubra* saplings in Wisconsin forest openings. Physiol. Plant. 89:741–750.

Kruger, E. L., and P. B. Reich. 1997a. Responses of hardwood regeneration to fire in mesic forest openings. I. Post-fire community dynamics. Canadian Journal of Forest Research 27:1822–1831.

Kruger, E. L., and P. B. Reich. 1997b. Responses of hardwood regeneration to fire in mesic forest openings. III. Whole-plant growth, biomass distribution, and nitrogen and carbohydrate relations. Canadian Journal of Forest Research 27:1841–1850.

Kruse, W. H., G. J. Gottfried, D. A. Bennett, and H. Mata-Manqueros. 1996. The role of fire in Madrean encinal oak and pinyon-juniper woodlands development. Pages 99–106 in: Effects of Fire on Madrean Province Ecosystems: A Symposium Proceedings (P. F. Ffolliott, L. F. DeBano, M. B. Baker, Jr., G. Solis-Garza, C. B. Edminster, D. G. Neary, L. S. Allen, and R. H. Hamre, technical coordinators). U.S. Forest Service General Technical Report RM-289.

Kubiske, M. E., and M. D. Abrams. 1992. Photosynthesis, water relations and leaf morphology of xeric versus mesic *Quercus rubra* ecotypes in central Pennsylvania in relation to moisture stress. Canadian Journal of Forest Research 22:1402–1407.

Kuchler, A. W. 1964. Potential natural vegetation of the conterminous United States. American Geographical Society Special Publication 36.

Kuchler, A. W. 1988. The map of natural vegetation of California. In: Terrestrial vegetation of California (M. G. Barbour and J. Major, eds.). California Native Plant Society, Special Publication 9.

Kurzejeski, E. W., and J. B. Lewis. 1990. Home ranges, movements, and habitat use of wild turkey hens in northern Missouri. Proceedings of the National Wild Turkey Symposium 6:67–71.

Lafon, A., and S. D. Schemnitz. 1995. Distribution, habitat use, and limiting factors of Gould's turkey in Chihuahua, Mexico. Proceedings of the National Wild Turkey Symposium 7:185–191.

Laing, C. 1994. Vegetation and fire history of the dwarf pine ridges, Shawangunk Mountains, New York. Final Report to the Nature Conservancy, Eastern New York Region.

Lalonde, R. G., and B. D. Roitberg. 1992. On the evolution of masting behavior in trees: Predation or weather? American Naturalist 139:1293–1304.

Landers, J. L., R. J. Hamilton, A. S. Johnson, and R. L. Marchington. 1979. Foods and habitat of black bears in southeastern North Carolina. Journal of Wildlife Management 43:143–153.

Lane, R. S., J. Piesman, and W. Burgdorfer. 1991. *Lyme borreliosis:* Relation of its causative agents to its vectors and hosts in North America and Europe. Annual Review of Entomology 36:587–609.

Langenau, E. E., Jr., S. R. Kellert, and J. E. Applegate. 1984. Values in management. Pages 699–720 in: White-tailed Deer: Ecology and Management (L. K. Halls, ed.). Stackpole Books, Harrisburg, PA.

Lanham, J. D., P. D. Keyser, P. H. Brose, and D. H. Van Lear. In press. Management options for neotropical migratory songbirds in oak shelterwood-burns. Forest Ecology and Management.

Larsen, D. R., and P. S. Johnson. 1998. Linking the ecology of natural oak regeneration to silviculture. Forest Ecology and Management 106:1–7.

Larsen, D. R., M. A. Metzger, and P. S. Johnson. 1997. Oak regeneration and overstory density in the Missouri Ozarks. Canadian Journal of Forest Resources 27:869–875.

Latham, R. M. 1956. Complete Book of the Wild Turkey. Stackpole Books, Harrisburg, PA.

Laudenslager, S. L., and L. D. Flake. 1987. Fall food habits of wild turkeys in south central South Dakota. Prairie Naturalist 19:37–40.

Lavenex, P., M. W. Shiflett, R. K. Lee, and L. F. Jacobs. 1998. Spatial versus nonspatial relational learning in free-ranging fox squirrels (*Sciurus niger*). Journal of Comparative Psychology 112:1–10.

Lawson, D. M. 1993. The effects of fire on stand structure of mixed *Quercus agrifolia* and *Q. engelmannii* woodlands. Master's thesis, San Diego State University.

Lawton, J. H. 1997. The science and non-science of conservation biology. Oikos 79:3–5.

Le Corre, V., N. Machon, R. J. Petit, and A. Kremer. 1997. Colonization with long-distance seed dispersal and genetic structure of maternally inherited genes in forest trees: A simulation study. Genetical Research 69:117–125.

LeCount, A. L. 1982. Characteristics of a central Arizona black bear population. Journal of Wildlife Management 46:861–868.

LeCount, A. L. 1983. Denning ecology of black bears in central Arizona. International Conference on Bear Research and Management 5:71–78.

Leduc, D. J. 1987. A comparative analysis of the reduced major axis technique of fitting lines to bivariate data. Canadian Journal of Forest Resources 17:654–659.

Lee, K. E. 1995. Earthworms and sustainable land use. Pages 215–234 in: Earthworm Ecology and Biogeography in North America (P. F. Hendrix, ed.). Lewis Publishers, Boca Raton, FL.

Leimgruber, P., W. J. McShea, and J. H. Rappole. 1994. Predation on artificial nests in large forest blocks. Journal of Wildlife Management 58:254–260.

Leitner, L. A., C. P. Dunn, G. R. Guntenspergen, F. Stearns, and D. M. Sharpe. 1991. Effects of site, landscape features, and fire regime on vegetation patterns in presettlement southern Wisconsin. Landscape Ecology 5:203–217.

Lentz, R. J., D. M. Sims, and P. J. Ince. 1989. Are our traditional attitudes restricting forestry management options? Pages 20–24 in: Proceedings of pine-hardwood mixtures: A symposium on management and ecology of the type (T. A. Waldrop, ed.). U.S. Forest Service General Technical Report SE-58.

Leopold, A. S. 1959. Wildlife of Mexico: The Game Birds and Mammals. University of California Press, Berkeley.

Leopold, B. D., G. H. Weaver, J. D. Cutler, and R. C. Warren. 1989. Pine-hardwood forests in north-central Mississippi: An ecological and economic perspective. Pages 211–222 in: Proceedings of Pine-Hardwood Mixtures: A Symposium on Management and Ecology of the Type. (T. A. Waldrop, ed.). U.S. Forest Service General Technical Report SE-58.

Levine, J. R., M. L. Wilson, and A. Spielman. 1985. Mice as reservoirs of the Lyme disease spirochete. American Journal of Tropical Medicine and Hygiene 34:355–360.

Lewis, A. R. 1980. Patch use by gray squirrels and optimal foraging. Ecology 61:1371–1379.

Lewis, A. R. 1982. Selection of nuts by gray squirrels and optimal foraging theory. American Midland Naturalist 107:250–257.

Lewis, J. B. 1992. Eastern turkey in midwestern oak-hickory forests. Pages 286–305 in: The Wild Turkey: Biology and Management (J. G. Dickson, ed.). Stackpole Books, Harrisburg, PA.

Lewis, J. B., and R. P. Breitenbach. 1966. Breeding potential of subadult wild turkey gobblers. Journal of Wildlife Management 30:618–622.

Lewis, J. B., and E. W. Kurzejeski. 1984. Wild turkey productivity and poult mortality in north central Missouri. Missouri Department of Conservation Final Report P-R Project W-13-R-38., Study 51, Job 1.

Lewis, J. C. 1962. Wild turkeys in Allegany County, Michigan. Master's thesis, Michigan State University, East Lansing.

Lewis, V. 1991. The temporal and spatial distribution of filbert weevil infested acorns in an oak woodland in Marin County, California. Pages 156–160 in: Proceedings of a Symposium on Oak Woodlands and Hardwood Rangeland Management. U.S. Forest Service General Technical Report PSW-126.

Liebhold, A. M., and J. S. Elkinton. 1989a. Characterizing spatial patterns of gypsy moth defoliation. Forest Science Monographs 35:557–568.

Liebhold, A. M., and J. S. Elkinton. 1989b. Elevated parasitism in artificially augmented populations of *Lymantria dispar* (Lepidoptera: Lymantriidae). Environmental Entomology 18:986–995.

Liebhold, A. M., J. S. Elkinton, D. Williams, and R. M. Muzika. In press. Dynamics of North American gypsy moth populations viewed from multiple trophic levels and spatial scales. Researches in Population Ecology.

Liebhold, A. M., K. W. Gottschalk, R. Muzika, M. E. Montgomery, R. Young, K. O'Day, and B. Kelley. 1995. Suitability of North American tree species to the gypsy moth: A summary of field and laboratory tests. U.S. Forest Service General Technical Report NE-211.

Liebhold, A. M., and M. L. McManus. 1991. Does larval dispersal cause the expansion of gypsy moth outbreaks? National Journal of American Foresters 8:95–99.

Ligon, J. S. 1946. History and management of Merriam's wild turkey. In: Biology 1. University of New Mexico Press.

Lindzey, F. G., and E. C. Meslow. 1976. Winter dormancy in black bears in southwestern Washington. Journal of Wildlife Management 40:408–415.

Lint, J. R., B. D. Leopold, and G. A. Hurst. 1995. Comparison of abundance indexes and population size estimates for wild turkey gobblers. Wildlife Society Bulletin 23:164–168.

Lippke, B., and J. Bishop. 1999. The economic perspective. Pages 597–638 in: Maintaining Biodiversity in Forest Ecosystems. (M. L. Hunter, Jr., ed.). Cambridge University Press, New York.

Little, E. L., Jr. 1950. Southwestern trees: A guide to the native species of New Mexico and Arizona. U.S. Department of Agriculture Handbook 9.

Little, E. L., Jr. 1971. Atlas of United States Trees. Vol 1. U.S. Forest Service Miscellaneous Publication 1146.

Little, E. L., Jr. 1979. Checklist of United States trees (native and naturalized). U.S. Forest Service Agriculture Handbook 541.

Little, S. 1974. Effects of fire on temperate forests: Northeastern United States. Pages 225–250 in: Fire and Ecosystems. (T. T. Koslowski and E. E. Ahlegren, eds.). Academic Press, New York.

Little, T. W. 1980. Wild turkey restoration in "marginal" Iowa habitat. Proceedings of the National Wild Turkey Symposium 4:45–60.

Litton, G. W. 1977. Food habits of the Rio Grande turkey in the Permian Basin of Texas. Texas Parks and Wildlife Department Technical Series 18.

Litvaitis, J. A., and D. M. Kane. 1994. Relationship of hunting technique and hunter selectivity to composition of black bear harvest. Wildlife Society Bulletin 22:604–606.

Loeb, R. E. 1989. Lake pollen records of the past century in northern New Jersey and southeastern New York, USA. Palynology 13:3–19.

Loeb, S. C., and M. R. Lennartz. 1989. The fox squirrel (*Sciurus niger*) in southeastern pine-hardwood forests. Pages 142–148 in: Proceedings of Pine-Hardwood Mixtures: A Symposium on Management and Ecology of the Type. (T. A. Waldrop, ed.). U.S. Forest Service General Technical Report SE-58.

Loewenstein, E. F., H. E. Garrett, P. S. Johnson, and J. P. Dwyer. 1995. Changes in a Missouri Ozark oak-hickory forest during 40 years of uneven-aged management. Pages 159–164 in: Proceedings of the Tenth Central Hardwood

Forest Conference (K. W. Gottschalk and S. L. C. Fosbroke, eds.). U.S. Forest Service General Technical Report NE-197.

Loewenstein, E. F., and M. S. Golden. 1995. Establishment of water oak is not dependent on advanced reproduction. Pages 443–446 in: 8th Biennial Southern Silvicultural Research Conference. U.S. Forest Service General Technical Report SRS-1.

Loftis, D. L. 1983. Regenerating southern Appalachian mixed hardwood stands with the shelterwood method. Southern Journal of Applied Forestry 7:212–217.

Loftis, D. L. 1990a. Predicting post-harvest performance of advance red oak reproduction in the southern Appalachians. Forest Science 36:908–916.

Loftis, D. L. 1990b. A shelterwood method for regenerating red oak in the southern Appalachians. Forest Science 36:917–929.

Loftis, D. L., and McGee, C. E., eds. 1993. Oak regeneration: Serious problems, practical recommendations. U.S. Forest Service General Technical Report SE-84.

Long, W. H. 1914. The death of chestnuts and oaks due to *Armillaria mellea*. U.S. Bureau of Plant Industry Bulletin 89.

Loomis, R. M. 1973. Estimating fire-caused mortality and injury in oak-hickory forests. U.S. Forest Service Research Paper NC-94.

Loomis, R. M. 1975. Annual changes in forest floor weights under a southeast Missouri oak stand. U.S. Forest Service Research Note NC-RN 184.

Lopes, V. L., and P. F. Ffolliott. 1992. Hydrology and watershed management of oak woodlands in southeastern Arizona. Pages 71–77 in: Ecology and management of oak and associated woodlands: Perspectives in the southwestern United States and northern Mexico (P. F. Ffolliott, G. J. Gottfried, D. A. Bennett, V. M. Hernandez C., A. Ortega-Rubio, and R. H. Hamre, technical coordinators.). U.S. Forest Service General Technical Report RM-218.

Lorimer, C. G. 1977. The presettlement forest and natural disturbance cycle of northeastern Maine. Ecology 58:139–148.

Lorimer, C. G. 1980. Age structure and disturbance history of a southern Appalachian virgin forest. Ecology 6:1169–1184.

Lorimer, C. G. 1983. Eighty-year development of northern red oak after partial cutting in a mixed-species Wisconsin forest. Forest Science 29:371–383.

Lorimer, C. G. 1984. Development of the red maple understory in northeastern oak forests. Forest Science 30:3–22.

Lorimer, C. G. 1985. The role of fire in the perpetuation of oak forests. Pages 8–25 in: Proceedings of a Symposium on Challenges in Oak Management and Utilization (J. E. Johnson, ed.). University of Wisconsin, Madison.

Lorimer, C. G. 1989. The oak regeneration problem: New evidence on causes and possible solutions. Forest Resources Analyses 8, University of Wisconsin Publishing R3484.

Lorimer, C. G. 1993. Causes of the oak regeneration problem. Pages 14–39 in:

Oak regeneration: Serious problems, practical recommendations (D. L. Loftis and C. E. McGee, eds.). U.S. Forest Service General Technical Report SE-84.

Lorimer, C. G., J. W. Chapman, and W. D. Lambert. 1994. Tall understory vegetation as a factor in the poor development of oak seedlings beneath mature stands. Journal of Ecology 82:227–237.

Lorimer, C. G., and L. E. Frelich. 1994. Natural disturbance regimes in old-growth northern hardwoods. Journal of Forestry 92:33–38.

MacClintock, D. 1970. Squirrels of North America. Van Nostrand Reinhold, New York.

MacDonald, J. E., and G. R. Powell. 1985. First growing period development of *Acer saccharum* stump sprouts arising after different dates of cut. Canadian Journal of Botany 63:819–828.

Mackey, D. L., and R. J. Jonas. 1982. Seasonal habitat use and food habits of Merriam's turkeys in southcentral Washington. Proceedings of the Western Wild Turkey Workshop 1:99–110.

Madansky, A. 1959. The fitting of straight lines when both variables are subject to error. Journal of American Statisticians Association 54:173–205.

Maehr, D. S., and J. R. Brady. 1984. Food habits of Florida black bears. Journal of Wildlife Management 48:230–235.

Maenza-Gmelch, T. E. 1997. Holocene vegetation, climate, and fire history of the Hudson Highlands, southeastern New York, USA. The Holocene 7:25–37.

Magnarelli, L. A., J. F. Anderson, K. E. Hyland, D. Fish and J. B McAninch. 1988. Serologic analyses of *Peromyscus leucopus*, a rodent reservoir for *Borrelia burgdorferi*, in northeastern United States. Journal of Clinical Microbiology 26:1138–1140.

Maguire, L. 1999. Social perspectives. Pages 639–666 in: Maintaining Biodiversity in Forest Ecosystems. (M. L. Hunter Jr., ed.). Cambridge University Press, New York.

Maingi, J. K., and P. F. Ffolliott. 1992. Specific gravity and estimated physical properties of Emory oak in southeastern Arizona. Pages 147–149 in: Ecology and management of oak and associated woodlands: Perspectives in the southwestern United States and northern Mexico (P. F. Ffolliott, G. J. Gottfried, D. A. Bennett, V. M. Hernandez C., A. Ortega-Rubio, and R. H. Hamre, technical coordinators). U.S. Forest Service General Technical Report RM-218.

Manion, P. D. 1991. Tree Disease Concepts. 2nd ed. Prentice-Hall, Englewood Cliffs, NJ.

Manville, A. M. 1987. Den selection and use by black bears in Michigan's northern lower peninsula. International Conference for Bear Research and Management 7:317–322.

Markley, M. H. 1967. Limiting factors. Pages 199–243 in: The Wild Turkey and Its Management. Wildlife Society, Washington, DC.

Marks, P. L., S. Gardescu, and F. K. Seischab. 1992. Late eighteenth century vegetation of central and western New York on the basis of original land surveys. State University of New York, Albany. New York State Museum Bulletin 484.

Marquis, D. A. 1965. Controlling light in small clearcuttings. U.S. Forest Service Research Paper NE-39.

Marquis, D. A. 1988. Guidelines for regenerating cherry-maple stands. Pages 167–188 in: Proceedings of a Symposium on Guidelines for Regenerating Appalachian Hardwood Stands (H. C. Smith, A. W. Perkey, and W. E. Kidd, Jr., eds.). Morgantown, WV. Society of American Foresters Publication 88-03.

Marquis, D. A., and R. Brenneman. 1981. The impact of deer on forest vegetation in Pennsylvania. U.S. Forest Service General Technical Report NE-65.

Marquis, D. A., P. L. Eckert, and B. A. Roach. 1976. Acorn weevils, rodents, and deer all contribute to oak-regeneration difficulties in Pennsylvania. U.S. Forest Service Research Paper NE-356.

Marquis, D. A., R. L. Ernst, and S. L. Stout. 1992. Prescribing silvicultural treatments in hardwood stands of the Alleghenies. Revised ed. U.S. Forest Service General Technical Report NE-96.

Marshall, J. T. 1957. Birds of the pine-oak woodland in southern Arizona and adjacent Mexico. Pacific Coast Avifauna 32:1–125.

Martin, A. C., F. H. May, and T. E. Clarke. 1939. Early winter food preferences of the wild turkey on the George Washington National Forest. Transactions of the North American Wildlife Conference 4:570–578.

Martin, A. C., H. S. Zim, and A. L. Nelson. 1961. American Wildlife and Plants: A Guide to Wildlife Food Habits. Dover Publishers, New York.

Martin, D., and D. Steffen. 1999. Virginia black bear status report-1999. In-house Report for the Virginia Department of Game and Inland Fisheries.

Martin, T. E. 1993. Nest predation among vegetation layers and habitat types: Revising the dogmas. American Naturalist 141:897–913.

Martin, U., S. G. Pallardy, and Z. A. Bahari. 1987. Dehydration tolerance of leaf tissues of six woody angiosperm species. Physiologia Plantarum 69:182–186.

Martin, W. H., and S. G. Boyce. 1993. Introduction: The southeastern setting. Pages 1–46 in: Biodiversity of the Southeastern United States Lowland Terrestrial Communities (W. H. Martin, S. G. Boyce, and A. C. Echternacht, eds.). John Wiley and Sons, New York.

Mather, T. N. 1993. The dynamics of spirochete transmission between ticks and vertebrates. Pages 43–62 in: Ecology and Management of Lyme Disease (H. Ginsberg, ed.). Rutgers University Press, New Brunswick, NJ.

Mather, T. N., and H. S. Ginsberg. 1994. Vector-host-pathogen relationships: Transmission dynamics of tick-borne infections. Pages 68–90 in: Ecological Dynamics of Tick-Borne Zoonoses (D. E. Sonenshine and T. N. Mather, eds.). Oxford University Press, New York.

Mather, T. N., M. L. Wilson, S. I. Moore, J. M. C. Ribeiro, and A. Spielman. 1989. Comparing the relative potential of rodents as reservoirs of the Lyme disease spirochete (*Borelia burgdorferi*). America Journal of Epidemiology 130:143–150.

Matson, J. R. 1946. Notes on the dormancy in the black bear. Journal of Mammalogy 27:203–212.

Matthews, J. D. 1963. Factors affecting the production of seed by forest trees. Forestry Abstracts 24:1–13.

Mautz, W. W. 1978. Sledding on a bushy hillside: The fat cycle in deer. Wildlife Society Bulletin 6:88–90.

Maxson, S. J., and L. W. Oring. 1978. Mice as a source of egg loss among ground-nesting birds. The Auk 95:582–584.

Mayer, K. E., and W. F. Laudenslayer, Jr., eds. 1988. A guide to wildlife habitats of California. California Department of Forestry and Fire Protection, Sacramento.

Mayer, K. E., P. C. Passof, C. Bolsinger, W. W. Grenfall, and H. Slack. 1985. Status of the hardwood resource of California: Report to Board of Forestry. California Department of Forestry and Fire Protection, Sacramento.

McCabe, R. E., and T. R. McCabe. 1984. Of slings and arrows: An historical retrospection. Pages 19–72 in: White-tailed Deer: Ecology and Management (L. K. Halls, ed.). Stackpole Books, Harrisburg, PA.

McCarthy, E. F. 1933. Yellow-poplar characteristics, growth, and management. U.S. Department of Agriculture Technical Bulletin 356.

McCarthy, J. J., and J. O. Dawson. 1990. Growth and water use efficiency of *Quercus alba*, *Q. bicolor*, *Q. imbricata*, and *Q. palustris* seedlings under conditions of reduced soil water availability and solar irradiance. Transactions of the Illinois State Academy of Science 83:128–148.

McClaran, M. P., L. S. Allen, and G. B. Ruyle. 1992. Livestock production and grazing management in the encinal oak woodlands of Arizona. Pages 57–64 in: Ecology and management of oak and associated woodlands: Perspectives in the southwestern United States and northern Mexico (P. F. Ffolliott, G. J. Gottfried, D. A. Bennett, V. M. Hernandez C., A. Ortega-Rubio, and R. H. Hamre, technical coordinators). U.S. Forest Service General Technical Report RM-218.

McClaren, M. P., and J. W. Bartolome. 1989a. Effect of *Quercus douglasii* (Fagaceae) on herbaceous understory along a rainfall gradient. Madrono 36:141–153.

McClaren, M. P., and J. W. Bartolome. 1989b. Fire-related recruitment in stagnant *Quercus douglasii* populations. Canadian Journal of Forest Research 19: 580–585.

McClaran, M. P., and G. R. McPherson. 1999. Oak savanna in the American Southwest. Pages 275–287 in: Savannas, Barrens, and Rock Outcrop Plant Communities of North America (R. Anderson, R. C., J. S. Fralish, and J. M. Baskin, eds.). Cambridge University Press, New York.

McComb, W. C., and R. N. Muller. 1983. Snag densities in old-growth and second-growth Appalachian forests. Journal of Wildlife Management 47:376–382.

McConnell, S. P. 1988. Effects of gypsy moth defoliation on acorn production and viability, litterfall, and litter layer depth and biomass in north-central Virginia and western Maryland. Master's thesis, Virginia Polytechnic Institute and State University, Blacksburg.

McCreary, D. D., W. D. Tietje, R. H. Schmidt, R. Gross, W. H. Willoughby, B. L. Weitkamp, and F. L. Bell. 1991. Stump sprouting of blue oaks. Page 64 in: Proceedings of a Symposium on Oak Woodlands and Hardwood Rangeland Management. U.S. Forest Service General Technical Report PSW-126.

McDonald, J. E., Jr., D. P. Fuller, T. K. Fuller, and J. E. Cardoza. 1994. The influence of food abundance on success of Massachusetts black bear hunters. Northeast Wildlife 51:55–60.

McGee, C. E. 1979. Fire and other factors related to oak regeneration. Pages 75–81 in: Proceedings of John Wright Forest Conference. (H. A. Holt and B. C. Fischer, eds.). Purdue University, West Lafayette, IN.

McGraw, J. B., K. W. Gottschalk, M. C. Vavrek, and A. L. Chester. 1990. Interactive effects of resource availabilities and defoliation on photosynthesis, growth and mortality of red oak seedlings. Tree Physiology 7:247–254.

McLaughlin, C. R., G. J. Matula, Jr., and R. J. O'Conner. 1994. Synchronous reproduction by Maine black bears. International Conference on Bear Research and Management 9:471–479.

McPherson, G. R. 1992. Ecology of oak woodlands in Arizona. Pages 24–33 in: Ecology and management of oak and associated woodlands: Perspectives in the southwestern United States and northern Mexico (P. F. Ffolliott, G. J. Gottfried, D. A. Bennett, V. M. Hernandez C., A. Ortega-Rubio, and R. H. Hamre, technical coordinators). U.S. Forest Service General Technical Report RM-218.

McPherson, G. R. 1994. Response of annual plants and communities to tilling in a semi-arid temperate savanna. Journal of Vegetation Science 5:415–420.

McPherson, G. R. 1997. Ecology and Management of North American Savannas. University of Arizona Press, Tucson.

McQuade, D. B., E. H. Williams, and H. Eichenbaum. 1986. Cues used for localizing food by the grey squirrel (*Sciurus carolinensis*). Ethology 72:22–30.

McShea, W. J. 2000. The influence of acorn crops on annual variation in rodent and bird populations. Ecology 81:228–238.

McShea, W. J., and J. H. Rappole. 1997. Herbivores and the ecology of forest understory birds. Pp. 298–309 in: The Science of Overabundance: Deer Ecology and Population Management (W. J. McShea, H. B. Underwood, and J. H. Rappole, eds.). Smithsonian Institution Press, Washington, DC.

McShea, W. J., and G. Schwede. 1993. Variable acorn crops: Responses of white-tailed deer and other mast consumers. Journal of Mammalogy 74:999–1006.

McShea, W. J., H. B. Underwood, J. H. Rappole. 1997. Deer management and the concept of overabundance. Pages 1–10 in: The Science of Overabundance: Deer Ecology and Population Management (W. J. McShea, H. B. Underwood, and J. H. Rappole, eds.). Smithsonian Institution Press, Washington, DC.

McWilliams, W. H., S. L. Arner, and C. J. Barnett. 1997. Summary of mortality statistics and forest health monitoring results for the northeastern United States. Pages 59–75 in: Proceedings of the 11th Central Hardwood Forest Conference. (S. G. Pallardy, R. A. Cecich, H. G. Garret, and P. S. Johnson, eds.). U.S. Forest Service General Technical Report NC-188.

McWilliams, W. H., S. L. Stout, T. W. Bowersox, and L. H. McCormick. 1995. Adequacy of advance tree-seedling regeneration in Pennsylvania's forests. Northern Journal of Applied Forestry 12:187–191.

Medina, A. L. 1987. Woodland communities and soils of Fort Bayard, southwestern New Mexico. Journal of the Arizona-Nevada Academy of Science 21:99–112.

Menzel, K. E. 1975. Population and harvest data for Merriam's turkeys in Nebraska. Proceedings of the National Wildlife Turkey Symposium 3:184–188.

Merkel, H. W. 1905. A deadly fungus on the American chestnut. Pages 97–103 in: New York Zoological Society Tenth Annual Report.

Merz, R. W., and S. G. Boyce. 1956. Age of oak "seedlings." Journal of Forestry 54:774–775.

Metropolitan District Commission. 1995. Quabbin Watershed: MDC Land Management Plan 1995–2004. Commonwealth of Massachusetts, Boston.

Meyer, J. 1986. Management of old growth forests in Missouri. Missouri Department of Conservation and U.S. Forest Service. Habitat Management Series, no. 3.

Mikan, C. J., D. A. Orwig, and M. D. Abrams. 1994. Age structure and successional dynamics of a presettlement-origin chestnut oak forest in the Pennsylvania Piedmont. Bulletin of the Torrey Botanical Club 121:13–23.

Miller, D. A., G. A. Hurst, and B. D. Leopold. 1997. Factors affecting gobbling activity of wild turkeys in central Mississippi. Proceedings of the Annual

Conference of the Southeastern Association of Fish and Wildlife Agencies 51:352–361.

Miller, H. A., and S. H. Lamb. 1985. Oaks of North America. Naturegraph Publishers, Happy Camp, CA.

Miller, S. G., S. P. Bratton, and J. Hadidian. 1992. Impacts of white-tailed deer on endangered and threatened vascular plants. Natural Areas Journal 12:67–74.

Millers, I., D. S. Shriner, and D. Rizzo. 1989. History of hardwood decline in the eastern United States. U.S. Forest Service General Technical Report NE-126.

Minckler, L. S., J. D. Woerheide, and R. C. Schlesinger. 1973. Light, soil moisture and tree reproduction in hardwood forest openings. U.S. Forest Service Research Paper NC-89.

Minser, W. G., T. Allen, B. Ellsperman, and S. E. Schlarbaum. 1995. Feeding response of wild turkeys to chestnuts and other hard mast. Proceedings of the Annual Conference Southeastern Association Fish and Wildlife Agencies 49:488–497.

Miyaki, M., and K. Kikuzawa. 1988. Dispersal of *Quercus mongolica* acorns in a broadleaved deciduous forest. 2. Scatterhoarding by mice. Forest Ecology and Management 25:9–16.

Mladenhoff, D. J., M. A. White, T. R. Crow, and J. Pastor. 1994. Applying principles of landscape design and management to integrate old-growth forest enhancement and commodity use. Conservation Biology 8:752–762.

Mohler, C. L. 1990. Co-occurrence of oak subgenera: Implications for niche differentiation. Bulletin of the Torrey Botanical Club 117:247–255.

Monsandl, R., and A. Kleinert. 1998. Development of oaks (*Quercus petraea* (Matt.) Liebl.) emerged from bird-dispersed seeds under old-growth pine (*Pinus silvestris* L.) stands. Forest Ecology and Management 106:35–44.

Montgomery, M. E., and W. E. Wallner. 1988. The gypsy moth, a westward migrant. Pages 353–375 in: Dynamics of Forest Insect Populations (A. A. Berryman, ed.). Plenum Publishing, New York.

Mooney, H. A., ed. 1977. Convergent evolution in Chile and California Mediterranean climate ecosystems. Dowden, Hutchinson and Ross, Stroudsberg, PA.

Moore, T. L. 1964. The summer food habits of the wild turkey on the high plateau. Master's thesis, St. Bonaventure University, Olean, NY.

Morrison, M. L. 1999. Bird habitat relationships in desert grasslands. Pages 122–124 in: Toward integrated research, land management, and ecosystem protection in the Malpai Borderlands: Conference summary (G. J. Gottfried, L. G. Eskew, C. G. Curtin, and C. B. Edminster, compilers). U.S. Forest Service Proceedings RMRS-P-10.

Mosby, H. S., and C. O. Handley. 1943. The wild turkey in Virginia: Its status, life history and management. Virginia Division of Game, Commission of Game and Inland Fisheries, Richmond. Pittman-Robertson Projects.

Moser, W. K., M. J. Ducey, and P. M. S. Ashton. 1996. Effects of fire intensity on competitive dynamics between red and black oaks and mountain laurel. Northern Journal of Applied Forestry 13:119–123.

Motroni, R. S., D. A. Airola, R. K. Marose, and N. D. Tosta. 1991. Using wildlife species richness to identify land protection priorities in California's hardwood rangelands. Pages 110–119 in: Proceedings of a Symposium on Oak Woodlands and Hardwood Rangeland Management. U.S. Forest Service General Technical Report PSW-126.

Mueller-Dombois, D., J. E. Canfield, R. A. Holt, and G. P. Buelow. 1983. Tree-group death in North American and Hawaiian forests: A pathological problem or a new problem for vegetation ecology? Phytocoenologia 11:117–137.

Muick, P. C., and J. R. Bartolome. 1987. An assessment of natural regeneration of oaks in California. Report submitted to the Forest and Rangeland Assessment Program, California Department of Forestry and Fires Protection, Sacramento.

Muller, R. N., and Y. Liu. 1991. Coarse woody debris in an old-growth deciduous forest on the Cumberland Plateau, southeastern Kentucky. Canadian Journal of Forest Research 21:1567–1572.

Munz, P. A., and D. D. Keck. 1973. California Flora: A Supplement. University of California Press, Berkeley.

Murie, A. 1946. The Merriam turkey on the San Carlos Indian reservation. Journal of Wildlife Management 10:329–333.

Muzika, R. M., and M. J. Twery. 1995. Regeneration in defoliated and thinned hardwood stands of north-central West Virginia. Pages 326–340 in: Proceedings of the Tenth Central Hardwood Forest Conference (K. W. Gottschalk and S. L. C. Fosbroke, eds.). U.S. Forest Service General Technical Report NE-197.

Myers, R. K., and D. H. Van Lear. 1998. Hurricane-fire interactions in coastal forests of the south: A review and hypothesis. Forest Ecology and Management 103:265–276.

Myers, R. L. 1985. Fire and the dynamic relationship between Florida sandhill and sand pine scrub vegetation. Bulletin of the Torrey Botanical Club 112:241–252.

Neilson, R. P., and L. H. Wullstein. 1980. Catkin freezing and acorn production in Gambel oak in Utah, 1978. American Journal of Botany 67:426–428.

Nelson, R. A., G. E. Folk, Jr., E. W. Pfeiffer, J. J. Craighead, C. J. Jonkel, and D. L. Steiger. 1983. Behavior, biochemistry, and hibernation in black, grizzly, and polar bears. International Conference on Bear Research and Management 5:284–290.

Nelson, R. M., I. H. Sims, and M. S. Abell. 1933. Basal fire wounds on some southern Appalachian hardwoods. Journal of Forestry 31:829–837.

Nelson, T. C. 1957. The original forest of the Georgia Piedmont. Ecology 38:390–397.

Nichols, J. O. 1968. Oak mortality in Pennsylvania: A ten-year study. Journal of Forestry 66:681–694.

Nielsen, G. A., and F. D. Hole. 1964. Earthworms and the development of co-progeneous A_1 horizons on forest soils of Wisconsin. Soil Science Society of America Proceedings 28:426–430.

Niering, W. A., R. H. Goodwin, and S. Taylor. 1970. Prescribed burning in southern New England: Introduction to long-range studies. Proceedings of the Tall Timbers Fire Ecology Conference 10:267–286.

Niering, W. A., and C. H. Lowe. 1984. Vegetation of the Santa Catalina Mountains: Community types and dynamics. Vegetatio 58:3–28.

Nix, L. E. 1989. Early release of bottomland oak enrichment plantings appears promising in South Carolina. Pages 379–383 in: Proceedings of the Fifth Biennial Southern Silvicultural Research Conference (J. Miller, ed.). U.S. Forest Service General Technical Report SO-GTR 74.

Nixon, C. M., and L. P. Hansen. 1987. Managing forests to maintain populations of gray and fox squirrels. Illinois Department of Conservation Technical Bulletin 5.

Nixon, C. M., L. P. Hansen, P. A. Brewer, and J. E. Chelsvig. 1991. Ecology of white-tailed deer in an intensively farmed region of Illinois. Wildlife Monographs 118:1–77.

Nixon, C. M., D. M. Worley, and M. W. McClain. 1968. Food habits of squirrels in southeast Ohio. Journal of Wildlife Management 32:294–305.

Nixon, W. 1995. As the worm turns. American Forests 101:34–36.

Norman, G. W., J. C. Pack, C. I. Taylor, D. E. Steffen, and K. H. Pollock. 2001. Reproductive ecology of the eastern wild turkey in Virginia and West Virginia. Journal of Wildlife Management 65:1–9.

Norton, D. A., and D. Kelly. 1988. Mast seeding over 33 years by *Dacrydium cupressinum* Lamb. (rimu) (Podocarpaceae) in New Zealand: The importance of economies of scale. Functional Ecology 2:399–408.

Novick, H. J., J. M. Siperek, and G. R. Stewart. 1981. Denning characteristics of black bears (*Ursus americanus*) in the San Bernardino Mountains of southern California. California Fish and Game 67:52–61.

Novick, H. J., and G. R. Stewart. 1982. Home range and habitat preference of black bears in the San Bernardino Mountains of southern California. California Fish and Game 68:21–35.

Nowacki, G. J., and M. D. Abrams. 1992. Community, edaphic and historical analysis of mixed-oak forests of the Ridge and Valley Province in central Pennsylvania. Canadian Journal of Forest Research 22:790–800.

Nowacki, G. J., and M. D. Abrams. 1997. Radial-growth averaging criteria for reconstructing disturbance histories from presettlement-origin oaks. Ecological Monographs 67:225–249.

Nowacki, G. J. , M. D. Abrams, and C. G. Lorimer. 1990. Composition, structure, and historical development of northern red oak stands along an edaphic gradient in northcentral Wisconsin. Forest Science 36:276–292.

Noyce, K. V., and D. L. Garshelis. 1997. Influence of natural food abundance on black bear harvests in Minnesota. Journal of Wildlife Management 61:1067–1074.

Nudds, N. T. 1996. Modification of ecosystems by ungulates. Journal of Wildlife Management 60:695–713.

Nupp, T. E., and R. K. Swihart. 1996. Effect of forest patch area on population attributes of white-footed mice (*Peromyscus leucopus*) in fragmented landscapes. Canadian Journal of Zoology 74:467–472.

Nupp, T. E., and R. K. Swihart. 1998. Effects of forest fragmentation on population attributes of white-footed mice and eastern chipmunks. Journal of Mammology 79:1234–1243.

Nuzzo, V. A. 1986. Extent and status of the mid-west oak savanna: Presettlement and 1985. Natural Areas Journal 6:6–36.

Nyandiga, C. O., and G. R. McPherson. 1992. Germination of two warm-temperature oaks, *Quercus emoryi* and *Quercus arizonica*. Canadian Journal of Forest Research 22:1395–1401.

Oak, S. W., D. A. Starkey, and J. M. Dabney. 1988. Oak decline alters habitat in southern upland forests. Proceedings of the Annual Conference of the Southeastern Association of Fish and Wildlife Agencies 42: 491–501. Hilton Head, SC.

Oak, S. W., and P. M. Croll. 1995. Evaluation of oak decline risk rating using the CISC database on the Cherokee National Forest. U.S. Forest Service Field Office Report 95-1-22.

Oak, S., F. Tainter, J. Williams, and D. Starkey. 1996. Oak decline risk rating for the southeastern United States. Annales des Sciences Forestieres Science. Forestry 53:721–730.

Ofcarcik, R. P., and E. E. Burns. 1971. Physical and chemical properties of selected acorns. Journal of Food Science 36:576–578.

Oliver, C. D. 1978. The development of northern red oak in mixed stands in central New England. Yale University School of Forestry Bulletin 91.

Oliver, C. D., and B. C. Larson. 1990. Forest Stand Dynamics. McGraw-Hill, New York.

Olson, D. F., Jr., and S. G. Boyce. 1971. Factors affecting acorn production and germination and early growth of seedlings and seedling sprouts. Pages 44–48 in: Oak Symposium Proceedings. U.S. Forest Service, Northeastern Forest Experiment Station, Broomall, PA.

O'Pezio, J., S. H. Clarke, and C. Hackford. 1983. Chronology of denning by black bears (*Ursus americanus*) in the Catskill region of New York (USA). New York Fish and Game Journal 30:1–11.

Orwig, D. A., and M. D. Abrams. 1994. Land-use history (1720–1992), composition, and dynamics of oak-pine forests within the Piedmont and Coastal Plain of northern Virginia. Canadian Journal of Forest Resources 24:1216–1225.

Ostfeld, R. S. 1994. The fence effect reconsidered. Oikos 70:340–348.

Ostfeld, R. S. 1997. The ecology of Lyme-disease risk. American Scientist 85:338–346.

Ostfeld, R. S., K. R. Hazler, and O. M. Cepeda. 1996. Temporal and spatial dynamics of *Ixodes scapularis* (Acari: Ixodidae) in a rural landscape. Journal of Medical Entomology 33:90–95.

Ostfeld, R. S., C. G. Jones, and J. O. Wolff. 1996. Of mice and mast, ecological connections in eastern deciduous forests. BioScience 46:323–330.

Ostfeld, R. S., and F. Keesing. 2000a. Biodiversity and disease risk: The case of Lyme disease. Conservation Biology 14:722–729.

Ostfeld, R. S., and F. Keesing. 2000b. Pulsed resources and community dynamics of consumers in terrestrial ecosystems. Trends in Ecology and Evolution 15:232–237.

Ostfeld, R. S., F. Keesing, C. G. Jones, C. D. Canham, and G. M. Lovett. 1998. Integrative ecology and the dynamics of species in oak forests. Integrative Biology 1:178–186.

Ostfeld, R. S., R. H. Manson, and C. D. Canham. 1997. Effects of rodents on survival of tree seeds and seedlings invading old fields. Ecology 78:1531–1542.

Pacific Meridian Resources. 1994. California hardwood rangeland monitoring final report. Unpublished report for Strategic Planning Program, California Department of Forestry and Fire Protection, Sacramento.

Pack, J. C., H. S. Mosby, and P. B. Siegel. 1967. Influence of social hierarchy on gray squirrel behavior. Journal of Wildlife Management 31:720–728.

Pack, J. C., G. W. Norman, C. I. Taylor, D. E. Steffen, D. A. Swanson, K. H. Pollock, and R. Alpizar-Jara. 1999. Effects of fall hunting on wild turkey populations in Virginia and West Virginia. Journal of Wildlife Management 63:964–975.

Paine, R. T. 1995. A conversation on refining the concept of keystone species. Conservation Biology 9:962–964.

Palik, B., and R. T. Engstrom. 1999. Species composition. Pages 65–94 in: Maintaining Biodiversity in Forest Ecosystems (M. L. Hunter, Jr., ed.). Cambridge University Press, New York.

Pallardy, S. G., and J. L. Rhoads. 1993. Morphological adaptations to drought in seedlings of deciduous angiosperms. Canadian Journal of Forest Resources 23:1766–1774.

Pallardy, S. G., T. A. Nigh, and H. E. Garrett. 1988. Changes in forest composition in central Missouri, 1968–1982. American Midland Naturalist 120:380–390.

Palmer, W. E., S. R. Priest, R. S. Seiss, P. S. Phalen, and G. A. Hurst. 1993. Reproductive effort and success in a declining wild turkey population. Proceedings of the Annual Conference of the Southeastern Association of Fish and Wildlife Agencies 47:138–147.

Parker, G. R. 1989. Old-growth forests of the Central Hardwood Region. Natural Areas Journal 9:5–11.

Pase, C. P. 1969. Survival of *Quercus turbinella* and *Q. emoryi* in an Arizona chaparral community. Southwestern Natural 14:149–156.

Pastor, J., and R. J. Naiman. 1992. Selective foraging and ecosystem processes in boreal forests. American Naturalist 139:690–705.

Pattee, O. H., and S. L. Beasom. 1979. Supplemental feeding to increase wild turkey productivity. Journal of Wildlife Management 43:512–516.

Patterson, W. A., III and K. E. Sassaman. 1988. Indian fires in the prehistory of New England. Pages 107–135 in: Holocene Human Ecology in Northeastern North America (G. P. Nicholas, ed.). Plenum Press, New York.

Peet, R. K., and O. L. Loucks. 1977. A gradient analysis of southern Wisconsin forests. Ecology 58:485–499.

Pekins, P. J., and W. W. Mautz. 1987. Acorn usage by deer: Significance of oak management. Northern Journal of Applied Forestry 4:124–128.

Pekins, P. J., and W. W. Mautz. 1988. Digestibility and nutritional value of autumn diets of deer. Journal of Wildlife Management 52:328–332.

Pekins, P. J., K. S. Smith, and W. W. Mautz. 1998. The energy cost of gestation in white-tailed deer. Canadian Journal of Zoology 76:1091–1097.

Pelham, P. H., and J. G. Dickson. 1992. Physical characteristics. Pages 32–45 in: The Wild Turkey: Biology and Management (J. G. Dickson, ed.). Stackpole Books, Harrisburg, PA.

Pelton, M. R. 1989. The impacts of oak mast on black bears in the southern Appalachians. Pages 7–11 in: Proceedings of the Workshop: Southern Appalachian Mast Management (C. E. McGee ed.). University of Tennessee, Knoxville.

Pelton, M. R. 1996. The importance of old growth to carnivores in eastern deciduous forests. In: Eastern old-growth forests (M. B. Davis, ed.). Island Press, Washington, DC.

Pelton, M. R., C. Bennett, J. Clark, and K. Johnson. 1986. Assessment of bear population census techniques. Proceedings of the Eastern Black Bear Workshop 8:208–233.

Peters, R. H. 1983. The Ecological Implications of Body Size. Cambridge University Press, Cambridge.

Peterson, D. W. 1998. Fire effects on oak savanna and woodland vegetation. Ph.D. diss., University of Minnesota, St. Paul.

Peterson, D. W., and P. B. Reich. 2001. Fire in oak savanna: Fire frequency effects on stand structure and dynamics. Ecological Applications.

Peterson, L. E., and A. H. Richardson. 1973. Merriam's wild turkey in the Black Hills of South Dakota. Pages 3–9 in: Wild turkey Management: Current Problems and Programs (G. C. Sanderson and H. C. Schultz, eds.). Missouri Chapter of the Wildlife Society and University of Missouri Press, Columbia.

Peterson, R. S., and C. S. Boyd. 1998. Ecology and management of sand shinnery communities: A literature review. U.S. Forest Service General Technical Report RMRS-16.

Phillips, F. J. 1912. Emory oak in southern Arizona. U.S. Forest Service Circular 201.

Pickett, S. T. A., R. S. Ostfeld, M. Shachak, and G. E. Likens. 1997. The Ecological Basis of Conservation: Heterogeneity, Ecosystems, and Biodiversity. Chapman and Hall, New York.

Piesman, J., and J. S. Gray. 1994. Lyme disease (*Lyme borreliosis*). Pages 327–350 in: Ecological Dynamics of Tick-borne Zoonoses (D. E. Sonenshine and T. N. Mather, eds.). Oxford University Press, New York.

Piesman, J., J. G. Donahue, T. N. Mather, and A. Spielman. 1986. Transovarially acquired Lyme disease spirochetes (*Borrelia burgdorferi*) in field-collected larval *Ixodes dammini* (Acari: Ixodidae). Journal of Medical Entomology 23:219.

Pigott, C. D., A. C. Newton, and S. Zammitt. 1991. Predation of acorns and oak seedlings by gray squirrels. Quarterly Journal of Forestry 85:173–178.

Pimm, S. L. 1991. The Balance of Nature? Ecological Issues in the Conservation of Species and Communities. University of Chicago Press, Chicago.

Poelker, R. J., and H. D. Hartwell. 1973. Black bear of Washington. Washington State Game Department Biological Bulletin 14.

Pollisco, R. R., P. F. Ffolliott, and G. J. Gottfried. 1995. Ecological conditions in the overlap areas of the pinyon-juniper and encinal woodlands. Pages 594–596 in: Biodiversity and management of the Madrean archipelago: The sky islands of the southwestern United States and northwestern Mexico (L. F. DeBano. P. F. Ffolliott, A. Ortega-Rubio, G. J. Gottfried, R. H. Hamre, and C. B. Edminster, technical coordinators.) U.S. Forest Service General Technical Report RM-265.

Pollock, K. H., R. Alpizar-Jara, and K. Tsai. 1997. Final report on the Virginia and West Virginia joint wild turkey survival analysis (1989–1994). Department of Statistics, North Carolina State University, Raleigh.

Porter, W. F. 1992. Habitat requirements. Pages 202–213 in: The Wild Turkey: Biology and Management (J. G. Dickson, ed.). Stackpole Books, Harrisburg, PA.

Porter, W. F., R. D. Tangen, G. C. Nelson, and D. A. Hamilton. 1980. Effect of corn food plots on wild turkeys in the upper Mississippi Valley. Journal of Wildlife Management 44:456–462.

Porter, W. F., G. C. Nelson, and K. Mattson. 1983. Effects of winter conditions on reproduction in a northern wild turkey population. Journal of Wildlife Management 47:281–290.

Portmann, A. 1961. Sensory organs: Skins, taste, and olfaction. Pages 37–48 in: Biology and Comparative Physiology of Birds. Part 1 (A. J. Marshall, ed.). Academic Press, New York.

Potter, T. D., S. D. Schemnitz, and W. D. Zeedyk. 1985. Status and ecology of

Gould's turkey in the Peloncillo Mountains of New Mexico. Proceedings of the National Wild Turkey Symposium 5:1–24.

Poulson, T. L., and W. J. Platt. 1989. Gap light regimes influence canopy tree diversity. Ecology 70:553–555.

Powell, D. S., J. L. Faulkner, D. R. Darr, Z. Zhu, and D. W. MacCleery. 1993. Forest resources of the United States, 1992. U.S. Forest Service General Technical Report RM-234.

Powell, D. S., and E. H. Tryon. 1979. Sprouting ability of advance growth in undisturbed hardwood stands. Canadian Journal of Forest Resources 9:116–120.

Powell, J. A. 1967. Management of the Florida turkey and eastern turkey in Georgia and Alabama. Pages 409–451 in: The Wild Turkey and Its Management (O. H. Hewitt, ed.). Wildlife Society, Washington, DC.

Powell, R. A., J. W. Zimmerman, and D. E. Seaman. 1997. Ecology and Behavior of North American Black Bears. Chapman and Hall, London.

Power, M. E., D. Tilman, J. A. Estes, B. A. Menge, W. J. Bond, L. S. Mills, G. Daily, J. C. Castilla, J. Lubchenko, and R. T. Paine. 1996. Challenges in the quest for keystones. BioScience 46:609–620.

Pozzanghera, S. A. 1990. The reproductive biology, winter dormancy, and denning physiology of black bears in Great Smoky Mountains National Park. Master's thesis, University of Tennessee, Knoxville.

Price, M. V., and S. H. Jenkins. 1986. Rodents as seed consumers and dispersers. Pages 191–235 in: Seed Dispersal (D. R. Murray, ed.). Academic Press, Sydney, Australia.

Pucek, Z., W. Jędrzejewski, B. Jędrzejewska, and M. Pucek. 1993. Rodent population dynamics in a primeval deciduous forest (Białowieża National Park) in relation to weather, seed crop, and predation. Acta Theriologica 38:199–232.

Pyne, S. J. 1982. Fire in America. Princeton University Press, Princeton, NJ.

Pyne, S. J. 1983. Indian fires. Natural History 2:6–11.

Pyne, S. J., P. L. Andrews, and R. D. Laven. 1996. Introduction to Wildland Fire. 2nd ed. John Wiley and Sons, New York.

Quinton, D. A., and A. K. Montei. 1977. Preliminary study of the diet of Rio Grande turkeys in north-central Texas. Southwest Naturalist 22:550–553.

Raine, R. M., and J. L. Kansas. 1990. Black bear seasonal food habits and distribution by elevation in Banff National Park, Alberta. International Conference on Bear Research and Management 8:297–304.

Randolph, S. E., and N. G. Craine. 1995. General framework for comparative quantitative studies on transmission of tick-borne diseases using *Lyme borreliosis* in Europe as an example. Journal of Medical Entomology 32:765–777.

Ratliff, R. D., D. Duncan, and S. E. Westfall. 1991. California oak-woodland overstory species affect herbage understory: Management implications. Journal of Range Management 44:306–310.

Rauscher, H. M. 1999. Ecosystem management decision support for federal forests in the United States: A review. Forest Ecology and Management. 114:173–197.

Rayner, J. M. V. 1985. Linear relations in biomechanics: The statistics of scaling functions. Journal of the Zoological Society of London 206:415–439.

Record, S. J. 1910. Forest conditions of the Ozark region of Missouri. University of Missouri, Agricultural Experimental Station Bulletin 89.

Reeves, R. H., and W. G. Swank. 1955. Food habits of Merriam's turkey. Arizona Pittman-Robertson Project W-49-R-3.

Regelbrugge, J. C., and D. W. Smith. 1994. Postfire tree mortality in relation to wildfire severity in mixed oak forests in the Blue Ridge of Virginia. Northern Journal of Applied Forestry 11:90–97.

Reich, P. B., and T. M. Hinckley. 1980. Water relations, soil fertility, and plant nutrient composition of a pygmy oak (*Quercus*) ecosystem in southeast Missouri. Ecology 61:400–416.

Reich, P. B., and T. M. Hinckley. 1989. Influence of pre-dawn water potential and soil-to-leaf hydraulic conductance on maximum daily leaf diffusive conductance in two oak species. Functional Ecology 3:719–726.

Reich, P. B., M. D. Abrams, D. S. Ellsworth, E. L. Kruger, and T. J. Tabone. 1990. Fire affects ecophysiology and community dynamics of central Wisconsin oak forest regeneration. Ecology 71:2179–2190.

Reisfield, A. S. 1995. Central Texas plant ecology and oak wilt. Pages 133–138 in: Oak Wilt Perspectives: Proceedings of the National Oak Wilt Symposium (D. N. Appel and R. F. Billings, eds.). Information Development, Inc., Houston.

Repenning, R. W., and R. F. Labisky. 1985. Effects of even-aged timber management on bird communities of the longleaf pine forest in northern Florida. Journal of Wildlife Management 48:895–911.

Reynolds, P. S. 1993. Effects of body size and fur on heat loss of collared lemmings, *Dicrostonx groenlandicus*. Journal of Mammalogy 74:291–303.

Reynolds, D. G., and J. J. Beecham. 1980. Home range activities and reproduction of black bears in west-central Idaho. International Conference on Bear Research and Management 4:181–190.

Ribbens, E., J. A. Silander, and S. W. Pacala. 1994. Seedling recruitment in forests: Calibration models to predict patterns of tree seedling dispersion. Ecology 75:1794–1806.

Rice, E. L., and W. T. Penfound. 1959. The upland forests of Oklahoma. Ecology 40:593–608.

Richter, D. D., C. W. Ralston, and W. R. Harms. 1982. Prescribed fire: Effects on water quality and nutrient cycling. Science 258:1099–1100.

Ricker, W. E. 1973. Linear regressions in fishery research. Journal of Fishery Research 30:409–434.

Ricker, W. E. 1984. Computation and uses of central trend lines. Canadian Journal of Forest Resources 62:1897–1905.

Rivers, D. C. 1940. A study of the food habits of the wild turkey of Virginia. Master's thesis, Virginia Polytechnic Institute, Blacksburg.

Roach, B. A., and S. F. Gingrich. 1968. Even-aged silviculture for upland central hardwoods. U.S. Forest Service Agricultural Handbook 335.

Robbins, C. T., A. E. Hagerman, P. J. Austin, C. McArthur, and T. A. Hanley 1991. Variation in mammalian physiological responses to a condensed tannin and its ecological implications. Journal of Mammalogy 72:480–486.

Robbins, C. T., S. Mole, A. E. Hagerman, and T. A. Hanley. 1987. Role of tannins in defending plants against ruminants: Reduction in protein availability. Ecology 68:98–107.

Roberts, S. D., and W. F. Porter. 1995. Importance of demographic parameters to annual changes in wild turkey abundance. Proceedings of the National Wild Turkey Symposium 7:15–20.

Robinson, S. K., F. R. Thompson, T. M. Donovan, D. Whitehead, and J. Faaborg. 1995. Regional forest fragmentation and the nesting success of migratory birds. Science 267:1987–1990.

Rodgers, C. S., and R. C. Anderson. 1979. Presettlement vegetation of two prairie counties. Botanical Gazette 140:232–240.

Rogers, L. L. 1976. Effects of mast and berry crop failures on survival, growth, and reproductive success of black bears. Transactions of the North American Wildlife and Natural Resources Conference 41:431–438.

Rogers, L. L. 1981. A bear in its lair. Natural History 90:64–70.

Rogers, L. L. 1983. Effects of food supply, predation, cannibalism, and other health problems on black bear populations. Pages 194–211 in: Symposium on Natural Regulation of Wildlife Populations. Forest Wildlife and Range Experiment Station (F. L. Bunnell, D. S. Eastman, and J. M. Peek, eds.). University of Idaho, Moscow.

Rogers, L. L. 1987. Effects of food supply and kinship on social behavior, movements, and population growth of black bears in northeastern Minnesota. Wildlife Monograph 97:1–72.

Rogers, M. J., L. K. Halls, and J. G. Dickson. 1990. Deer habitat in the Ozark forests of Arkansas. U.S. Forest Service Research Paper SO-259.

Romme, W. H., and W. H. Martin. 1982. Natural disturbance by tree falls in old-growth mixed mesophytic forest: Lilley Cornett Woods, Kentucky. Pages 367–383 in: Proceedings of the 4th Central Hardwood Forest Conference (R. N. Muller, ed.). University of Kentucky Press, Lexington.

Ross, A. S., and G. A. Wunz, 1990. Habitats used by wild turkey hens during the summer in oak forests in Pennsylvania. Proceedings of the National Wild Turkey Symposium 6:39–43.

Rossiter, M. C. 1994. Maternal effects hypothesis of herbivore outbreak. BioScience 44:752–763.

Rossiter, M. C., J. C. Schultz, and I. T. Baldwin. 1988. Relationships among defoliation, red oak phenolics, and gypsy moth growth and reproduction. Ecology 69:267–277.

Roth, E. R., and G. H. Hepting 1943. Origin and development of oak stump sprouts as affecting their likelihood to decay. Journal of Forestry 41:27–36.

Rumble, M. A., and R. A. Hodorff. 1993. Nesting ecology of Merriam's turkeys in the Black Hills, South Dakota. Journal of Wildlife Management 57:789–801.

Runkle, J. R. 1982. Patterns of disturbance in some old-growth mesic forests of eastern North America. Ecology 63:1533–1546.

Runkle, J. R. 1985. Disturbance regimes in temperate forests. Pages 17–33 in: The Ecology of Natural Disturbance and Patch Dynamics (S. T. A. Pickett and P. S. White, eds.). Academic Press, San Diego.

Runkle, J. R. 1990. Gap dynamics in an Ohio *Acer-Fagus* forest and speculations on the geography of disturbance. Canadian Journal of Forest Resources 20:632–641.

Russell, E. W. B. 1980. Vegetation changes in northern New Jersey from precolonization to the present: A paleological interpretation. Bulletin of the Torrey Botanical Club 107:432–446.

Russell, E. W. B. 1981. Vegetation of northern New Jersey before European settlement. American Midland Naturalist 105:1–12.

Russell, E. W. B. 1983. Indian-set fires in the forests of the northeastern United States. Ecology 64:78–88.

Rusterholz, K. A. 1991. Oaks and old growth. Pages 100–106 in: The Oak Resource in the Upper Midwest: Implications for Management (S. B. Laursen and J. F. DeBoe, eds.). Minnesota Extension Service, University of Minnesota, St. Paul.

Ryan, L. A., and A. B. Carey. 1995. Biology and management of the western gray squirrel and Oregon white oak woodland with emphasis on the Puget Trough. U.S. Forest Service Research Paper PNR-348.

Saether, B. E., and H. Haagenrud. 1983. Life history of the moose (*Alces alces*): Fecundity rates in relation to age and carcass weight. Journal of Mammalogy 64:226–232.

Salwasser, H. 1988. Managing ecosystems for viable populations of vertebrates: A focus on biodiversity. Pages 87–104 in: Ecosystem Management for Parks and Wilderness (J. K. Agee and D. R. Johnson, eds.). University of Washington Press, Seattle.

Sampson, A. W. 1944. Plant succession on burned chaparral lands in northern California. California Agricultural Experiment Station Bulletin 685.

Sanchini, P. J. 1981. Population structure and fecundity patterns in *Quercus emoryi* and *Q. arizonica* in southeastern Arizona. Ph.D. diss., University of Colorado, Boulder.

Sander, I. L. 1971. Height growth of new oak sprouts depends on size of advance reproduction. Journal of Forestry 69:809–811.

Sander, I. L. 1979. Regenerating oaks with the shelterwood system. Pages 54–60 in: The 1979 John S. Wright Forestry Conference (H. A. Holt and B. C. Fischer, eds.). Purdue University, West Lafayette, IN.

Sander, I. L., P. S. Johnson, and R. Rogers. 1992. Evaluating oak advance reproduction in the Missouri Ozarks. U.S. Forest Service Research Paper NC-251.

Sander, I. L., C. E. McGee, K. G. Day, and R. E. Willard. 1983. Oak-hickory. Pages 116–120 in: Silvicultural Systems of the Major Forest Types of the United States (R. M. Burns and B. H. Honkala, eds.). U.S. Forest Service Agriculture Handbook 445.

SAS Institute. 1989. SAS/STAT user's guide. Version 6. 4th ed. Vol. 1. SAS Institute, Cary, NC.

Scanlon, J. J. 1992. Managing forests to enhance wildlife diversity in Massachusetts. Northeast Wildlife 49:1–9.

Scarlett, T. L., and K. G. Smith. 1991. Acorns preference of urban blue jays (*Cyanocitta cristata*) during fall and spring in northwestern Arkansas. Condor 93:438–442.

Schaberg, R. H., T. P. Holmes, K. J. Lee, and R. C. Abt. 1999. Ascribing value to ecological processes: An economic view of environmental change. Forest Ecology and Management 114:329–338.

Schafale, M. P., and P. A. Harcombe. 1983. Presettlement vegetation of Hardin County, Texas. American Midland Naturalist 109:355–366.

Schemnitz, S. D. 1956. Wild turkey food habits in Florida. Journal of Wildlife Management 20:132–137.

Schemnitz, S. D., D. E. Fligert, and R. C. Willging. 1990. Ecology and management of Gould's turkeys in southwestern New Mexico. Proceedings of the National Wild Turkey Symposium 6:72–83.

Schemnitz, S. D., D. L. Goerndt, and K. H. Jones. 1985. Habitat needs and management of Merriam's turkey in southcentral New Mexico. Proceedings of the National Wild Turkey Symposium 5:199–231.

Schemnitz, S. D., and W. D. Zeedyk. 1992. Gould's turkey. Pages 350–360 in: The Wild Turkey: Biology and Management (J. G. Jackson, ed.). Stackpole Books, Harrisburg, PA.

Schemnitz, S. D., and M. L. Zornes. 1995. Management practices to benefit Gould's turkey in the Peloncillo, Mountains, New Mexico. Pages 461–464 in: Biodiversity and management of the Madrean archipelago: The sky islands of the southwestern United States and northwestern Mexico (L. F. DeBano, P. F. Ffolliott, A. Ortega-Rubio, G. J. Gottfried, R. H. Hamre, and C. B. Edminster, technical coordinators). U.S. Forest Service General Technical Report RM-265.

Schlesinger, R. C., I. L. Sander, and K. R. Davidson. 1993. Oak regeneration potential increased by shelterwood treatments. Northern Journal of Applied Forestry 10:149–153.

Schmidt, K. A., and R. S. Ostfeld. 2001. Biodiversity and the dilution effect in disease ecology. Ecology 82:609–619.

Schmidt-Nielson, K. 1984. Scaling: Why Is Animal Size So Important? Cambridge University Press, Cambridge.

Schoener, T. W. 1983. Simple models of optimal feeding-territory size: A reconciliation. American Naturalist 121:608–629.

Schooley, R. L., C. R. McLaughlin, G. J. Matula, Jr., and W. B. Krohn. 1994. Denning chronology of female black bears: Effects of food, weather, and reproduction. Journal of Mammalogy 75:466–477.

Schopmeyer, T. 1974. Seeds of Woody Plants in the United States. U.S. Forest Service Agricultural Handbook 450.

Schorger, A. W. 1949. Squirrels in early Wisconsin. Transactions of the Wisconsin Academy of Science, Arts, and Letters 39:195–247.

Schorger, A. W. 1955. The Passenger Pigeon: Its Natural History and Extinction. University of Wisconsin Press, Madison.

Schorger, A. W. 1966. The Wild Turkey: Its History and Domestication. University of Oklahoma Press, Norman.

Schrage, M. S. 1994. Influence of gypsy moth–induced oak mortality on a black bear population. Master's thesis, Virginia Polytechnic Institute and State University, Blacksburg.

Schrage, M. S., and M. R. Vaughan. 1998. Population responses of black bears following oak mortality induced by gypsy moths. Ursus 10:49–54.

Schroeder, R. L. 1985. Habitat suitability index models: Eastern wild turkey. U.S. Department of the Interior, Fish and Wildlife Service Biological Report 82.

Schroeder, R. L., and L. D. Vangilder. 1997. Tests of wildlife habitat models to evaluate oak-mast production. Wildlife Society Bulletin 25:639–646.

Schuler, T. M., and G. W. Miller. 1995. Shelterwood treatments fail to establish oak reproduction on mesic forest sites in West Virginia: 10-year results. Pages 375–387 in: Proceedings of the Tenth Central Hardwood Forest Conference (K. W. Gottschalk and S. L. C. Fosbroke, eds.). U.S. Forest Service General Technical Report NE-197.

Schwartz, C. C., S. D. Miller, and A. W. Franzmann. 1987. Denning ecology of three black bear populations in Alaska. International Conference on Bear Research and Management 7:281–291.

Scott, C. T. 1998. Sampling methods for estimating change in forest resources. Ecological Applications 8:228–233.

Scott, T. A., and N. Pratini. 1997. Edge effects and recreational impacts on a population of woodland birds. In: Proceedings of a Symposium on Oak Woodlands: Ecology, Management, and Urban Interface Issues. U.S. Forest Service Research Paper PSW-GTR-160.

Scott, V. E., and E. L. Boeker. 1973. Seasonal food habits of Merriam's turkeys on the Fort Apache Indian Reservation. Pages 151–157 in: Wild Turkey Management: Current Problems and Programs (G. C. Sanderson and H. C. Schultz, eds.). Missouri Chapter of the Wildlife Society and University of Missouri Press, Columbia.

Seidel, K. W. 1972. Drought resistance and internal water balance of oak seedlings. Forest Science 18:34–40.

Seischab, F. K. 1990. Presettlement forests of the Phelps and Gorham purchase in western New York. Bulletin of the Torrey Botanical Club 117:27–38.

Seiss, R. S. 1989. Reproductive parameters and survival rates of wild turkey hens in east-central Mississippi. Master's thesis, Mississippi State University, Starkville.

Servello, F. A., and R. L. Kirkpatrick. 1987. Regional variation in the nutritional ecology of ruffed grouse. Journal of Wildlife Management 51:749–770.

Servello, F. A., and R. L. Kirkpatrick. 1988. Nutrition and condition of ruffed grouse during the breeding season in southwestern Virginia. Condor 90:836–842.

Seymour, R. S., and M. L. Hunter Jr. 1999. Principles of ecological forestry. Pages 22–61 in: Maintaining Biodiversity in Forest Ecosystems (M. L. Hunter, Jr., ed.). Cambridge University Press, New York.

Sharitz, R. R., and W. J. Mitsch. 1993. Southern floodplain forests. Pages 311–372 in: Biodiversity of the southeastern United States lowland terrestrial communities (W. H. Martin, S. G. Boyce, and A. C. Echternacht, eds.). John Wiley and Sons, New York.

Sharman J. W., and P. F. Ffolliott. 1992. Structural diversity in oak woodlands of southeastern Arizona. Pages 132–136 in: Ecology and management of oak and associated woodlands: Perspectives in the southwestern United States and northern Mexico (P. F. Ffolliott, G. J. Gottfried, D. A. Bennett, V. M. Hernandez C., A. Ortega-Rubio, and R. H. Hamre, technical coordinators). U.S. Forest Service General Technical Report RM-218.

Sharp, W. M. 1958. Evaluating mast yields in the oaks. Pennsylvania State University, Agricultural Experimental Station Bulletin 635, University Park.

Sharp, W. M., and V. G. Sprague. 1967. Flowering and fruiting in the white oaks: Pistillate flowering, acorn development, weather, and yields. Ecology 48:243–251.

Shaw, S. P. 1971. Wildlife and oak management. Pages 84–89 in: Proceedings of the Oak Symposium. U.S. Forest Service, Northeastern Forest Experiment Station, Upper Darby, PA.

Shaw, H. G., and C. Mollohan. 1992. Merriam's turkey. Pages 331–349 in: The Wild Turkey: Biology and Management (J. G. Dickson, ed.). Stackpole Books, Harrisburg, PA.

Shaw, M. W. 1968a. Factors affecting the natural regeneration of sessile oak (Quercus petraea) in north Wales. Journal of Ecology 56:565–583.

Shaw, M. W. 1968b. Factors affecting the natural regeneration of sessile oak (Quercus petraea) in north Wales: II. Acorn losses and germination under field conditions. Journal of Ecology 56:647–660.

Shearin, A. T., M. H. Bruner, and N. B. Goebel. 1972. Prescribed burning stimulates natural regeneration of yellow poplar. Journal of Forestry 70:482–484.

Shiflet, T. N, ed. 1994. Rangeland Cover Types of the United States. Society for Rangeland Management, Denver.

Shigo, A. L. 1979. Tree decay: An expanded concept. U.S. Forest Service Information Bulletin Number 419.

Shimazu, M., and R. S. Soper. 1986. Pathogenicity and sporulation of *Entomophaga maimaiga* Humber, Shimazu, Soper & Hajek (Entomophthorales: Entomophthoraceae) on larvae of the gypsy moth, *Lymantria dispar* L. (Lepidoptera: Lymantriidae). Applied Entomology Zoology 21:589–596.

Short, H. L. 1976. Composition and squirrel use of acorns of black and white oak groups. Journal of Wildlife Management 40:479–483.

Short, H. L., and E. A. Epps, Jr. 1976. Nutrient quality and digestibility of seeds and fruits from southern forests. Journal of Wildlife Management 40:283–289.

Silvertown, J. W. 1980. The evolutionary ecology of mast seeding in trees. Biological Journal of the Linnean Society 14:235–250.

Simard, A. J., J. Eenigenburg, and R. Blank. 1986. Predicting injury and mortality to trees from prescribed burning. Pages 65–72 in: Proceedings from Prescribed Burning in the Midwest: State-of-the-Art Symposium (A. L. Koonce, ed.). Stevens Point, WI.

Sinclair, A. R. E. 1997. Carrying capacity and the overabundance of deer: A framework for management. Pages 380–394 in: The Science of Overabundance, Deer Ecology and Management (W. J. McShea, H. B. Underwood, and J. H. Rappole, eds.). Smithsonian Institution Press, Washington, DC.

Sinclair, W. A., H. H. Lyon, and W. T. Johnson. 1987. Diseases of Trees and Shrubs. Cornell University Press, Ithaca, NY.

Singer, D. K., S. T. Jackson, B. J. Madsen, and D. A. Wilcox. 1996. Differentiating climatic and successional influences on long-term development of a marsh. Ecology 77:1765–1778.

Sisojevic, P. 1975. Population dynamics of tachinid parasites of the gypsy moth (*Lymantria dispar* L.) during a gradation period (in Serbo-Croatian). Zasitia Bilja 26:97–170.

Skeen, J. N., P. D. Doerr, and D. H. van Lear. 1993. Oak-hickory-pine forests. Pages 1–33 in: Biodiversity of the Southeastern United States: Upland Terrestrial Communities (W. H. Martin, S. G. Boyce, and A. C. Echternacht, eds.). John Wiley and Sons, New York.

Smallwood, P. D. 1992. Temporal and spatial scales in foraging ecology: testing hypotheses with spiders and squirrels. Ph.D. diss., University of Arizona, Tucson.

Smallwood, P. D., and W. D. Peters. 1986. Grey squirrel food preferences: The effects of tannin and fat concentration. Ecology 67:168–174.

Smallwood, P. D., M. A. Steele, E. Ribbens, and W. McShea. 1998. Detecting the effects of seed hoarders on the distribution of seedlings of tree species: Gray squirrels (*Sciurus carolinensis*) and oaks (*Quercus* spp.) as a model system. Pages 211–222 in: Ecology and Evolutionary Biology of Tree Squirrels (M. A. Steele, J. F. Merritt, and D. A. Zegers, eds.). Virginia Museum of Natural History Special Publication 6.

Smith, C. C. 1968. The adaptive nature of social organization in the genus of tree squirrels (*Tamiasciurus*). Ecological Monographs 38:31–63.

Smith, C. C. 1970. The coevolution of pine squirrels *(Tamiasciurus)* and conifers. Ecological Monographs. 40:349–371.

Smith, C. C. 1981. The indivisible niche of *Tamiasciurus:* An example of non-partitioning of resources. Ecological Monographs. 51:343–363.

Smith, C. C. 1995. The niche of diurnal tree squirrels. Pages 209–225 in: Storm over a Mountain Island: Conservation Biology and the Mt. Graham Affair (C. A. Istock and R. S. Hoffmann, eds.). University of Arizona Press, Tucson.

Smith, C. C. 1998. The evolution of reproduction in trees: Its effect on squirrel ecology and behavior. Pages 203–210 in: Ecology and Evolutionary Biology of Tree Squirrels (M. A. Steele, J. F. Merritt, and D. A. Zegers, eds.). Special Publication 6, Virginia Museum of Natural History.

Smith, C. C., and R. P. Balda. 1979. Competition between insect, bird and mammals for conifer seeds. American Zoologist 19:1065–1083.

Smith, C. C., and D. Follmer. 1972. Food preferences of squirrels. Ecology 53:82–91.

Smith, C. C., J. L. Hamrick, and C. L. Kramer. 1990. The advantage of mast years for wind pollination. American Naturalist 136:154–166.

Smith, C. C., and O. J. Reichman. 1984. The evolution of food caching by birds and mammals. Annual Review of Ecology and Systematics 15:329–351.

Smith, D. W. 1993. Oak regeneration: The scope of the problem. Pages 40–52 in: Oak Regeneration: Serious Problems, Practical Recommendations (D. L. Loftis and C. E. McGee, eds.). U.S. Forest Service General Technical Report SE-84.

Smith, D. W., and W. E. Linnartz. 1980. The southern hardwoods region. Pages 145–320 in: Regional Silviculture of the United States (J. W. Barrett, ed.). John Wiley and Sons, New York.

Smith, H. C. 1981. Diameters of clearcut openings influence central Appalachian hardwood stem development: A 10-year study. U.S. Forest Service Research Paper NE-476.

Smith, H. R. 1985. Wildlife and the gypsy moth. Wildlife Society Bulletin 13:166–174.

Smith, H. R. 1989. Predation: Its influence on population dynamics and adaptive changes in morphology and behavior of the *Lymantriidae*. Pages 469–488 in: Lymantriidae: A comparison of features of new and old world tussock moths. U.S. Forest Service General Technical Report NE-123.

Smith, H. R., and R. A. Lautenschlager. 1981. Gypsy moth. Pages 96–124 in: The gypsy moth predators: Research toward integrated pest management (C. C. Doane and M. L. McManus, eds.). U.S. Department of Agriculture Technical Bulletin 1584.

Smith, K. G., and Scarlett, T. 1987. Mast production and winter populations of red-headed woodpeckers and blue jays. Journal of Wildlife Management 51:459–467.

Smith, N. S., and R. G. Anthony. 1992. Coues white-tailed deer and the oak woodlands. Pages 50–56 in: Ecology and management of oak and associated woodlands: Perspectives in the southwestern United States and northern Mexico (P. F. Ffolliott, G. J. Gottfried, D. A. Bennett, V. M. Hernandez C., A. Ortega-Rubio, and R. H. Hamre, technical coordinators). U.S. Forest Service General Technical Report RM-218.

Smith, T. M., and M. R. Pelton. 1994. Home ranges and movements of black bears in a bottomland hardwood forest in Arkansas. International Conference on Bear Research and Management 8:213–218.

Smith, W. A., and B. Browning. 1967. Wild turkey food habits in San Luis Obispo County, California. California Fish and Game 53:246–253.

Smitley, D. R., L. S. Bauer, A. E. Hajek, F. J. Sapio, and R. A. Humber. 1995. Introduction and establishment of *Entomophaga maimaiga*, a fungal pathogen of gypsy moth (Lepidoptera: Lymantriidae) in Michigan. Environmental Entomology 24:1685–1695.

Society of American Foresters. 1991. Task force report on biological diversity in forest ecosystems. Society of American Foresters, Bethesda, MD.

Sokal, R. R., and F. J. Rohlf. 1981. Biometry: The Principles and Practice of Statistics in Biological Research. 2nd ed. W. H. Freeman, San Francisco.

Soloman, J. D. 1995. Guide to insect borers of North American broadleaf trees and shrubs. U.S. Forest Service Agricultural Handbook 706.

Solomon, D. S., and B. M. Blum. 1967. Stump sprouting of four northern hardwoods. U.S. Forest Service Research Paper NE-59.

Sonesson, K. L. 1994. Growth and survival after cotyledon removal in *Quercus robur* seedlings, grown in different natural soil types. Oikos 69:65–70.

Sork, V. L. 1984. Examination of seed dispersal and survival in red oak, *Quercus rubra* (Fagaceae), using metal-tagged acorns. Ecology 65:1020–1022.

Sork, V. L. 1993. Evolutionary ecology of mast-seeding in temperate and tropical oaks (*Quercus* spp.) Vegetatio 107–108:133–147.

Sork, V. L., and J. E. Bramble. 1993. Prediction of acorn crops in three species of North American oaks: *Quercus alba, Q. rubra* and *Q. velutina*. Annals Science Forestry 50:128s–136s.

Sork, V. L., J. Bramble, and O. Sexton. 1993. Ecology of mast-fruiting in three species of North American deciduous oaks. Ecology 74:528–541.

Southwood, T. R. E., and H. N. Comins. 1976. A synoptic population model. Journal of Animal Ecology 45:949–965.

Speake, D. W., T. E. Lynch, W. J. Fleming, G. A. Wright, and W. J. Hamrick. 1975.

Habitat use and seasonal movements of wild turkeys in the Southeast. Proceedings of the National Wild Turkey Symposium 3:122–130.

Speare, A. T., and R. H. Colley. 1912. The artificial use of the brown-tail fungus in Massachusetts, and a brief note on a fungus disease of the gypsy caterpillar. Wright and Potter, Boston.

Spetich, M. A., S. R. Shifley, and G. R. Parker. 1999. Regional distribution and dynamics of coarse woody debris in midwestern old-growth forests. Forest Science 45:302–313.

Spies, T. A., J. F. Franklin, and T. B. Thomas. 1988. Coarse woody debris in Douglas-fir forests of western Oregon and Washington. Ecology 69:1689–1702.

Spurr, S. H. 1951. George Washington, surveyor and ecological observer. Ecology 32:544–549.

Stahle, D. W., and P. L. Chaney. 1994. A predictive model for the location of ancient forests. Natural Areas Journal 14:151–158.

Stalheim-Smith, A. 1984. Comparative study of forelimbs of the semi-fossorial prairie dog, *Cynomys gunnisoni*, and the scansorial fox squirrel, *Sciurus niger*. Journal of Morphology 180:55–68.

Standiford, R. B., D. McCreary, S. Gaertner, and L. Forero. 1996. Impact of firewood harvesting on hardwood rangelands varies with region. California Agriculture 50:7–12.

Standiford, R. B., N. K. McDougald, R. Phillips, and A. Nelson. 1991. South Sierra oak regeneration survey. California Agriculture 45:12–14.

Standiford, R. B., and P. Tinnin. 1996. Guidelines for managing California's hardwood rangelands. University of California Division of Agriculture and Natural Resources Leaflet 3368.

Stapanian, M. A., and D. L. Cassell 1999. Regional frequencies of tree species associated with anthropogenic disturbance in three forest types. Forest Ecology Management 117:241–252.

Stapanian, M. A., D. L. Cassell, and S. P. Cline. 1997. Regional patterns of local diversity of trees: Associations with anthropogenic disturbance. Forest Ecology and Management 93:33–44.

Stapanian, M. A., and C. C. Smith. 1978. A model for seed scatterhoarding: Coevolution of fox squirrels and black walnuts. Ecology 59:884–896.

Stapanian, M. A., and C. C. Smith. 1984. Density dependent survival of scatterhoarded nuts: An experimental approach. Ecology 65:1387–1396.

Stapanian, M. A., and C. C. Smith. 1986. How fox squirrels influence the invasion of prairies by nut-bearing trees. Journal of Mammalogy 67:326–332.

Stapanian, M. A., S. D. Sundberg, G. A. Baumgardner, and A. Liston. 1999. Alien plant species composition and associations with anthropogenic disturbance in North American forests. Plant Ecology 139:49–62.

Starkey, D. A., S. W. Oak, G. Ryan, F. H. Tainter, C. Redmond, and H. D. Brown. 1989. Evaluation of oak decline areas in the South. U.S. Forest Service Protection Report R8 PR-17.

Steele, M. A., K. Gavel, and W. Bachman. 1998. Dispersal of half-eaten acorns by gray squirrels: Effects of physical and chemical seed characteristics. Pages 223–231 in: Ecology and Evolutionary Biology of Tree Squirrels (M. A. Steele, J. F. Merritt, and D. A. Zegers, eds.). Virginia Museum of Natural History Special Publication 6.

Steele, M. A., L. Z. Hadj-Chikh, and J. Hazeltine. 1996. Caching and feeding behavior of gray squirrels: Responses to weevil infested acorns. Journal of Mammalogy 77:305–314.

Steele, M., T. Knowles, K. Bridle, and E. Simms. 1993. Tannins and partial consumption of acorns: Implications for dispersal of oaks by seed predators. American Midland Naturalist 130:229–238.

Steele, M. A., and P. D. Smallwood. 1994. What are squirrels hiding? Natural History 103:40–45.

Steele, M. A., G. Turner, P. D. Smallwood, J. O. Wolff, and J. Radillo. 2001. Cache management by tree squirrels: Experimental evidence for the significance of acorn embryo excision. Journal of Mammalogy 82:000–000.

Steiner, K. C. 1996. Autumn predation of northern red oak seed crops. Pages 489–494 in: 10th Central Hardwood Forest Conference (K. W. Gottschalk and S. L. C. Fosbroke, eds.). U.S. Forest Service General Technical Report NE-197.

Stephens, S. L. 1997. Fire history of a mixed oak-pine forest in the foothills of the Sierra Nevada, El Dorado County, California. Pages 191–198 in: Proceedings of a Symposium on Oak Woodlands: Ecology, Management, and Urban Interface Issues. U.S. Forest Service General Technical Report PSW-GTR-160.

Stephenson, S. L. 1986. Changes in a former chestnut-dominated forest after a half-century of succession. American Midland Naturalist 116:173–179.

Stickel, P. W. 1935. Forest fire damage studies in the Northeast II. First-year mortality in burned-over oak stands. Journal of Forestry 33:595–598.

Stringer, J. W., and L. Taylor. 1999. Effects of leaf litter depth on acorn germination. Pages 289–290 in: 12th Central Hardwood Forest Conference (J. W. Stringer and D. L. Loftis, eds.). U.S. Forest Service General Technical Report SRS-24.

Strole, T. A., and R. C. Anderson. 1992. White-tailed deer browsing: Species preferences and implications for central Illinois forests. Natural Areas Journal 12:139–144.

Stromayer, K. A. K., and R. J. Warren. 1997. Are overabundant deer herds in the eastern United States creating alternate stable states in forest plant communities? Wildlife Society Bulletin 25:227–234.

Sutherland, E. K. 1997. History of fire in a southern Ohio second-growth mixed-oak forest. Pages 172–183 in: Proceedings of the 11th Central Hardwood Forest Conference (S. G. Pallardy, R. A. Cecich, H. G. Garrett, and P. S. Johnson, eds.). U.S. Forest Service General Technical Report NC-188.

Swan, F. R., Jr. 1970. Post-fire response of four plant communities in south central New York state. Ecology 51:1074–1082.

Swetnam, T. W., C. H. Baisan, P. M. Brown, and A. C. Capio. 1989. Fire history of Rhyolite Canyon, Chiricahua National Monument. U.S. National Park Service Cooperative Park Studies Unit Technical Report 31.

Swiecki, T. J., E. A. Bernhardt, and C. Drake. 1997. Factors affecting blue oak sapling recruitment. Pages 157–168 in: Proceedings of a Symposium on Oak Woodlands: Ecology, Management, and Urban Interface Issues. U.S. Forest Service General Technical Report PSW-GTR-160.

Swindel, B. F., L. F. Conde, and J. E. Smith. 1984. Species diversity: Concept, measurement and response to clearcutting and site preparation. Forest Ecology and Management 8:11–22.

Szeicz, J. M., and G. M. MacDonald. 1991. Postglacial vegetation history of oak savanna in southern Ontario. Canadian Journal of Botany 69:1507–1519.

Tabatabai, F. R., and M. L. Kennedy. 1984. Spring food habits of the eastern wild turkey in southwestern Tennessee. Journal of the Tennessee Academy of Science 59:74–76.

Tainter, F. H., and F. A. Baker. 1996. Principles of Forest Pathology. John Wiley and Sons, New York.

Tainter, F. H., W. A. Retzlaff, D. A. Starkey, and S. W. Oak. 1990. Decline of radial growth in red oaks is associated with short-term changes in climate. European Journal of Forest Pathology 20:95–105.

Taitt, J. J. 1981. The effect of extra food on small rodent populations: I. Deermice (*Peromyscus maniculatus*). Journal of Animal Ecology 50:111–124.

Tappeiner, J. C. 1969. Effect of cone production on branch, needle, and xylem ring growth of Sierra Nevada Douglas-fir. Forest Science 15:171–174.

Teskey, R. O., and R. B. Shrestha. 1985. A relationship between carbon dioxide, photosynthetic efficiency and shade tolerance. Physiologia Plantarum 63:126–132.

Texas Parks and Wildlife Department. 1988. The black-capped vireo in Texas. Texas Parks and Wildlife Department PWD-BR-300-2-4/88.

Thor, E., and G. M. Nichols. 1974. Some effects of fire on litter, soil, and hardwood regeneration. Pages 455–482 in: Proceedings Tall Timbers Fire Ecology Conference 12.

Thoreau, H. D. [1860] 1993. Faith in a Seed: The Dispersion of Seeds and other Late Natural History Writings. Island Press, Washington, DC.

Tietje, W. D., R. H. Barrett, E. B. Kleinfelder, and B. T. Carre. 1991. Wildlife diversity in valley-foothill riparian habitat: North central versus central coast California. Pages 120–125 in: Proceedings of a Symposium on Oak Woodlands and Hardwood Rangeland Management. U.S. Forest Service General Technical Report PSW-100.

Tietje, W. D., and R. L. Ruff. 1980. Denning behavior of black bears in boreal forest of Alberta. Journal of Wildlife Management 44:858–870.

Tilghman, N. G. 1989. Impacts of white-tailed deer on forest regeneration in northwestern Pennsylvania. Journal of Wildlife Management 53:524–532.

Tisch, E. L. 1961. Seasonal food habits of the black bear in the White Fish Range of northwestern Montana. Master's thesis, Montana State University, Missoula.

Touchan R., and P. F. Ffolliott. 1999. Thinning of Emory oak coppice: Effects on growth, yield, and harvesting cycles. Southwestern Naturalist 44:1–5.

Tripathi, R. S., and M. L. Khan. 1990. Effects of seed weight and microsite characteristics on germination and seedling fitness in two species of *Quercus* in a subtropical wet hill forest. Oikos 57:289–296.

Tryon, E. H., and D. S. Powell. 1984. Root ages of advanced hardwood reproduction. Forest Ecology and Management 8:293–298.

Turchin, P. 1990. Rarity of density dependence or population regulation with lags? Nature 344:660–663.

Tyree, M. T., and H. Cochard. 1996. Summer and winter embolism in oak: Impact on water relations. Annuals Science Forestry 53:173–180.

Tyrrell, L. E., and T. R. Crow. 1994. Dynamics of dead wood in old-growth hemlock-hardwood forests of northern Wisconsin and northern Michigan. Canadian Journal of Forest Research 24:1672–1683.

Uhlig, H. G., and R. W. Bailey. 1952. Wild turkey in West Virginia. Journal of Wildlife Management 16:24–32.

Uhlig, H. G., and H. L. Wilson. 1952. A method of evaluating an annual mast index. Journal of Wildlife Management 16:338–343.

U.S. Forest Service. 1987. Forest and woodland habitat types (plant associations) of Arizona south of the Mogollon Rim and southwestern New Mexico. Southwestern Region, U.S. Forest Service, Albuquerque, New Mexico.

U.S. Forest Service. 1988a. Forest health through silviculture and integrated pest management-strategic plan. U.S. Forest Service Report 1988-576-488.

U.S. Forest Service. 1988b. The South's fourth forest: Alternatives for the future. U.S. Forest Service Forest Resource Report 24.

U.S. Forest Service. 1992. Ecosystem management of the national forests and grasslands. U.S. Forest Service Memorandum 1330-1.

U.S. Natural Resources Conservation Service. 1998. The Plants National Database. U.S. Department of Agriculture, Natural Resources Conservation Service, Washington, DC.

Van Buskirk, J., and R. S. Ostfeld. 1995. Controlling Lyme disease by modifying the density and species composition of tick hosts. Ecological Applications 5:1133–1140.

Van Buskirk, J., and R. S. Ostfeld. 1998. Habitat heterogeneity, dispersal, and local risk of exposure to Lyme disease. Ecological Applications 8:365–378.

Vander Haegan, W. M., W. E. Dodge, and M. W. Sayre. 1988. Factors affecting productivity in a northern wild turkey population. Journal of Wildlife Management 52:127–133.

Vander Haegan, W. M., M. W. Sayre, and W. E. Dodge. 1989. Winter use of agricultural habitats by wild turkeys in Massachusetts. Journal of Wildlife Management 53:30–33.

van der Pijl, L. 1972. Principles of Dispersal in Higher Plants. Springer Verlag, New York.

Van Dersal, W. R. 1940. Utilization of oaks by birds and mammals. Journal of Wildlife Management 4:404–428.

Vander Wall, S. B. 1990. Food hoarding in Animals. University of Chicago Press, Chicago.

Van Hooser, D. D., R. A. O'Brien, and D. C. Collins. 1990. New Mexico's forest resources. U.S. Forest Service Resource Bulletin INT-79.

Van Lear, D. H. 1991. Fire and oak regeneration in the southern Appalachians. Pages 15–21 in: Fire and the environment: Ecological and cultural perspectives (S. C. Nodvin and T. A. Waldrop, eds.). U.S. Forest Service General Technical Report SE-69.

Van Lear, D. H. 1993. The role of fire in oak regeneration. Pages 66–78 in: Proceedings of a Symposium on Oak Regeneration: Serious Problems, Practical Recommendations (D. Loftis and C. E. McGee, eds.). U.S. Forest Service General Technical Report SE-84.

Van Lear, D. H., and V. J. Johnson. 1983. Effects of prescribed burning in southern Appalachian and Piedmont forests: A review. Clemson University Department of Forestry, Forestry Bulletin 36.

Van Lear, D. H., and T. A. Waldrop. 1988. Effects of fire on natural regeneration in the Appalachian Mountains. Pages 56–70 in: Guidelines for regenerating Appalachian hardwood stands (H. C. Smith, A. W. Perkey and W. E. Kidd Jr., eds.). Morgantown, WV. SAF Publications 88-03.

Van Lear, D. H., and T. A. Waldrop. 1989. History, uses, and effects of fire in the Appalachians. U.S. Forest Service General Technical Report SE-54.

Van Lear, D. H., and J. M. Watt. 1993. The role of fire in oak regeneration. Pages 66–78 in: Proceedings of the Symposium on Oak Regeneration: Serious Problem, Practical Recommendations (D. L. Loftis and C. E. McGee, eds.). U.S. Forest Service General Technical Report SE-84.

Vangilder, L. D. 1995. Survival and cause-specific mortality of wild turkeys in the Missouri Ozarks. Proceedings of the National Wild Turkey Symposium 7:21–31.

Verme, L. J. 1967. Influence of experimental diets on white-tailed deer reproduction. Transactions of the North American Wildlife and Natural Resources Conference 32:405–420.

Verme, L. J. 1969. Reproductive patterns of white-tailed deer related to nutritional plane. Journal of Wildlife Management 33:881–887.

Verme, L. J., and J. J. Ozoga. 1980. Influence of protein-energy intake on deer fawns in autumn. Journal of Wildlife Management 44:305–314.

Verner, J. 1980. Birds of California oak habitats: Management implications. Pages 246–264 in: Proceedings of the Symposium on the Ecology, Management, and Utilization of California Oaks. U.S. Forest Service General Technical Report PSW-44.

Vessey, S. H. 1987. Long-term population trends in white-footed mice and the impact of supplemental food and shelter. American Zoologist 27:879–890.

Vose, J. M., W. T. Swank, B. D. Clinton, J. D. Knoepp, and L. W. Swift. 1999. Using stand replacement fires to restore southern Appalachian pine-hardwood ecosystems: Effects on mass, carbon, and nutrient pools. Forest Ecology and Management 114:215–226.

Vozzo, J. A. 1984. Insects and fungi associated with acorns of *Quercus* spp. Pages 40–43 in: Proceedings of the Cone and Seed Insects Working Party Conference. U.S. Forest Service Research Station, Asheville, NC.

Wahl, R., and D. D. Diamond. 1990. The golden-cheeked warbler: A status review. U.S. Fish and Wildlife Service, Fort Worth, TX.

Wainio, W. W., and E. B. Forbes. 1941. The chemical composition of forest fruits and nuts from Pennsylvania. Journal of Agricultural Research 62:627–635.

Waite, R. K. 1985. Food caching and recovery by farmland corvids. Bird Study 32:45–49.

Wakeling, B. F., and T. D. Rogers. 1995. Winter diet and habitat selection by Merriam's turkeys in north-central Arizona. Proceedings of the National Wild Turkey Symposium 7:175–184.

Waldrop, T. A., E. R. Buckner, and J. A. Muncy. 1985. Cultural treatments in low-quality hardwood stands for wildlife and timber production. Pages 493–500 in: Proceedings of the Third Biennial Southern Silvicultural Research Conference (E. Shoulders, ed.). U.S. Forest Service General Technical Report SO-54.

Waldrop, T. A., and F. T. Lloyd. 1991. Forty years of prescribed burning on the Santee fire plots: Effects on overstory and midstory vegetation. Pages 45–50 in: Fire and the environment: Ecological and cultural perspectives (S. C. Nodvin and T. A. Waldrop, eds.). U.S. Forest Service General Technical Report SE-69.

Waldrop, T. A., D. H. Van Lear, F. T. Lloyd, and W. H. Harms. 1987. Long-term studies of prescribed burning in loblolly pine forests of the southeastern Coastal Plain. U.S. Forest Service General Technical Report SE-45.

Walker, E. A. 1941. The wild turkey and its management in the Edwards Plateau. Pages 21–23 in: Texas Game, Fish, and Oyster Commission Quarterly Progress Report Oct.–Dec., P-R Project 1-R, Austin.

Walker, L. C. 1980. The southern pine region. Pages 231–276 in: Regional Silviculture of the United States (J. W. Barrett ed.). Wiley-Interscience, New York.

Waller, D. M., and W. S. Alverson. 1997. The white-tailed deer: A keystone herbivore. Wildlife Society Bulletin 25:217–226.

Wallmo, O. C. 1955. Vegetation of the Huachuca Mts., Arizona. American Midland Naturalist 54:466–480.

Walters, M. B., E. L. Kruger, and P. B. Reich. 1993. Relative growth rate in relation to physiological and morphological traits for northern hardwood tree seedlings: Species, light environment and ontogenetic considerations. Oecologia 96:219–231.

Walters, C. J. 1998. Improving links between ecosystem scientists and managers. Pages 272–286 in: Successes, Limitations, and Frontiers in Ecosystem Science (M. L. Pace and P. M. Groffman, eds.). Springer-Verlag, Berlin.

Ward, J. S., and E. Gluck. 1999. Using prescribed burning to release oak seedlings from shrub competition in southern Connecticut. Page 283 in: Proceedings of the 12th Central Hardwood Conference (J. W. Stringer and Loftis, D. L., eds.). U.S. General Technical Report SRS-24.

Ward, J. S., and G. R. Stephens. 1989. Long-term effects of a 1932 surface fire on stand structure in a Connecticut mixed-hardwood forest. Page 267–273 in: Proceedings of the Seventh Central Hardwood Forest Conference. (G. Rink and C. Budelsky, eds.). U.S. Forest Service General Technical Report NC-132.

Wargo, P. M. 1977. *Armillaria mellea* and *Agrilus bilineatus* and mortality of defoliated oak trees. Forest Science 23:485–492.

Wargo, P. M., and D. R. Houston. 1974. Infection of defoliated sugar maple trees by *Armillaria mellea*. Phytopathology 64:817–822.

Warren, R. J. 1997. The challenge of deer overabundance in the 21st century. Wildlife Society Bulletin 25:213–214.

Warren, R. J., and R. L. Kirkpatrick. 1982. Evaluating nutritional status of white-tailed deer using fat indices. Pages 463–472 in: Proceedings of the Annual Conference of the Southeast Association of Fish and Wildlife Agencies 36.

Wathen, W. G. 1983. Reproduction and denning of black bears in the Great Smoky Mountains. Master's thesis, University of Tennessee, Knoxville.

Wathen, W. G., K. G. Johnson, and M. R. Pelton. 1986. Characteristics of black bear dens in the southern Appalachian region. International Conference on Bear Research and Management 6:119–127.

Watts, C. H. S. 1969. The regulation of wood mouse (*Apodemus sylvaticus*) numbers in Wytham woods, Berkshire. Journal of Animal Ecology 38:285–304.

Watts, W. A. 1979. Late Quaternary vegetation of the central Appalachian and New Jersey Coastal Plain. Ecological Monographs 49:427–469.

Wauters, L. A., and P. Casale. 1996. Long-term scatterhoarding by Eurasian red squirrels (*Sciurus vulgaris*). Journal of Zoology, London 238:195–207.

Webb, T., III. 1988. Glacial and Holocene vegetation history: Eastern North America. Pages 385–414 in: Vegetation History (B. Huntley and T. Webb, eds.). Kluwer Academic, Amsterdam.

Webb, S. L. 1986. Potential role of passenger pigeons and other vertebrates in the rapid Holocene migrations of nut trees. Quaternary Research 26:367–375.

Webb, L. G. 1941. Spring and winter foods of the wild turkey in Alabama. Master's thesis, Auburn University, Auburn, AL.

Weigel, D. R., and G. R. Parker. 1997. Tree regeneration response to the group selection method in southern Indiana. Northern Journal of Applied Forestry 14:90–94.

Weigl, P. D., M. A. Steele, L. J. Sherman, J. C. Ha, and T. L. Sharpe. 1989. The ecology of the fox squirrel (*Sciurus niger*) in North Carolina: Implications for survival in the southeast. Bulletin of Tall Timbers Research Station, Tallahassee, FL.

Weinstein, M., B. D. Leopold, and G. A. Hurst. 1995. Evaluation of wild turkey population estimation techniques. Proceedings of the Annual Conference of the Southeastern Association of Fish and Wildlife Agencies 49:467–487.

Welker, J. M., and J. W. Menke. 1990. The influence of simulated browsing on tissue water relations, growth and survival of *Quercus douglasii* (Hook and Arn.) seedlings under slow and rapid rates of soil drought. Functional Ecology 1990 (4): 807–817.

Welsh, S. L., N. D. Atwood, S. Goodrich, and L. C. Higgens. 1993. A Utah Flora. Brigham Young University, Monte L. Bean Life Science Museum, Provo, UT.

Welsh, C. J. E., and W. M. Healy. 1993. Effect of even-aged timber management on bird species diversity and composition in northern hardwoods of New Hampshire. Wildlife Society Bulletin 21:143–154.

Wendel, G. W. 1975. Stump sprout growth and quality of several Appalachian hardwood species after clearcutting. U.S. Forest Service Research Paper NE-329.

Wendel, G. W., and H. C. Smith. 1986. Effects of prescribed fire in a central Appalachian oak-hickory stand. U.S. Forest Service Research Paper NE-594.

Wentworth, J. M., A. S. Johnson, and P. E. Hale. 1990. Influence of acorn use on nutritional status and reproduction of deer in the southern Appalachians. Pages in 142–154 in: Proceedings of the Annual Conference of the Southeastern Association of Fish and Wildlife Agencies 44.

Wentworth, J. M., Johnson, A. S., and Hale, P. E. 1992. Relationships of acorn abundance and deer herd characteristics in the southern Appalachians. Southern Journal of Applied Forestry 16:5–8.

Westin, S. 1992. Wildfire in Missouri. Missouri Department of Conservation. Jefferson City.

Wheeler, R. J., Jr. 1948. The wild turkey in Alabama. Alabama Department of Conservation Bulletin 12.

Whelan, R. J. 1995. The Ecology of Fire. Cambridge University Press, Cambridge.

White, D. L., and F. T. Lloyd. 1995. Defining old growth: Implications for management. Pages 51–62 in: Proceedings of the Eighth Biennial Southern Silvicultural Conference (M. B. Edwards, compiler). U.S. Forest Service, Southern Research Station, Asheville, NC.

White, T. H., Jr., H. A. Jacobson, and B. D. Leopold. 1996. Black bears and timber management in forested wetlands of the Mississippi alluvial valley. Habitat 13:3–5.

Whitehead, C. J. 1969. Oak mast yields on wildlife management areas in Tennessee. Tennessee Wildlife Research Agency, Nashville.

Whitehead, C. J. 1980. A quantitative study of mast production in the southern Appalachian region of Georgia, North Carolina, and Tennessee (the tri-state black bear study region). Tennessee Wildlife Research Agency Technical Report 80-2.

Whitney, G. G. 1984. Fifty years of change in the arboreal vegetation of Heart's Content, an old-growth hemlock–white pine–northern hardwood stand. Ecology 65:403–408.

Whitney, G. G. 1986. Relation of Michigan's presettlement pine forests to substrate and disturbance history. Ecology 67:1548–1559.

Whitney, G. G. 1990. The history and status of the hemlock-hardwood forests of the Allegheny Plateau. Journal of Ecology 78:443–458.

Whitney, G. G. 1994. From Coastal Wilderness to Fruited Plain: A History of Environmental Change in Temperate North America from 1500 to the Present. Cambridge University Press, Cambridge.

Whitney, G. G., and W. C. Davis. 1986. Thoreau and the forest history of Concord, Massachusetts. Journal of Forest History 30:70–81.

Whittaker, R. H., and W. A. Niering. 1964. Vegetation of the Santa Catalina Mountains, Arizona. I. Ecological classification and distribution of species. Journal of the Arizona Academy of Science 3:9–34.

Whittaker, R. H., and W. A. Niering. 1965. Vegetation of the Santa Catalina Mountains, Arizona. II. A gradient analysis of the south slope. Ecology 46:429–452.

Wigley, T. B., R. L. Willett, M. E. Garner, and J. B. Baker. 1989. Wildlife habitat quality in varying mixtures of pine and hardwood. Pages 131–136 in: Proceedings of pine-hardwood mixtures: A symposium on management and ecology of the type (T. A. Waldrop ed.). U.S. Forest Service General Technical Report SE-58.

Will-Wolf, S. 1991. Role of fire in maintaining oaks in mesic oak maple forests. Pages 27–33 in: The oak resource in the Upper Midwest: Implications for Management (S. B. Laursen and J. F. DeBoe, eds.). Minnesota Extension Service, University of Minnesota, St. Paul.

Willey, C. H. 1978. The Vermont Black Bear. Vermont Fish and Game Department, Montpelier.

Williams, C. E. 1989. Checklist of North American nut-infesting insects and host plants. Journal of Entomological Science 24:550–562.

Williams, D. W., R. W. Fuester, W. W. Balaam, R. J. Chianese, and R. C. Reardon. 1992. Incidence and ecological relationships of parasitism in larval populations of *Lymantria dispar.* Biological Control 2:35–43.

Williams, D. W., and A. M. Liebhold. 1995. Influence of weather on the synchrony of gypsy moth (Lepidoptera: Lymantriidae) outbreaks in New England. Environmental Entomology 24:987–995.

Williams, L. E., Jr. 1992. Florida turkey. Pages 214–231 in: The Wild Turkey: Biology and Management (J. G. Dickson, ed.). Stackpole Books, Harrisburg, PA.

Williams, M. 1989. Americans and Their Forests. Cambridge University Press, New York.

Wilson, B. F., and M. J. Kelty. 1994. Shoot growth from the bud bank in black oak. Canadian Journal of Forest Resources 24:149–154.

Wilson, M. L., S. R. Telford, J. Piesman, and A. Spielman. 1988. Reduced abundance of immature Ixodes dammini (Acari: Ixodidae) following elimination of deer. Journal of Medical Entomology 25:224–228.

Wilson, M. L., A. M. Ducey, T. S. Litwin, T. A. Gavin, and A. Spielman. 1990. Microgeographic distribution of immature Ixodes dammini ticks correlated with that of deer. Medical and Veterinary Entomology 4:151–159.

Wilson, M. L., and J. E. Childs. 1997. Vertebrate abundance and the epidemiology of zoonotic diseases. Pages 224–248 in: The Science of Overabundance: Deer Ecology and Population Management (W. J. McShea, H. B. Underwood, and J. H. Rappole, eds.). Smithsonian Institution Press, Washington, DC.

Wimsatt, W. A. 1963. Delayed implantation in the Ursidae, with particular reference to the black bear (Ursus americanus Pallas). In: Delayed Implantation (A. E. Enders, ed.). University of Chicago Press, Chicago.

Winchcombe, R. J. 1993. Controlled access hunting for deer population management: A case study. Northeast Wildlife 50:1–9.

Winkler, M. G., A. M. Swain, and J. E. Kutzback. 1986. Middle Holocene dry period in the northern midwestern United States: Lake levels and pollen stratigraphy. Quarternary Research 25:235–250.

Wolf, W. W., Jr. 1988. Shelterwood cutting to regenerate oaks: The Glatfelter experience. Pages 210–218 in: Proceedings for a Symposium on Guidelines for Regenerating Appalachian Hardwood Stands (H. C. Smith, A. W. Perkey, and W. E. Kidd, Jr., eds.). West Virginia University Books, Morgantown.

Wolff, J. O. 1986. Life history strategies of white-footed mice (Peromyscus leucopus). Virginia Journal of Science 37:208–220.

Wolff, J. O. 1996. Population fluctuations of mast-eating rodents are correlated with production of acorns. Journal of Mammalogy 77:850–856.

Wolgast, L. J., and B. B. Stout. 1977. Effects of age, stand density, and fertilizer application on bear oak reproduction. Journal of Wildlife Management 41:685–691.

Woods, F. W., and R. E. Shanks. 1959. Natural replacement of chestnut by other species in the Great Smoky Mountains National Park. Ecology 40:349–361.

Woods, S. A., and J. S. Elkinton. 1987. Bimodal patterns of mortality from nuclear polyhedrosis virus in gypsy moth (*Lymantria dispar*) populations. Journal of Invertebrate Patholology 50:151–157.

Woods, S. A., J. S. Elkinton, K. D. Murray, A. M. Liebhold, J. R. Gould, and J. D. Podgwaite. 1991. Transmission dynamics of a nuclear polyhedrosis virus and predicting mortality in gypsy moth (Lepidoptera: Lymantriidae) populations. Journal of Economic Entomology 84:423–430.

Woodward, A., D. G. Silsbee, E. G. Schreiner, and J. E. Means. 1994. Influence of climate on radial growth and cone production in subalpine fir (*Abies lasiocarpa*) and mountain hemlock (*Tsuga mertensiana*). Canadian Journal of Forest Research 24:1133–1143.

Wootton, J. T. 1993. Indirect effects and habitat use in an intertidal community: Interaction chains and interaction modifications. American Naturalist 141:71–89.

Woudenberg, S. W., and T. O. Farrenkopf. 1995. The Westwide Forest Inventory Database: User's manual. General Technical Report INT-317. U.S. Department of Agriculture, Forest Service, Intermountain Research Station, Ogden, UT.

Wright, H. A., and A. W. Bailey. 1982. Fire Ecology: United States and Southern Canada. John Wiley and Sons, New York.

Wright, H. E., Jr. 1964. Aspects of the early postglacial forest succession in the Great Lakes region. Ecology 45:439–448.

Wright, H. E., Jr., T. C. Winter, and H. L. Patten. 1963. Two pollen diagrams from southeastern Minnesota: Problems in the regional late- and postglacial vegetational history. Geological Society of America Bulletin 74:1371–1396.

Wright, S. L. 1986. Prescribed burning as a technique to manage insect pests of oak regeneration. Pages 91–96 in: Proceedings from Prescribed Burning in the Midwest: State-of-the-Art Symposium. Stevens Point, WI.

Wunz, G. A. 1979. Wild turkey study. Pennsylvania Game Commission, Harrisburg. P-R Project W-46-R-25, Final Report.

Wunz, G. A., and J. C. Pack. 1992. Eastern turkey in eastern oak-hickory and northern hardwood forests. Pages 233–264 in: The Wild Turkey: Biology and Management (J. G. Dickson, ed.). Stackpole Books, Harrisburg, PA.

Wyant, J. G., R. J. Alig, and W. A. Bechtold, 1991. Physiographic position, disturbance and species composition in North Carolina coastal plain forests. Forest Ecology and Management 41:1–19.

Wykoff, M. W. 1991. Black walnut on Iroquoian landscapes. Northeast Indian Quarterly, Summer Issue:4–17.

Yaffee, S. L. 1999. Three faces of ecosystem management. Conservation Biology 13:713–725.

Yahner, R. H., and H. R. Smith. 1991. Small mammal abundance and habitat relationships on deciduous forest sites with different susceptibility to gypsy moth defoliation. Environmental Management 15:113–120.

Yasuda, M., N. Nakagoshi, and F. Takahashi. 1991. Examination of spool-and-line method as quantitative technique to investigate seed dispersal by rodents. Japanese Journal of Ecology 41:257–262.

Young, B. F., and R. L. Ruff. 1982. Population dynamics and movements of black bears in east central Alberta. Journal of Wildlife Management 46:845–860.

Zhang, Q. F., K. S. Pregitzer, and D. D. Reed. 1999. Catastrophic disturbance in the presettlement forests of the Upper Peninsula of Michigan. Canadian Journal of Forest Resources 29:106–114.

Zar, J. H. 1984. Biostatistical Analysis. Prentice-Hall, Englewood Cliffs, NJ.

Index

Abies balsamea (balsam fir), 25

Acer (maple), 52, 66, 70, 123–24, 126, 257

Acer rubrum (red maple): and chestnut blight, 84; in eastern forests, 22; and ecosystem management, 338; and fire, 44; and gypsy moths, 110; in old-growth forests, 117–18, 331; and shelterwood-burn method, 275, 278; as succession species, 320; and wildlife, 221

Acer saccharum (sugar maple): in eastern forests, 22; and European settlement, 41; and fire, 44, 121–22, 125; in Holocene period, 38–40; in old-growth forests, 117, 120; and seed dispersal, 123; and shelterwood-burn method, 279; as succession species, 65, 222, 320; and wildlife, 124, 221

acorn nutrition: and black bears, 173, 226–35; chemical composition of, 173–81, 190–91, 248; and deer, 215–16; and mice, 102–3; and native diseases and insects, 86; and oak regeneration methods, 322; and old-growth oak forests, 115; and squirrels, 173, 256, 259–61, 263–64; and wildlife, 69, 129, 152, 173–81; and wild turkeys, 177, 179, 241–44, 248–49, 254

acorn production: and black bears, 234–35, 238–39; and climate, 5, 137–39; and deer, 215–24; and dispersal, 69, 123, 182–94, 261, 263; and ecosystem management, 4–9, 61, 149–50, 152, 166–67, 170–72, 328, 336–38, 340; estimating, 70, 149–50, 152–56, 163–70; and even-aged stands, 322, 325; by evergreen oak, 308; failure of, 132, 142–43, 147, 156–57, 228, 230, 232, 235, 237–39, 245, 249–51, 253, 263–64; and fire, 270, 299–300; and gypsy moths, 100–101, 103–8, 111–12, 149, 204; and Lyme disease, 149, 202; and mice, 100–102, 149, 206; and native diseases and insects, 86, 89–90, 93–97, 99; and oak regeneration, 68, 149, 278, 293, 322, 325–26; patterns in, 132–35, 156–58; and resource switching, 143–45, 148; synchrony in, 139–43, 145, 147–48; and tree variation, 159–61; variability in, 144–47; and weather, 137–39; and wildlife, 6–7, 85, 317, 322–26; and wild turkeys, 245–53, 311

acorn weevils (Curculio), 95, 99, 188, 192, 245, 299–300

acorn woodpecker (Melanerpes formicivorus), 143

agriculture: and forest landscape, 2, 27, 29, 43, 46, 53–57, 59, 76, 281, 287; and native diseases and insects, 83; and old-growth forests, 116, 118, 123, 126, 330; and wildlife, 210, 215, 225, 245, 250, 264, 318

alder, 36, 38

alligator juniper (Juniperus deppeana), 304, 311, 313

Ambrosia (ragweed), 38–39

American basswood (Tilia americana), 221

American beech. See Fagus

American chestnut. See Castanea